Orthorectification of High-Resolution Remotely Sensed Images

Orthorectification in remote sensing is used to geometrically correct satellite or aerial image distortions caused by terrain elevation, sensor lens, etc. This book explains the orthorectification for different imaging methods, such as frame, linear, and video imaging; for different imaged areas, such as Greenland without ground features or urban areas with high buildings; for different types of sensors, such as videos and images; and for different requirements of processing speeds, such as real-time, onboard, and PC-based processing. It starts with the definition of orthorectification and provides a step-by-step introduction to the models, the algorithms, the methods, and Digital Orthophoto Map (DOM) quality control.

Key Features:

- Focuses on a specific topic that is crucial for applications like land-use mapping, urban planning, and environmental monitoring.
- Clearly explains the theories and the applications through many illustrations and detailed examples.
- Addresses orthorectification systematically and comprehensively for all levels of readers.
- Explains the challenges faced during constructing models, algorithms, and methods for orthorectification.
- Includes the lessons learned from the author's longstanding research and experience on the topic.

This book is an excellent reference for a complete understanding of orthorectification. It's for senior undergraduate students and graduate students in remote sensing, information technology, computer science and engineering, electronic engineering, and geography. It is also valuable for researchers, professionals, and those who work in academia, governmental agencies, and commercial sectors interested in satellite remote sensing image processing.

Guoqing Zhou is a professor at the Guilin University of Technology, Guilin, China. He received his first PhD from Wuhan University, Wuhan, China, in 1994 and his second PhD from Virginia Tech at Blacksburg, Virginia, USA, in 2001. He was a visiting scholar at the Department of Computer Science and Technology, Tsinghua University, Beijing, China, and a postdoctoral researcher at the Institute of Information Science, Beijing Jiaotong University, Beijing, China, from 1994 to 1996. He continued his research as an Alexander von Humboldt Fellow at the Technical University of Berlin, Berlin, Germany, from 1997 to 1998 and afterward became a postdoctoral researcher at the Ohio State University, Columbus, OH, USA, from 1998 to 2000. Later, he worked at Old Dominion University, Norfolk, VA, USA, as an assistant professor, an associate professor, and a full professor in 2000, 2005, and 2010, respectively. He has authored 14 books and has published more than 326 papers in peer-reviewed journals and 332 papers in conference proceedings.

Orthorectification of High-Resolution Remotely Sensed Images

Guoqing Zhou

CRC Press
Taylor & Francis Group
Boca Raton London New York

CRC Press is an imprint of the
Taylor & Francis Group, an **informa** business

Designed cover image: © Guoqing Zhou

First edition published 2027
by CRC Press
2385 NW Executive Center Drive, Suite 320, Boca Raton FL 33431

and by CRC Press
4 Park Square, Milton Park, Abingdon, Oxon, OX14 4RN

CRC Press is an imprint of Taylor & Francis Group, LLC

ISBN: 978-1-041-05272-2 (hbk)
ISBN: 978-1-041-05273-9 (pbk)
ISBN: 978-1-003-63210-8 (ebk)

DOI: 10.1201/9781003632108

Typeset in Times
by codeMantra

Contents

Preface

Digital orthophoto map (DOM) (also called digital orthoimage map) is a critical component of the National Spatial Data Infrastructure. The computational process for the generation of a DOM is called orthorectification (sometimes unstrictly called *geometric correction*). Theoretically, a DOM should be a spatially accurate image with ground features represented in their correct, upright, and true planimetric positions with all distortion removals, including lens distortion; terrain relief displacements caused by hills, valleys, buildings, etc.; aircraft attitude; and atmospheric distortion. Therefore, a DOM contains both the image characteristics of a photograph and the geometric qualities of a map, which makes the distances, areas, and angles directly measurable and the detailed portions of ground features directly visible that are typically omitted or generalized on traditional maps.

With the emergence of different types of imaging sensor technologies, such as frame, push-broom linear, and video imaging; different types of imaged regions, such as urban areas with high buildings and Greenland without ground features; and different requirements of processing speeds, such as real-time generation of and post-processing generation of DOM, these factors present many challenges during constructing models, algorithms, and methods for orthorectification. Therefore, the objective of this book is to advance readers' theories, models, algorithms, and methods on orthorectification of high-resolution remotely sensed images for DOM generation through systemically describing principles, theories, models, and methods of orthorectification under different imaging modes, different types of imaged areas, different types of sensor's physical models, and different requirements of processing speeds.

Several of the apparent features of this book that I believe can enhance its value as a textbook and as a reference book are as follows: First, this book only concenters on a special topic, orthorectification for different imaging modes (frame and video imaging), differently imaged areas (Greenland without ground features and urban areas with high buildings), different types of sensor's physical models (unknown and known), and different requirements of processing speeds (real-time, onboard, and post-processing). In fact, none of the books have addressed all of these topics in a comprehensive, systemic, and integrated way. Second, this book emphasizes the principles, models, algorithms, and methods of the high-resolution remotely sensed image orthorectification. Each topic is clearly explained and illustrated by detailed examples. This book can therefore advance readers' knowledge and skills in geometric correction and orthorectification. Third, the content of this book is specially deliberated from orthorectification definition, to models, algorithms, and methods of orthorectification for different types of images, to DOM quality control. Such a deliberate chapter arrangement is to help the reader completely understand the principles of orthorectification step-by-step.

This book involves 12 chapters, which are based on my past 26-year research work. Parts of the chapter are from the published papers that are co-authored with my postdoctoral researcher and PhD students. They are Dr. Weilong Chen (Chapters 5 and 7), Dr. Wenhan Xie (Chapter 8), Dr. Rongting Zhang (Chapters 9 and 10), Dr. Tao Yue (Chapters 5 and 11), and Dr. Qingyang Wang (Chapter 12). The author has faithfully shared heartfelt thanks for the work they do!

This book is appropriate as a textbook for senior undergraduate and graduate students with one-semester, three-credit course, and for the references of instructors, researchers, scientists, engineers, and professionals in academia, governmental agencies, and industrial sectors.

Guoqing Zhou
October 2025

Acknowledgments

This book is based on my past 26-year research work on the topic of high-resolution remotely sensed image orthorectification. Many thanks to my postdoctoral researcher, Wenhan Xie; five PhD students, Ronglin Zhang, Jingjin Huang, Tao Yue, Qinyang Wang, and Deqian Liu; one intern, Weirong Chen; and two MS students, Linjun Jiang and Yujun Shi, for their contributions on the co-author for the articles published in journals. Special thanks go to research assistants Xia Wang, Wenyu Wang, and Minhui Jiang for their format edition.

Finally, I should very much like to thank my wife, Michelle M. Xie, and my two sons, Deqi Zhou and Mitchell Zhou. I thank them for those occasional times that I did dedicate to work on the manuscript rather than spending more time with them.

Guoqing Zhou
December 2025

1 Introduction

1.1 WHAT IS DIGITAL ORTHOPHOTOMAP OR ORTHOIMAGEMAP, OR ORTHOPHOTOGRAPH?

A digital *orthophotomap* (*DOM*) is a remotely sensed image, such as airborne aerial image (or photograph) and spaceborne image (or photograph), whose displacements and distortions caused by terrain and height relief, atmospheric refraction, flight errors, and camera distortion errors are orthorectified into a given map coordinate system (Wolf et al. 2004; McGlone et al. 2004).

In the early 1960s, film was a major carrier of the captured objects and is thus called photograph or photo. The photo is orthorectified by a rectifier, and the corresponding orthorectified photo is called orthophoto. With the development of digital imagery and computer software, such as photogrammetric workstations in the 1980s, a digital orthophoto was generated using computer software or a digital photogrammetric workstation (Kasser and Egels 2002). Thus, one called the digital orthophoto an "orthoimage" (Wolf et al. 2004; McGlone et al. 2004). The term orthophoto or orthophotograph corresponds to the orthorectified raw photo, and the orthoimage corresponds to the orthorectified image.

Since the columns and rows of a digital orthophoto or orthoimage are aligned with north and east of a geodetic mapping coordinate system, each pixel in the digital orthophoto or orthoimage not only retains the image gray information but also contains both image coordinates (*I*, *J*) and geodetic mapping coordinates (*X*, *Y*) with respect to a given geodetic coordinate system. Therefore, a digital orthophoto or digital orthoimage is called "digital orthophoto map", i.e., "DOM", or "digital orthoimage map", i.e., "DOM". This book will call orthoimage, or orthophoto, or DOM without confusion hereafter.

Digital orthophotomaps (DOMs) are a critical component of National Spatial Data Infrastructure (Federal Geographic Data Committee [FGDC] 1997; Liu et al. 2018; Maitra 1998; Yang et al. 2016; Zhou et al. 2005). DOMs (1) serve as a geospatial foundation upon which to add detail and attach attribute information; (2) provide a base on which to accurately register and compile other themes of data; and (3) orient and link the results of an application to the landscape (Zhou et al. 2022a; FGDC 1995). Especially, a highly detailed DOM is capable of serving as a source for locating the features to be mapped and measured (Jamil and Bayram 2017).

The process of generating a DOM is called *orthorectification*. This process is traditionally used to transform an image from a *perspective projection* (also called *central projection*) to an *orthogonal projection* (also called *parallel projection*). The mathematical models and methods used for the orthorectification range from simple affine transformation to rigorous collinearity (Figure 1.1).

1.2 THE CHARACTERISTICS OF AN ORTHOIMAGE

A DOM should theoretically be a spatially accurate image with ground features represented in brightness form and in their upright and correct planimetric positions of a given map coordinate system after the displacements and distortions caused by terrain relief, atmospheric refraction, and camera systems are rectified by photogrammetric techniques (Zhou et al. 2005; McGlone et al. 2004). Thus, a DOM contains both the image characteristics of a photograph and the geometric properties of a map. So, a DOM contains the characteristics below (Federal Geographic Data Committee 1995, 1997):

a. DOM is an accurate representation of the Earth's surface at a reduced uniform scale; thus, all ground features are shown in their correct ground positions. This makes it possible to measure ground-true distances and areas.

DOI: 10.1201/9781003632108-1

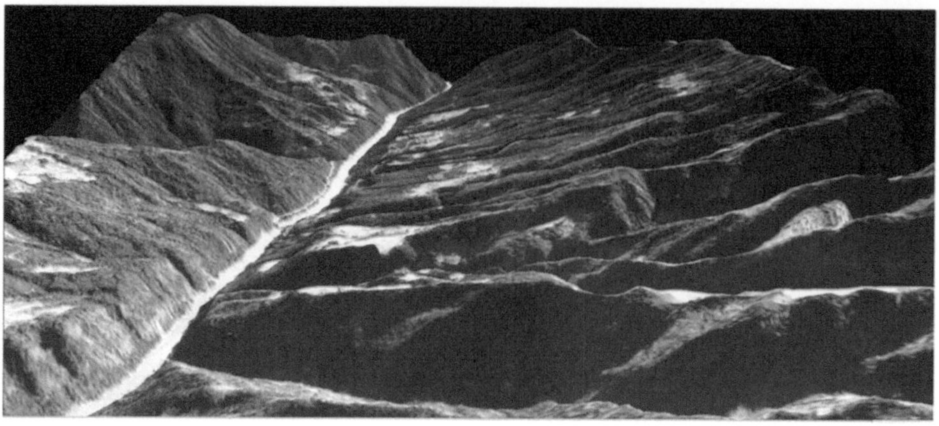

FIGURE 1.1 An example of DOM.

b. DOM should theoretically remove all distortions caused by camera lens, flight attitudes, atmospheric refraction, etc., and displacements caused by elevation differences.
c. DOM usually serves as a geospatial base on which an organization may add geospatial data details and attach geospatial attribute information.
d. DOM provides a base on which an organization can accurately register and compile other vector data and attribute data.

1.3 WHAT IS GEOREFERENCING?

Similar terminologies include *image geo-referencing*, *image geo-registration*, or *geo-correction*. Theoretically, a geo-correction majorly means to correct the geometric distortions, such as geometric displacement caused by lens distortion, atmospheric refraction, earth curvature, etc., but does not contain the relief displacements caused by object height and terrain. Therefore, strictly, geo-correction does not mean orthorectification. Conversely, since orthorectification can effectively correct both geometric displacement and relief displacement (a type of geometric displacement in photogrammetry), orthorectification thus belongs to one type of geo-correction.

Georeferencing (also called "geo-registration" or "ground registration") is a method with which an original image is referenced to a given geodetic mapping coordinate system so that the columns and rows of the resulting image are aligned with north and east of a given geodetic mapping coordinate system. Generally, the methods for georeferencing an image are usually a two-dimensional (2D) affine transformation, not rigorous orthorectification. Thus, this type of method is usually for low-accuracy georeferencing of a low-accuracy image. The process of georeferencing an image consists of two fundamental steps: (1) computing the parameters of 2D coordinate transformation from original digital image to the resulting georeferenced image, which is aligned in a given geodetic coordinate system, and (2) filling a gray for each pixel in the resulting georeferenced image, which is aligned with the given geodetic coordinate system, through gray interpolation methods, such as linear interpolation and bilinear interpolation.

1.4 THE DIFFERENCE BETWEEN THE ORIGINAL IMAGE AND DIGITAL ORTHOIMAGEMAP (DOM)

An original image is one with a perspective projection containing various types of distortions, such as camera lens distortion caused by nonuniform magnification from optical curvature and light bending, especially in wide-angle designs and cost-constrained lenses; relief displacements caused by the tilting of the camera and terrain relief (topography). With the existence of the relief displacement in an original image, it does not have a uniform scale. Usually, the tops of hills closer to the sensor have a larger scale than the bottoms of valleys that are further away. With various scales in an entire image, the geometric measurements, such as distances and areas on an aerial photograph, vary (McGlone et al. 2004; Wolf et al. 2004).

For a DOM, all distortions caused by relief displacement, camera lens, atmospheric refraction, and aircraft attitude changing, etc., are theoretically removed. Moreover, the DOM is georeferenced to a given geodetical mapping coordinate system. Thus, a DOM combines the image characteristics of an image with the geometric qualities of a map, and the scale of an entire orthoimage is uniform, so measurement of distances, areas, angles, and the detailed portions of ground features on the DOM are equivalent to those on the map. Thus, a DOM is usually called a *photographic map*, or simply *orthophotomap*, since it is like a map (McGlone et al. 2004; Wolf et al. 2004).

1.5 THE DIFFERENCE BETWEEN THE ORTHOIMAGEMAP (DOM) AND MAP

A map utilizes lines and symbols to represent the locations, relationships, and natural features and social features of objects on the Earth (or another area) at a reduced scale, while a DOM utilizes brightness to represent the natural features and social features of the Earth's surface (McGlone et al. 2004; Wolf et al. 2004). Both are a reproduction, at a reduced scale, of an orthographic projection of the terrain onto a reference datum plane, i.e., each point of the earth surface in both orthoimage and map is orthogonally projected onto its true planimetric position of a defined geodetic coordinate system. DOMs are therefore geometrically equivalent to conventional line and symbol planimetric maps, which also show true orthographic positions of objects on the Earth's surface (McGlone et al. 2004; Wolf et al. 2004).

The major difference between a DOM and a map is that a DOM is composed of images of features, whereas maps utilize lines and symbols plotted to scale to depict features. This means that DOMs offer the advantages of a conventional map product; but unlike the traditional map, the DOM is able to display actual ground features, is an accurate representation of the Earth's surface, and can be used to measure true geometric distance, area, and angle (McGlone et al. 2004; Wolf et al. 2004).

1.6 BRIEF HISTORY OF ORTHOIMAGE GENERATION

DOM generation, whose process is called *orthorectification*, is a classical photogrammetric topic. The development of DOM generation is heavily dependent upon the advances in computer techniques and optical--mechanical techniques. With the development of computer technology and software algorithms, the process of DOM generation in the different phases is different. It has evolved through the following stages.

1.6.1 Analogy Orthorectification

The earliest foundation of photo rectification emerged in 1898 with the development of zonal transformation. A patent for the "Photoperspektograph" was registered in 1903 (Hobbie 2010). In 1953, an orthorectifier, called "Orthophotoscope," was developed by Russell Bean (Hobbie 2010). With significant efforts by scientists from the USA and Germany, Carl Zeiss first announced the "Orthophotoskop Gigas-Zeiss" at the Photogrammetric Week in Karlsruhe in 1963 (Spennemann 1963) and the prototype of the GZ 1 orthoprojector in 1964 (Ahrend et al. 1964) (Figure 1.2).

1.6.2 Analytic Orthorectification

The emergence of computers marked the beginning of analytic photogrammetry and the development of analytic orthoprojectors. Wilson and Vlcek (1970) solved the coefficients of the projective transformation equation using analytical triangulation and adjustments for strips or blocks of

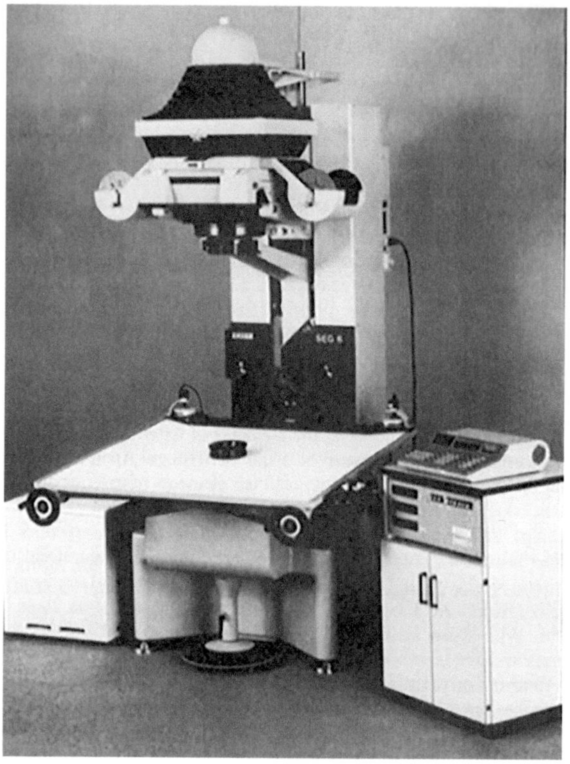

FIGURE 1.2 Carl Zeiss SEG 6 automatic rectifier. (Courtesy of Hobbie, 2010.)

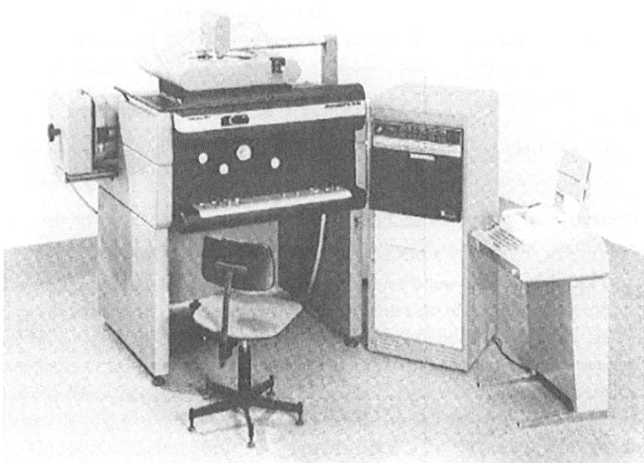

FIGURE 1.3 The Wild AVIOPLAN OR1 analytical orthoprojector. (Courtesy of Hobbie, 2010)

photographs. These coefficients could then be substituted into formulas that yield the settings of the rectifier. Formulae applicable to the Wild E4 Rectifier were also derived. Mugnier (1978) presented a technique that was operational in a production environment, enabling rectification by computing the coordinates of fiducials projected into the object space. Wang (1991) proposed a method that treats a small area as a rectification unit and rectifies each block by raising and lowering the photo-receiving surface according to changes in the elevation of the central point of the small area. This approach is called "differential rectification" or "orthoimages projection method" (Figure 1.3) (Wang 1991; McGlone et al. 2004).

1.6.3 Digital Orthorectification

With digital computers, photogrammetry entered the digital era, and digital orthorectification can be completely performed using a computer. Keating and Boston (1979) described a system using DEMs and aerial film negatives with a microdensitometer. Konecny (1979) showed that low-resolution aerial photos could also be rectified with DEMs, requiring a dedicated scanner and processor. Hossler (1980) introduced direct and indirect differential rectification methods for satellite, aerial, and digitized 2D images. Wiesel (1985) explored differential rectification using digital image processing. Mayr and Heipke (1988) proposed creating high-quality orthoimages by combining pixel-by-pixel methods with geomorphological data. Hood et al. (1989) developed a system that generates digital orthophotoquads by rectifying tile and relief distortions and mosaicking the corrected images. Novak (1992) compared three common rectification methods—polynomial, projective, and differential—using digitized aerial photos and video images on a workstation. Zheng et al. (1997) analyzed geometric distortions in shuttle imagery, derived a model based on incidence and azimuth angles, and developed an orthorectification method, successfully applied to two space photos, improving their compatibility with standard maps. Simard (1997) studied how DEM accuracy affects orthoimage accuracy, finding that even with a sevenfold scale difference between DEM and imagery, orthoimage accuracy did not significantly degrade.

1.6.4 True Orthophotomap (TOM) Generation

In the 1980s, due to challenges in performing orthorectification for tall buildings in cities (Zhou et al. 2004; Amhar et al. 1998; Biasion et al. 2004; Habib et al. 2007), traditional methods often resulted in ghost orthoimages (Mayr 2002; Nielsen 2004; Pala and Arbiol 2002; Rau et al. 2002), To address this issue, the so-called TOM generation was proposed by many researchers (Sheng et al. 2003; Siachalou 2004; Xie and Zhou 2006; Xie and Zhou 2009; Zhou 2009). Several studies have addressed the challenges posed by abrupt terrain elevation changes, such as those caused by buildings and bridges (Amhar et al. 1998; Schickier and Thorpe 1998; Xie and Zhou 2008; Zhou et al. 2017; Zhou et al. 2019). Jauregui et al. (2002) developed a method to orthorectify aerial imagery for the purpose of updating topographic surface maps. Applications of IKONOS satellite imagery include generating orthoimages for urban areas (Siachalou 2004) and monitoring volcanic activity on Nisyros Island in Greece (Vassilopoulou et al. 2002). Cameron et al. (2000) utilized ortho-corrected aerial photos to assess changes in Scottish pine populations, while Passini and Jacobsen (2004) focused on evaluating the geometric accuracy of high-resolution orthoimages. Additionally, Biasion et al. (2004) investigated approaches for the automated production of true orthoimages. Orthoimage generation generally includes occlusion and shadow detection and compensation (Zhou et al. 2020; Zhou et al. 2022a; Zhou et al. 2022b; Wang et al. 2022). Skarlatos (1999) showed that architectural occlusion reduces image quality and orthoimage accuracy. Zhou et al. (2016) proposed a method to detect occlusions in urban orthoimages by modeling the link between ghost images and building occlusion boundaries, using building displacements to identify occluded areas. Yoo and Lee (2016) developed a facet-based approach to generate realistic building orthoimages, detecting occlusions at the building surface level and recovering them using multiple images and detailed DBM data. Oliveira et al. (2018) introduced an occlusion detection method for true orthophoto mosaics using irregular point clouds to avoid interpolation errors. Sheng et al. (2003) adopted a model-driven approach to generate realistic forest orthoimages, addressing occlusion and distortion from trees. Zhou et al. (2005) filled occluded regions by merging nearby orthoimage blocks. Zhou et al. (2016) extended this by using overlapping "slave" orthoimages to fill occlusions in a "master" orthoimage, enabling complete and true orthoimages. Rau et al. (2002) proposed a technique to recover information in building shadows. Zhou and Sha (2020) introduced a method to detect rooftops and ground shadows simultaneously using a DBM. By calculating sun zenith and altitude angles from shadow boundaries, they mapped actual shaded areas on ghost images, independent of ground reflectivity and lighting. Leone and Distance (2007) improved shadow detection through better classification, segmentation, and localization. Makarau et al. (2011) developed a robust, adaptive shadow detection method usable across various sensors and lighting. Tiwari et al. (2015) proposed an algorithm that uses the light intensity-to-hue ratio and local thresholding for shadow detection and compensation, performing well especially on low- and medium-intensity images.

1.6.5 True² Orthoimages Map (T²OM) Generation

The generation of TOM does not consider building façade textures, which are widely applied in microenvironment analysis and the distribution of cell phone transmission stations. For this reason, Zhou et al. (2022b) first proposed a method for generating T²OM in 2022, which is completely different from traditional D/TOM. The basic idea for the T²OM generation of a single building is to orthorectify the DBM-based building roof from up to down, the building façade from front to back, from back to front, from left side to right side, and from right side to left side, as well as DTM-based T²OM, of which a superpixel is proposed to store building ID, texture ID, elevation of each pixel, and gray information. The details for T²OM generation can be referenced to Zhou et al. (2022b).

1.7 DIGITAL ORTHOPHOTO QUARTER QUADRANGLE (DOQQ)

The DOQQ program was carried out by the USGS for the creation of national digital orthophotos from aerial photographs. The primary DOQQ is a 1-m ground resolution, quarter-quadrangle (3.75 min of latitude by 3.75 min of longitude) image cast on the Universal Transverse Mercator Projection on the North American Datum of 1983 (NAD83) (McGlone et al. 2004). This material is based on the information excerpted from the USGS website.

The *U.S. Geological Survey National Mapping Division* is the primary federal agency responsible for producing *USGS Digital Orthophoto Quadrangles* (*DOQs*) at the national level. DOQs are coincident with the USGS 1:24,000-scale (1 in. = 2000 ft.) topographic quadrangle coverage, have been geometrically corrected to conform to a standard cartographic map projection, and possess a half × half meter ground spatial resolution. The statewide National Aerial Photography Program is the primary imagery source used to produce DOQs. DOQs depict physical and cultural features in far greater detail and are more current than published USGS quadrangle maps. DOQs can be readily combined with existing, older quadrangle map information in a GIS to produce an updated map of a desired geographic area.

DOQ data are widely used in government agencies, academia, and the private sector in many disciplines, such as urban planning, environmental assessment, crop science, agriculture, geology, soils science, watershed management, geography, landscape architecture, civil engineering, biology, and pollution prevention.

REFERENCES

Ahrend, M.; Brucklacher, W.; Meier, H.-K.; Utz, H., Der orthoprojektor gigas-zeiss. *Bildmessung Luftbildwesen*, 3, 1964, pp. 153–160.

Amhar, F.; Jansa, J.; Ries, C., The generation of true orthophotos using a 3D building model in conjunction with a conventional DTM. *The International Archives of the Photogrammetry, Remote Sensing and Spatial Information Sciences*, 32, 1998, pp. 16–22.

Biasion, A.; Dequal, S.; Lingua, A., A new procedure for the automatic production of true orthophotos. *The International Archives of the Photogrammetry, Remote Sensing and Spatial Information Sciences*, 35, 2004, pp. 1682–1777.

Cameron, A.; Miller, D.; Ramsay, F.; Nikolaou, I.; Clarke, G., Temporal measurement of the loss of native pinewood in Scotland through the analysis of orthorectified aerial photographs. *The Journal of Environmental Management*, 58, 2000, pp.33–43.

Federal Geographic Data Committee, *Development of a National Digital Geospatial Data Framework*. Federal Geographic Data Committee: Washington, DC, 1995.

Federal Geographic Data Committee, *Fact sheet: National digital geospatial data framework: A status report.* Federal Geographic Data Committee: Reston, VA, 1997.

Habib, A. F.; Kim, E.-M.; Kim, C.-J., New methodologies for true orthophoto generation. *Photogrammetric Engineering & Remote Sensing*, 73, 2007, pp. 25–36.

Hobbie, D., *The Development of Photogrammetric Instruments and Methods at Carl Zeiss in Oberkochen.* Verlag der Bayerischen Akademie der Wissenschaften: München, 2010.

Hood, J.; Ladner, L.; Champion, R., Image processing techniques for digital orthophotoquad production. *Photogrammetric Engineering & Remote Sensing*, 55, 1989, pp. 1323–1329.

Hossler, R., Differential rectification of digital or digitized imageries. *The International Archives of the Photogrammetry*, 23, 1980.

Jamil, A.; Bayram, B., Tree species extraction and land use/cover classification from high-resolution digital orthophoto maps. *IEEE Journal of Selected Topics in Applied Earth Observations and Remote Sensing*, 11(1), 2017, pp. 89–94.

Jauregui, M.; Vílchez, J.; Chacón, L., A procedure for map updating using digital mono-plotting. *Computers & Geosciences*, 28, 2002, pp. 513–523.

Kasser, M.; Egels, Y., *Digital Photogrammetry*. Taylor & Francis: London, 2002.

Keating, T. J.; Boston, D., Digital orthophoto production using scanning microdensitometers. *Photogrammetric Engineering & Remote Sensing*, 45, 1979, pp. 735–740.

Konecny, G., Methods and possibilities for digital differential rectification. *Photogrammetric Engineering & Remote Sensing*, 45, 1979, pp. 727–734.

Leone, A.; Distante, C., Shadow detection for moving objects based on texture analysis. *Pattern Recognition*, 40, 2007, pp. 1222–1233.

Liu, Y.; Zheng, X.; Ai, G.; Zhang, Y.; Zuo, Y., Generating a high-precision true digital orthophoto map based on UAV images. *The ISPRS International Journal of Geo-Information*, 7, 2018, p. 333.

Maitra, J.B., The national spatial data infrastructure in the United States: Standards; Metadata, Clearinghouse, and Data Access; *Federal Geographic Data Committee c/o US Geological Survey: Reston*, VA, USA, 1998

Makarau, A.; Richter, R.; Muller, R.; Reinartz, P., Adaptive shadow detection using a blackbody radiator model. *IEEE Transactions on Geoscience and Remote Sensing*, 49, 2011, pp. 2049–2059.

Mayr, W. Bemerkungen zum Thema, True orthoimage. *Photogrammetrie-Fernerkundung-Geoinformation*, 2002, pp. 237–244.

Mayr, W.; Heipke, C. A., contribution to digital orthophoto generation. *The International Archives of the Photogrammetry, Remote Sensing*, 27, 1988, pp. 430–439.

McGlone, J. C.; Mikhail, E. M.; Bethel, J. S.; Mullen, R., *Manual of Photogrammetry*, 6th ed.; American Society for Photogrammetry and Remote Sensing: Bethesda, MD, 2004.

Mugnier, J., Analytical rectification using artificial points. *Photogrammetric Engineering & Remote Sensing*, 44, 1978, pp. 579–583.

Nielsen, M. Ø., *True Orthophoto Generation*. Technical University of Denmark, DTU: Kgs. Lyngby, 2004.

Novak, K., Rectification of digital imagery. *Photogrammetric Engineering & Remote Sensing*, 58, 1992, pp. 339–339.

Oliveira, H. C.; Dal Poz, A. P.; Galo, M.; Habib, A. F., Surface gradient approach for occlusion detection based on triangulated irregular network for true orthophoto generation. *The IEEE Journal of Selected Topics in Applied Earth Observations and Remote Sensing*, 11, 2018, pp. 443–457.

Pala, V.; Arbiol, R., True orthoimage generation of urban areas. *GIM International*, 16, 2002, pp. 50–51.

Passini, R.; Jacobsen, K., Accuracy analysis of digital orthophotos from very high resolution imagery. *The International Archives of the Photogrammetry, Remote Sensing and Spatial Information Sciences*, 35, 2004, pp. 695–700.

Rau, J.-Y.; Chen, N.-Y.; Chen, L.-C., True orthophoto generation of built-up areas using multi-view images. *Photogrammetric Engineering & Remote Sensing*, 68, 2002, pp. 581–588.

Schickier, W.; Thorpe, A. Operational procedure for automatic true orthophoto generation. *International Archives of the Photogrammetry, Remote Sensing and Spatial Information Sciences*, 32, 1998, pp. 527–532.

Sheng, Y.; Gong, P.; Biging, G. S., True orthoimage production for forested areas from large-scale aerial photographs. *Photogrammetric Engineering & Remote Sensing*, 69, 2003, pp. 259–266.

Siachalou, S., Urban orthoimage analysis generated from IKONOS data. *The International Archives of the Photogrammetry, Remote Sensing and Spatial Information Sciences*, 14, 2004, pp. 12–23.

Simard, P. G., *Accuracy of Digital Orthophotos*. University of New Brunswick: Fredericton, New Brunswick, 1997.

Skarlatos, D., Orthophotograph production in urban areas. *Photogrammetric Record*, 16, 1999, pp. 643–650.

Spennemann, R., Bericht über die 9. Photogrammetrische Wochen 1963. *Bildmessung Luftbildwesen*, 4, 1963, pp. 206–209.

Tiwari, S.; Chauhan, K.; Kurmi, Y., Shadow detection and compensation in aerial images using MATLAB. *International Journal of Computer Applications*, 119, 2015.

Vassilopoulou, S.; Hurni, L.; Dietrich, V.; Baltsavias, E.; Pateraki, M.; Lagios, E.; et al., Orthophoto generation using IKONOS imagery and high-resolution DEM: a case study on volcanic hazard monitoring of Nisyros Island (Greece). *The ISPRS Journal of Photogrammetry and Remote Sensing*, 57, 2002, pp. 24–38.

Wang, Q.; Zhou, G.; Song, R.; Xie, Y.; Luo, M.; Yue, T., Continuous space ant colony algorithm for automatic selection of orthophoto mosaic seamline network. *The ISPRS Journal of Photogrammetry and Remote Sensing*, 186, 2022, pp. 201–217.

Wang, Z. a, *Principles of Photogrammetry*. Wuhan University Press: Wuhan, China, 1991.

Wiesel, J., Digital image processing for orthophoto generation. *Photogrammetria*, 40, 1985, pp. 69–76.

Wilson, K.; Vlcek, J., Analytical rectification. *Photogrammetric Engineering*, 36, 1970, pp. 570–575.

Wolf, P. R.; Dewitt, B. A.; Wilkinson, B. E., *Elements of Photogrammetry with Application in GIS*, 4th ed.; McGraw Hill: New York, 2004.

Xie, W.; Zhou, G., Experimental realization of urban large-scale true orthoimage generation. In: *Proceedings of the ISPRS Congress,* Beijing, China, 2008, pp. 3–11.

Xie, W.; Zhou, G., High-accuracy of orthorectification model with self-geometric constraint for high buildings in urban area. In: *2009 International Geoscience and Remote Sensing* Symposium, Cape Town, South Africa, 2009, pp. IV-394–IV-397.

Xie, W.; Zhou, G., Urban 3D building model applied to true orthoimage generation. In: *1st EARSeL Workshop on Urban Remote Sensing*, Beijing, China, 2006, pp. 2–3.

Yang, M.; Liu, J.; Zhang, Y.; Li, X. Design and construction of massive digital orthophoto map database in China. *International Archives of the Photogrammetry, Remote Sensing and Spatial Information Sciences,* 41, 2016, pp. 103–106.

Yoo, E. J.; Lee, D.-C., True orthoimage generation by mutual recovery of occlusion areas. *GIScience & Remote Sensing*, 53, 2016, pp. 227–246.

Zheng, Q.; Klemas, V.; Yan, X.-H.; Wang, Z.; Kagleder, K., Digital orthorectification of space shuttle coastal ocean photographs. *International Journal of Remote Sensing*, 18, 1997, pp. 197–211.

Zhou, G., Near real-time orthorectification and mosaic of small UAV video flow for time-critical event response. *IEEE Transactions on Geoscience and Remote Sensing*, 47, 2009, pp. 739–747.

Zhou, G.; Bao, X.; Ye, S.; Wang, H.; Yan, H., Selection of optimal building facade texture images from UAV-based multiple oblique image flows. *IEEE Transactions on Geoscience and Remote Sensing*, 59, 2020, pp.1534–1552.

Zhou, G.; Chen, W.; Kelmelis, J. A.; Zhang, D., A comprehensive study on urban true orthorectification. *IEEE Transactions on Geoscience and Remote Sensing*, 43, 2005, pp. 2138–2147.

Zhou, G.; Li, H.; Song, R.; Wang, Q.; Xu, J.; Song, B., Orthorectification of fisheye image under equidistant projection model. *Remote Sensing*, 14, 2022a, p. 4175.

Zhou, G.; Schickler, W.; Thorpe, A.; Song, P.; Chen, W.; Song, C., True orthoimage generation in urban areas with very tall buildings. *International Journal of Remote Sensing*, 25, 2004, pp. 5163–5180.

Zhou, G.; Sha, H., Building shadow detection on ghost images. *Remote Sensing*, 12, 2020, p. 679.

Zhou, G.; Wang, Q.; Huang, Y.; Tian, J.; Li, H.; Wang, Y., True^2 orthoimage map generation. *Remote Sensing*, 14, 2022b, p. 4396.

Zhou, G.; Wang, Y.; Yue, T.; Ye, S.; Wang, W., Building occlusion detection from ghost images. *IEEE Transactions on Geoscience and Remote Sensing*, 55, 2016, pp. 1074–1084.

Zhou, G.; Zhang, R.; Liu, N.; Huang, J.; Zhou, X., On-board ortho-rectification for images based on an FPGA. *Remote Sensing*, 9, 2017, p. 874.

Zhou, G.; Zhang, R.; Zhang, D.; Huang, J.; Baysal, O., Real-time ortho-rectification for remote-sensing images. *International Journal of Remote Sensing*, 40, 2019, pp. 2451–2465.

2 Error Sources of Images

2.1 INTRODUCTION

A digital orthophotomap (DOM) generation is a process of applying the mathematics of photogrammetry to remove all distortions caused by various types of errors. This processing is called "image orthorectification". These errors generally cause the point positions in the image plane to deviate from their right positions; thus, this type of error is sometimes called "geometric distortion" or "image geometric distortion". This book will not obviously distinguish the differences through text. In addition, for the generation of a high-quality DOM, it is necessary for people to first know what types of errors an image has and how these errors impact the distortions of the image, so that these errors can be effectively corrected or rectified by models.

Usually, various influences cause the targets of an image to be geometrically distorted. The sources of distortion come from Earth's curvature, film distortion, relief displacement, atmospheric refraction, camera lens, etc. These distortions are called systematic error in photogrammetry and can be modeled; therefore, they can be corrected using models. Additionally, most satellite images usually contain other kinds of geometric distortions due to orbit drift, attitude fluctuations of the satellite, flying velocity, and nonlinear motions of the scanning mirror. These errors can generally be considered minimal during short periods (Wolf et al. 2004). This chapter mainly describes a few classical errors that can be rectified.

2.2 IMAGERY GEOMETRIC DISTORTION

The imagery geometric distortion means that the target in the image plane is offset from its correct position. This section describes the imagery geometric distortion caused by camera lens distortion.

2.2.1 Lens Distortion

The existence of an ideal camera model is impossible. The deviation from the ideal model is defined as the *distortion* of the camera, which is modeled by (Zhou et al. 1998)

$$\Delta R = R' - R = f \cdot \left(\tan(\alpha') - \tan(\alpha)\right) \tag{2.1}$$

where R' denotes the distance measured from the principal point to the respective image point, f is the focal length of the camera, and α is the angle of incidence of a ray (see Figure 2.1). Equation (2.1) shows that the offset is a function of the focal length, the incidence angle, and the refracting angle. The following ratiocinations can be drawn up by analyzing equation (2.1):

1. If the incidence angle is equal to the refracting angle and the focal length is a constant, the offset of the image coordinate is zero. This is the so-called *distortion-free* imaging system or *pinhole model*.
2. If the focal length of the camera is a constant, the offset would vary with changes in the incidence angle. Obviously, the incidence angle of a wide-angle camera is bigger than that of a long focal length camera. This is why Weng et al. (1992) concluded that "wide-angle cameras (short focal length) are not beneficial for calibration accuracy improvement relative to long focal length cameras".

DOI: 10.1201/9781003632108-2

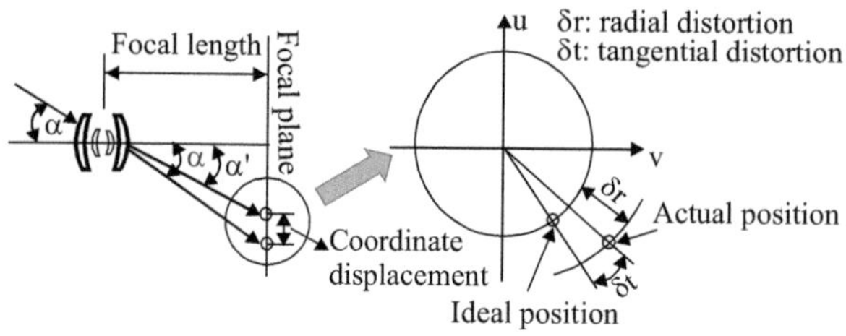

FIGURE 2.1 Radial and tangential distortion.

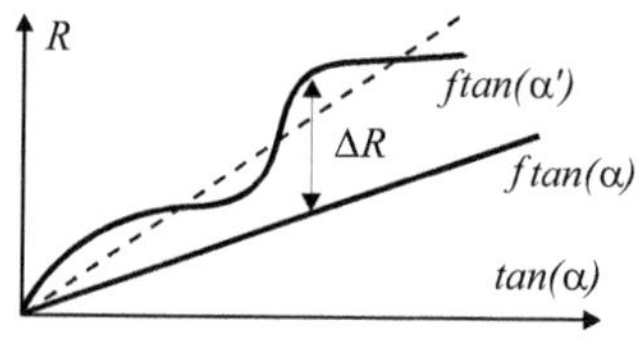

FIGURE 2.2 Distortion model choice.

3. Provided that both the focal length and the incidence angle are constants, the offset only depends on the refracting angle. This means that the thickness or thinness of the lens impacts the size of the geometric distortion of the imaging system.
4. If the incidence angle were equal to the refracting angle, the offset would vary with the change in the focal length of the camera. In other words, the offset will vary with different focal lengths in spite of the same camera. Thus, the identical camera with various zooms will result in various distortion models.

The choice of distortion model can be illustrated in a rectangular coordinate system in Figure 2.2, where the distortion values (i.e., deviations of the point locations from the ideal one in the image plane) are plotted along a vertical direction, and the abscissa represents the function of the incidence angle. It is not hard to discover from Figure 2.2 that the choice of distortion model is, in fact, to seek a function that can best fit the real curve.

Usually, three types of distortion models, radial, decentering, and thin prism, are considered. The components of offset in equation (2.1) can be expressed (Wei and Ma 1994; Weng et al. 1992) as

$$\delta_u(u,v)=\overbrace{\kappa_1 u\left(u^2+v^2\right)}^{radial}+\overbrace{\rho_1\left(3u^2+v^2\right)+2\rho_2 uv}^{decentering}+\overbrace{s_1\left(u^2+v^2\right)}^{thin\ prism}$$

$$\delta_v(u,v)=\overbrace{\kappa_1 v\left(u^2+v^2\right)}^{radial}+\overbrace{\rho_2\left(3u^2+v^2\right)+2\rho_1 uv}^{decentering}+\overbrace{s_2\left(u^2+v^2\right)}^{thin\ prism} \quad (2.2)$$

2.2.2 Radial Distortion

A type of distortion results in the actual coordinates in the image plane apparently departing from the ideal ones along the radial direction. This type of distortion is called *radial distortion*, which is caused by the flawed radial curvature of the lens elements. No matter whether the displacements are negative or positive, they are always symmetric about the optical axis. This type of distortion was modeled by

$$\delta r_r = k_1 u\left(u^2 + v^2\right) \quad (2.3)$$

$$\delta r_t = k_1 v\left(u^2 + v^2\right) \quad (2.4)$$

where k_1 is the coefficient of radial distortion. Almost all camera calibration methods first consider radial distortion. This is because the camera lens is a curved one whose thickness varies from the center of the lens to the edge of the lens. The refract rate also varies. Thus, the different rays of incidence (angle of incidence) have various rays of refraction. Thus, as a result of the radial component, rays of incidence at equal angles to the optic axis are associated with different distances in the image plane (measured from the intersection point of the optic axis to those of the respective imaging rays) depending on the azimuth. In fact, the radial component corresponds to an accidental tilt of the camera in taking the photograph of the test object (Abdel-Aziz 1975). Therefore, it should be able to be compensated by intentionally tilting the taking axis toward the object.

2.2.3 Decentering Model

The decentering model, also called *Conrady's model* (Abdel-Aziz 1975; Brown 1966), is caused by the optical centers of lens elements and is not strictly collinear. This is a type of asymmetrical lens distortion. Geometric optical computation based on the assumption of such asymmetric-point dislocation discloses the existence of a tangential component, δr_t, in addition to a radial one, δr_r. The components δr_r, δr_t are expressed as

$$\delta r_r = 3P_2 \sin(\varphi - \varphi_0) \quad (2.5)$$

$$\delta r_t = P_2 \cos(\varphi - \varphi_0) \quad (2.6)$$

$$P_2 = \left(j_1 r^2 + j_2 r^4 + \cdots\right) \quad (2.7)$$

Usually, we choose

$$P_2 = j_1 r^2 \quad (2.8)$$

Thus, equations (2.5) and (2.6) are rewritten as

$$\delta r_r = \rho_1\left(3u^2 + v^2\right) + 2\rho_2 uv \quad (2.9)$$

$$\delta r_t = \rho_2\left(u^2 + 3v^2\right) + 2\rho_1 uv \quad (2.10)$$

where $\rho_1 = -j_1 \sin(\varphi_0)$ and $\rho_2 = -j_1 \cos(\varphi_0)$. As a consequence, straight lines in object space may appear as curves in the image plane even if they pass the image center (see Figure 2.3). Thus, direct compensation of this tangential distortion effect by the operation of taking a photograph is impossible. In other words, only if an analytical method is employed, this type of distortion can be compensated.

2.2.4 Thin Prism

Scholer (1975) introduced the thin prism theory. The thin prism theory permits the derivation of a mathematical relationship between the radial and tangential components, in accordance with

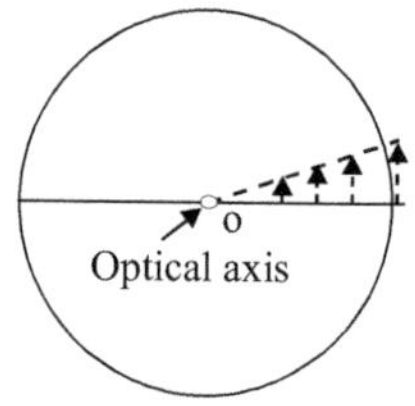

FIGURE 2.3 Tangential distortion.

practical experiments on optical systems. This distortion is also a type of asymmetrical lens distortion. The components δr_r and δr_t can be expressed as

$$\delta r_r = P_1 \sin(\varphi - \varphi_0) \quad (2.11)$$

$$\delta r_t = P_1 \cos(\varphi - \varphi_0) \quad (2.12)$$

$$P_1 = \left(i_1 r^2 + i_2 r^4 + \cdots\right) \quad (2.13)$$

where φ_0 is the angle between the positive u axis (see Figure 2.2) and the axis of maximum tangential distortion. i_1, i_2, and i are constants. δr_r is the radial component of asymmetrical lens distortion. δr_t is the tangential component of asymmetrical lens distortion. Usually, the equation can be rewritten as (McGlone et al. 2004)

$$\delta r_r = s_1 \left(u^2 + v^2\right) \quad (2.14)$$

$$\delta r_t = s_2 \left(u^2 + v^2\right) \quad (2.15)$$

where $s_{1=-}i_1 \sin(\varphi_0)$, $s_{2=-}i_1 \cos(\varphi_0)$, and u, v are the components of coordinates along the u, v axes.

As analyzed above, the decentering model and the thin prism model reveal one important fact that the radial component and the tangential component of asymmetrical lens distortion are correlated, and the tangential component should be smaller than the radial distortion.

Some researchers focused on calibration algorithms including computational processes and sensitivity analysis. It should be noted that the depth from the center of camera to control field, imaging patterns (such as intersection stereo pair and parallel stereo pair), data error of ground control point, illuminated condition and voltage stability, except for camera's type (focal length, field angle, and image resolution) (Weng et al. 1992), distributions of the control points (Wang 1991), and image coordinate error discussed in this paper also influence calibration accuracy to some extent. Additionally, this paper has not considered the choice of radiometric distortion.

2.3 EARTH CURVATURE DISTORTION AND CORRECTION

The primary problem with the earth curvature distortion is that imaged ground features are on the earth surface, which is referenced to an approximately spherical datum (i.e., mean sea level), whereas the rectified orthoimage and rectification mathematical model (i.e., mathematical equations) are referenced to a projected coordinate system, such as Universal Transverse Mercator (UTM), or State Plane Coordinate System, which assume that the projected frame is a horizontal plane (Wolf et al. 2004).

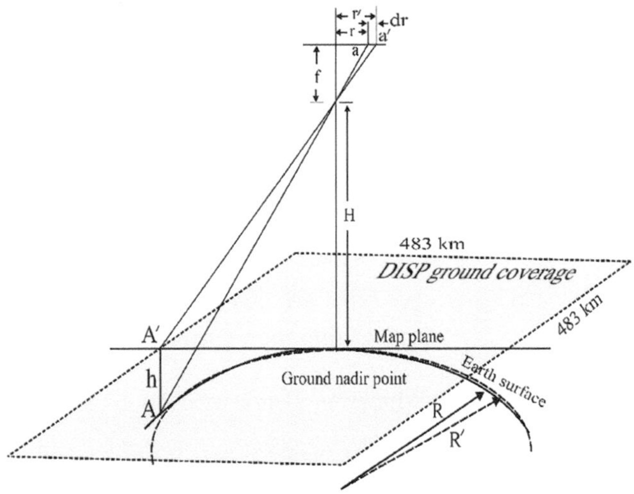

FIGURE 2.4 Model for Earth's curvature on imagery. (Zhou et al. 2002).

The mathematical model is commonly used to correct the measured image coordinates for compensation of the effects of the Earth's curvature. This method models the Earth's curvature distortion using flying height, focal length of the camera, radial distance from the principal point to the image point, and the Earth's radius.

Figure 2.4 illustrates the principle of this correction method. Point A is a ground point and A' is its corresponding position in a map plane, which is tangent to the earth at the ground nadir point. Points a and a' are the images of Points A and A', respectively. From Figure 2.4, the distortion, dr, for earth curvature is (Wolf et al. 2004)

$$dr = \frac{Hr^3}{2Rf^2} \tag{2.16}$$

where H is the flying height above ground, r is the radial distance from the principal point to the image point, R is Earth's radius, and f is the camera focal length.

It has widely been recognized that the practice of making earth curvature correction to measured image coordinates is not theoretically high, but it has been accepted that using the correction generally leads to more accurate coordinates than ignoring it (Wolf et al. 2004; Zhou et al. 2002).

However, the impact of earth curvature correction is more apparent in polar areas than in the equatorial area. For example, the early investigation on the orthorectification of the Declassified Intelligence Satellite Photography of Greenland, which appeared in the 1960s with 483 km by 483 km coverage, demonstrated that the Earth's curvature distortion is significant. The maximum distortion caused by Earth's curvature on corner points of the photograph achieves 2.2 m, which is equivalent to 9 km in ground distance, inward from its actual position (Zhou et al. 2002) because of the extremely small ratio of flying height to focal length (Figure 2.5).

On the other hand, the correction accuracy depends upon the map projection used. For example, when a UTM projection is used, the application of earth curvature corrections will yield more accurate elevation values and Y coordinate values, but will degrade X coordinate values since the UTM projection does not curve in the X direction. If the State Plane Coordinate System is used in object space, the axis system is also curved, although it is curved primarily in only one direction (either east-west or north-south). Thus, one sometimes employs a proper approach, such as a geocentric coordinate system, to avoid the need for any sort of earth curvature correction.

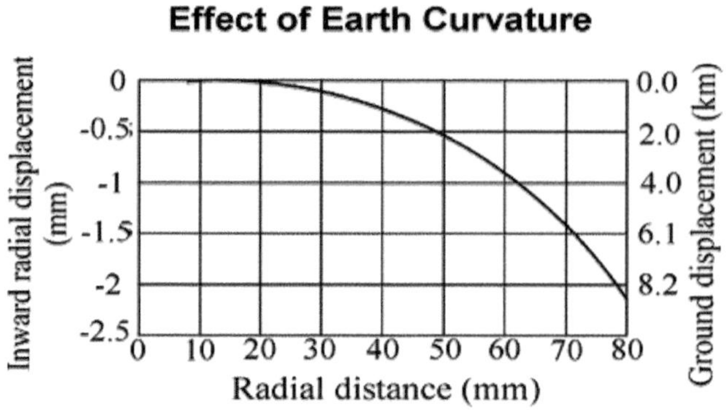

FIGURE 2.5 Effect of Earth's curvature on imagery. (Zhou et al. 2002).

2.4 ATMOSPHERIC DISTORTION

The primary problem of atmospheric distortion is that the density (and hence the index of refraction) of the atmosphere decreases with increasing altitude, so light rays do not travel in straight lines through the atmosphere, but rather they are bent according to Snell's law (Wolf et al. 2004). For example, as shown in Figure 2.1, the angular distortion due to atmospheric refraction is denoted θ, the relationship between the angular distortion θ and angle a, which is the angle between the vertical and the ray of light, is

$$\theta = K \tan\alpha \tag{2.17}$$

where K is a value that depends on the flying height above mean sea level and the elevation of the ground point of the Earth's surface. Different approaches are used to calculate the value of K. The Manual of Photogrammetry (McGlone et al. 2004) recommends the equation of computing K by

$$K = \left(7.4\times10^{-4}\right)(H-h)\left[1-0.02(2H-h)\right] \tag{2.18}$$

where H is the flying height of the aircraft above the mean sea level in kilometers and h is the elevation of the object point above mean sea level in kilometers; K is with units of degree.

Also from Figure 2.6, we have

$$\tan\alpha = \frac{r}{f} \tag{2.19}$$

where r is radial and f is the focal length of a camera. Substituting equation (2.19) into equation (2.17), to compute the refraction angle $\Delta\alpha$

$$\Delta\alpha = K\frac{r}{f} \tag{2.20}$$

The radial distance r' from the principal point to the corrected image location can then be computed by

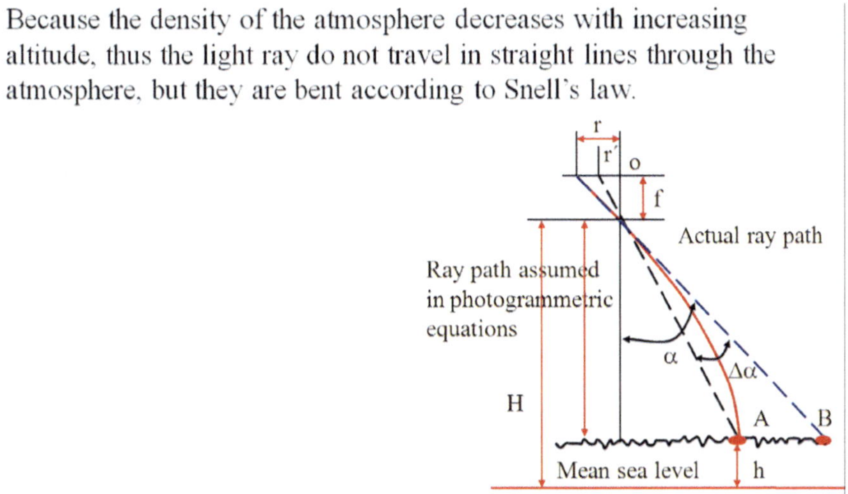

FIGURE 2.6 Atmospheric refraction and image coordinate displacement.

$$r' = f\tan(\alpha - \Delta\alpha) \tag{2.21}$$

The error in radial distance δ_r is then computed by

$$\delta r = r - r' \tag{2.22}$$

The x and y components of atmospheric refraction distortion corrections (δ_x and δ_y) can then be computed by

$$\delta x = x' \frac{\Delta r}{r} \tag{2.23}$$

$$\delta y = y' \frac{\Delta r}{r} \tag{2.24}$$

The early investigations have found that radial displacement on the photograph due to the effect of atmospheric refraction was smaller than 1.2 μm (about 7 m in the ground plane) because of the extremely small ratio of flying height to focal length (Wolf et al. 2004).

2.5 RELIEF DISPLACEMENT

Relief displacement, as implied by its name, is a shift or displacement of an imaged point offset its right position in the photographic position (Wolf et al. 2004). This relief displacement is mainly caused by the relief of the object, i.e., its elevation above or below a selected datum.

The geometric illustration for the concept of relief displacement is depicted in Figure 2.7, in which a vertical photograph is taken from a flying height of H above the given datum. For a vertical object AA' with a height of h_A, the ending point A' is located on the datum. The points A and A' are imaged on the image planes, and a', respectively. Draw a radial line from the principal point o, to the image point a, which passes through the image point a'. Denote r and r' as the distances from

4. It has also been shown that relief displacement occurs radically from the principal point.

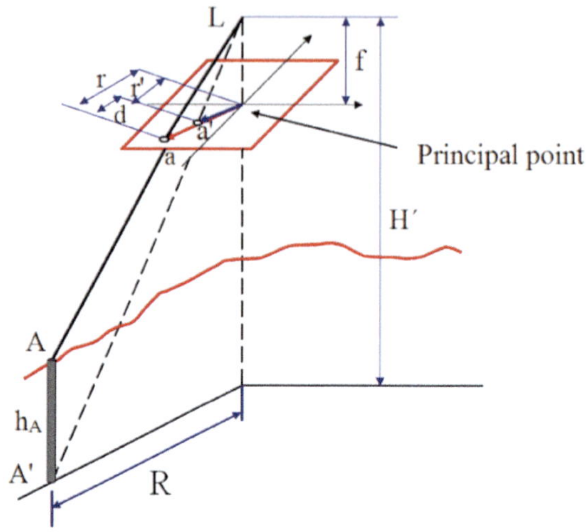

FIGURE 2.7 Computation of relief displacement above a datum.

the principal point o to the image points a and a', respectively. The distance difference between r and r' is the relief displacement, which is caused by the height of h_A. The mathematical model of calculating the relief displacement on a vertical photograph can be realized as follows.

As seen, trianglesLao and LAA_o are similar triangles, so we have

$$\frac{r}{R} = \frac{f}{H - h_A} \tag{2.25}$$

Rearranging equation (2.25) gives

$$r(H - h_A) = fR \tag{2.26}$$

Also, triangles$La'o$ and $LA'P$ are similar triangles, so we have

$$\frac{r'}{R} = \frac{f}{H} \tag{2.27}$$

Rearranging equation (2.27) gives

$$r'H = fR \tag{2.28}$$

Equating expressions (2.25) and (2.28) yields

$$r'H = r(H - h_A) \tag{2.29}$$

As noted above, $d = r - r'$. Rearranging the above equations, and substituting the symbol d for $r - r'$, gives

$$d = \frac{rh_A}{H} \tag{2.30}$$

where d is the relief displacement, h is the height above datum of the object point whose image is displaced, r represents the radial distance on the photograph from the principal point to the displaced image (the units of d and r must be the same), and H is the flying height above the same datum selected for measurement of h.

As analyzed from equation (2.30) and Figure 2.7, it can be found that the characteristics of relief displacement are as follows:

1. With respect to a datum, relief displacement is outward for points whose elevations are above datum or inward for points whose elevations are below datum.
2. Relief displacement increases with increasing radial distance to the image. Relief displacement occurs radically from the center of the photograph (principal point). This radial pattern is also readily apparent for the relief displacement of all the other vertical buildings in the photo (Figure 2.8). If a building in the center is one of the tallest images in the photo, its relief displacement is essentially close to zero due to its position near the principal point.
3. Relief displacement increases with increased elevation of the object point above datum (Figure 2.9). So, the vertical height of an object such as buildings appearing on aerial photographs can be calculated from relief displacements by $h = \frac{dH}{r}$.
4. Relief displacement decreases with increased flying height above datum (Figure 2.10). So, for the same height of building, relief displacement in spaceborne-based images is smaller than that in airborne-based images.

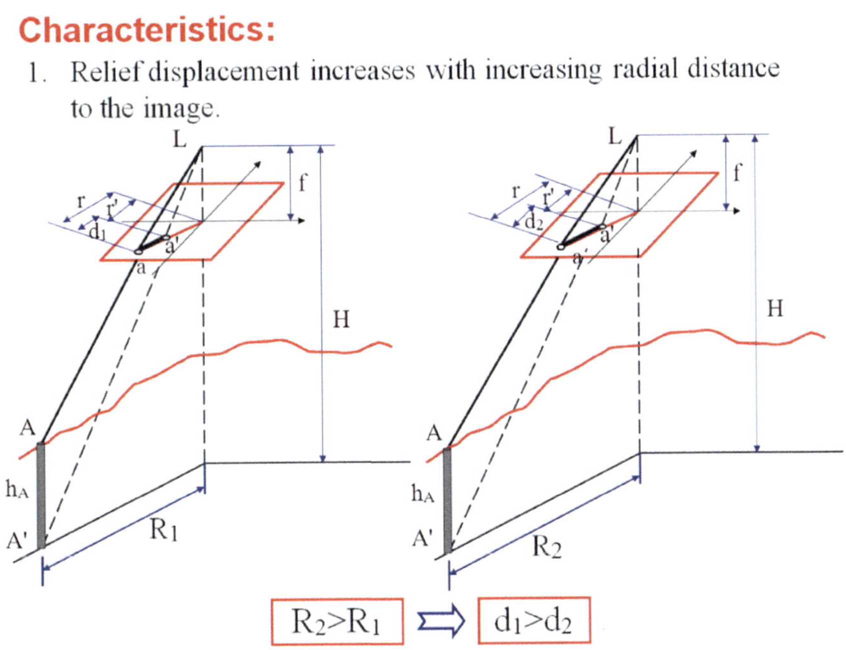

FIGURE 2.8 Various relief displacements with different distances away the ground nadir point above a datum.

2. It also increases with increased elevation of the object point above datum.

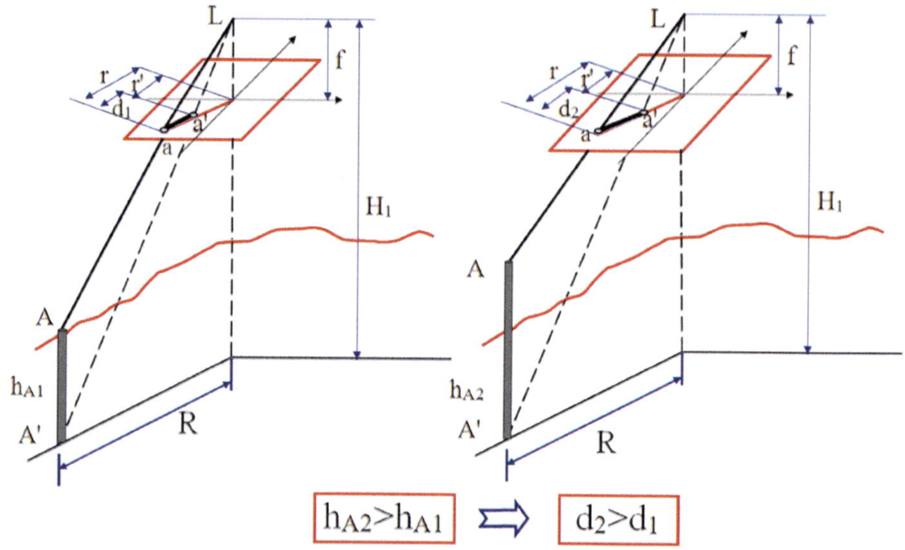

FIGURE 2.9 Various relief displacements with different building highs above a datum.

3. Relief displacement decreases with increased flying height above datum.

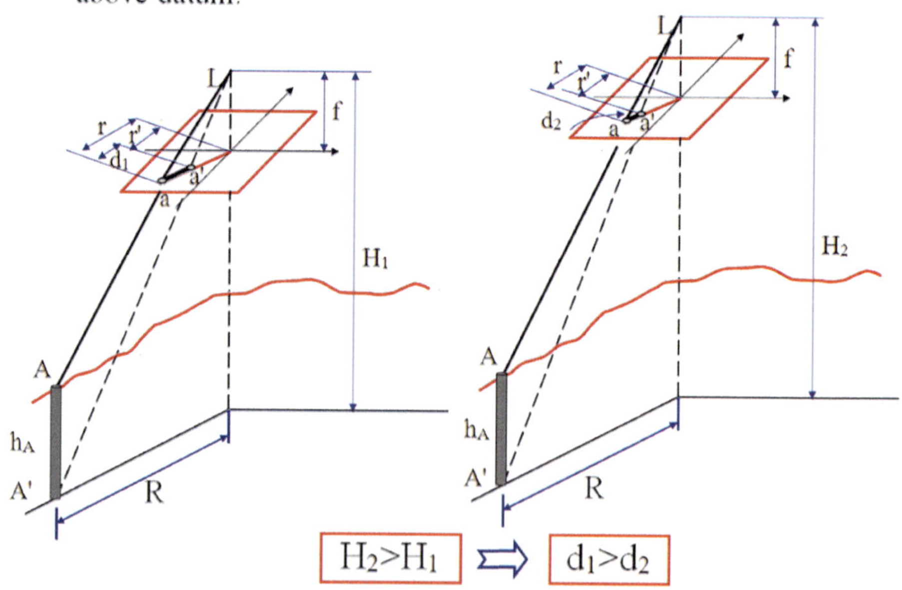

FIGURE 2.10 Relief displacement related to flying height.

REFERENCES

Abdel-Aziz, Y. L., Asymmetrical lens distortion. *Photogrammetric Engineering and Remote Sensing*, 41, 1975, 337.

Brown, D. C., Decentering distortion of lenses. *Photogrammetric Engineering*, 32, 1966, pp. 444–462.

McGlone, J. C.; Mikhail, E. M; Bethel, J. S; Mullen, R., *Manual of Photogrammetry*, 6th ed.; American Society for Photogrammetry and Remote Sensing: Bethesda, MD, 2004.

Scholer, H., On photogrammetric distortion. *Photogrammetric Engineering and Remote Sensing*, 41(6), 1975, pp. 721–728.

Wang, Z., *Principles of Photogrammetry*, English ed.; Wuhan University Press: Wuhan, 1991.

Wei, G. Q.; Ma, S. D., Implicit and explicit camera calibration: theory and experiments, *IEEE Transactions on Pattern Analysis and Machine Intelligence*, 16(5), May 1994, pp. 469–480.

Weng, J.; Cohen, P.; Herniou, M., Camera calibration with distortion models and accuracy evaluation. *IEEE Transactions on Pattern Analysis and Machine Intelligence*, 14, 1992, pp. 965–981.

Wolf, P.; Dewitt, B.; Wilkinson, B., *Elements of Photogrammetry with Application in GIS*, 4th ed.; McGraw Hill: New York, 2004.

Zhou, G.; Jezek, K.; Wright, W.; Rand, J.; Granger, J., Orthorectification of 1960s satellite photographs covering Greenland. *IEEE Transactions on Geoscience and Remote Sensing*, 40, 2002, pp. 1247–1259.

Zhou, G.; Uzi, E.; Feng, W.; Yuan, B., Charge-coupled device camera calibration based on natural landmark. *Pattern Recognition*, 31, 1998, pp. 1715–1724.

Zhou, G.; Yuan, B.; Tang, X., The intrinsic and extrinsic factor of charge-coupled device calibration. *Acta Electronica Sinica*, 24, 1996, pp. 12–17.

3 Principle of Orthorectification

3.1 INTRODUCTION

The orthorectification is, in fact, to transform a perspective projection to an orthogonal projection, meanwhile to remove the distortions that are inherent in aerial photography. This process is called *orthorectification*. Orthorectification models range from a simple 2D-to-2D affine transformation, which utilize linear or higher-order polynomials to model the distortions, to 3D-to-2D projective transformations, e.g., differential rectification for relief displacement. The early geometric correction methods can, in general, be divided into two categories (Wolf et al. 2004; McGlone et al. 2004):

1. The first approach is called the *parametric approach*. This method does not need to recover the sensor orientation when orthorectifying an image. In this method, ground control points (GCPs) are collected through identifying points that are coincident on both the image and the corresponding map. Once enough GCPs are collected, 2D-2D affine transformation model, which describe a map relationship between 2D image space and 2D map space, is built and is solved using the least square solution. Some models, such as second-order polynomial functions (e.g., Leckie 1980; De Leeuw et al. 1988), interpolative splines (e.g., Baker et al. 1975; Goshtasby 1988), and piecewise continuous surfaces (e.g., Craig and Green 1987; Devereux et al. 1990; Novak 1992; Roy 1997; Zheng et al. 1997), have been used. The advantages of this method are simple, no requirement for sensor's interior and exterior orientation parameters, and fast for orthorectifying image. One of the major disadvantages is low accuracy of orthorectified image (digital orthophotomap, DOM).
2. The second approach is called the *nonparametric approach*. This method needs to recover a rigorous sensor imaging model, i.e., the spatial relationships between an image pixel and its conjugate ground point are characterized by the imaging geometry, which is generally described by the collinearity equation in photogrammetry. Before applying this method, the orientation parameters should be recovered through the solution of collinearity equation on the basis of GCPs. The advantages of this method are high accuracy of orthorectified image (DOM), strict orthorectification process, recovery of sensor imaging mechanism, etc. One of the major disadvantages of this method is that the image space and object space must be reversible, but not all images are in practice reversible (Zhou et al. 2005).

Which approach is used depends upon several factors, such as DOM accuracy required, difficulty of collection of GCPs (e.g., homogeneous area in Greenland ice sheet) or unstructured scene content in ocean or jungle, and labor-intensive.

3.2 ORTHOGONAL PROJECTION VERSUS PERSPECTIVE PROJECTION

To deeply understand the orthorectification, this section first introduces the definitions of *orthogonal projection* and *perspective projection*. The orthogonal projection (also called *orthographic projection* or *parallel projection*) is that the points are projected onto a plane along directions perpendicular (90°) to the target surface (Figure 3.1a). For example, a map is a reproduction, at a reduced scale, of an orthographic projection of the terrain onto a reference datum plane. The major characteristics of an orthogonal project are as follows:

1. Orthogonal projection maps point onto a surface along perpendicular directions, preserving true scale without perspective effects.

DOI: 10.1201/9781003632108-3

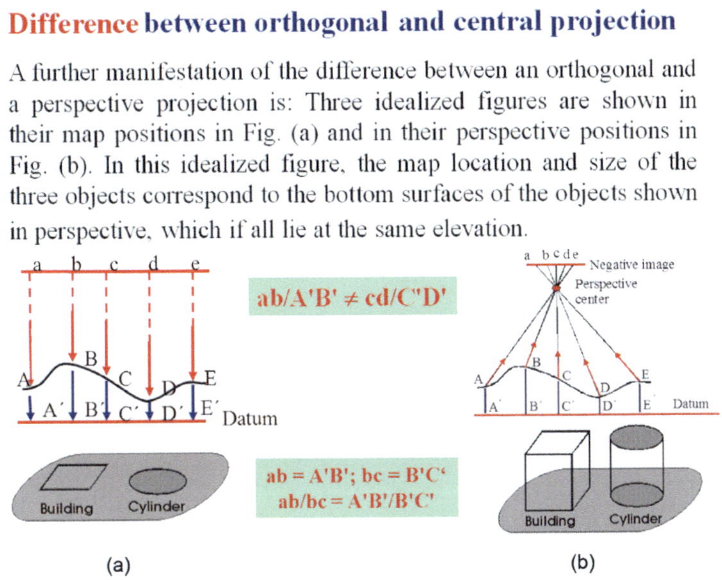

FIGURE 3.1 (a) Orthogonal projection versus (b) perspective projection.

2. Each point on the terrain is seen to be projected normal to the reference plane.
3. A distance measured between any two points on the map when multiplied by the scale of the map will be equal to the corresponding horizontal distance measured directly in the field.

Perspective projection (also called *central projection*) is the one in which all points are projected onto the reference plane through one point called the *perspective center* (Figure 3.1b). Its major characteristics (McGlone et al. 2004):

1. In photography, the sensitized film occupies the reference plane where the negative is formed behind the perspective center.
2. If all the projected points were to lie in one plane parallel to the projection plane, the resulting projection would be a scaled reproduction of the original points.
3. The points lie at different elevations, and then the projected points would not bear the same relation to one another as the original points.

A further manifestation of the difference between an orthogonal and a perspective projection is shown below. Three idealized figures are shown in their map positions in Figure 3.1a and in their perspective positions in Figure 3.1b. In this idealized figure, the map location and size of the three objects correspond to the bottom surfaces of the objects shown in perspective, which of all lie at the same elevation (Wolf et al. 2004).

3.3 2D-TO-2D AFFINE TRANSFORMATION FOR ORTHORECTIFICATION

This section focuses on the model and method of 2D–2D affine transformation, which has typically been applied in image-to-map geometric rectification for removing geometric distortion of remotely sensed image data. Two basic operations must be performed to geometrically rectify a remotely sensed image to a given map coordinate system (Wang 1991):

1. The geometric relationship between the input pixel coordinates (column and row) and the associated map coordinates of this same point (X, Y) must be established. A number of GCP pairs are used to establish the nature of the geometric coordinate transformation that must be applied to rectify every pixel in the output orthoimage from the surrounding

pixels of the original image. This process is called space interpolation (Wolf et al. 2004; McGlone et al. 2004).

2. Pixel brightness values must be determined. There is no direct one-to-one relationship between the pixel gray of input image position and the pixel gray of output image position. This means that a pixel in the rectified output image often requires a value from the input pixel grid that does not fall neatly on row- and column-coordinates. When this occurs, interpolation method must be used for determining the gray value to be assigned to the output rectified pixel. This process is called intensity interpolation (Wolf et al. 2004; McGlone et al. 2004).

The details for image-to-map rectification process involved the steps below:

1. Establishing the 2D–2D affine transformation model describing image-to-map relations. The simplest model is linear transformation, i.e.,

$$X = T_X + a_1 I + b_1 J \tag{3.1}$$

$$Y = T_Y + a_2 I + b_2 J \tag{3.2}$$

   where X and Y represent geodetic coordinates in a given map coordinate system and I and J represent image coordinates, *i.e.*, row and column of an image; (T_X, T_Y), (a_1, b_1), (a_2, b_2) are coefficients of the affine transformation equations, which should be solved using the least square solution on the basis of a sufficient number of GCPs. Once computed, these parameters are used to transform coordinates from the ground system to the image system. However, before applying the coefficients to create the rectified output image, it is important to determine how well the six coefficients derived from the least-squares regression of the initial GCPs account for the geometric distortion in the input image. The method used most often involves the computation of the root-mean-square error for each of the GCPs (Ton and Jain 1989).
2. Selecting GCP image pixel coordinates (row and column) with their corresponding map coordinate counterparts (e.g., meters northing and easting in a Universal Transverse Mercator map projection). A sufficient number of GCPs are needed. For example, for equation (3.1), at least three GCPs are needed for solution of six coefficients (T_X, T_Y; a_1, b_1; a_2, b_2).
3. Orthorectifying the imagery. The process of orthorectifying (sometimes called georeferencing) an image involves two fundamental steps: (1) inputting the parameters of a two-dimensional affine transformation, which are solved and (2) filling gray in the resulting image, which uses the appropriate digital numbers that brightness of the ground at the corresponding locations. The usual method for intensity interpolation, which in practice is commonly referred to gray resampling, has (Wang 1991)
   - Nearest neighbor,
   - Bilinear interpolation, and
   - Cubic convolution.

The details of the methods for gray interpolation are below.

Nearest-neighbor interpolation: Using nearest-neighbor interpolation, the brightness value closest to the specified row and column (I, J coordinate) is assigned to the output row and column (I', J') coordinate. However, there are nearby values at the integer grid intersections. Distances to neighboring pixels are computed using the Pythagorean theorem. A nearest-neighbor rule would assign the output pixel (I', J') value, which is the value found at the nearest input pixel (Wolf et al. 2000; McGlone et al. 2004).

Bilinear interpolation: Bilinear interpolation assigns output pixel values by interpolating gray values in two orthogonal directions in the input image. It basically fits a plane to the four-pixel values nearest to the desired position (I, J) in the input image and then computes a new gray value based on the weighted distances to these points. This method usually acts as a spatial moving filter that subdues extremes in brightness value throughout the output image (Wolf et al. 2000; McGlone et al. 2004).

Cubic convolution interpolation: This method assigns gray values to output pixels in much the same manner as bilinear interpolation, except that the weighted values of 16 input pixels surrounding the location of the desired (I, J) pixel are used to determine the value of the output pixel (I', J') (Wolf et al. 2000; McGlone et al. 2004).

Sometimes, it is difficult to gain an appreciation of exactly describing the relationship between image and map using the six coefficients in an affine (linear) transformation. This is because the six coefficients cannot completely describe all geometric distortion of an image, such as distortions caused by lens, relief displacements caused by elevation above a given datum, and distortions caused earth curvature. Therefore, it is usually for people to use a higher-order polynomial transformation to model these distortions; further, the solved coefficients in the higher-order polynomial transformation model are used to rectify the image. For example, a second-order (quadratic) polynomial equation is expressed by

$$X = T_X + a_1 I + b_1 J + c_1 IJ + d_1 I^2 + e_1 J^2 \tag{3.3}$$

$$Y = T_Y + a_2 I + b_2 J + c_2 IJ + d_2 I^2 + e_2 J^2 \tag{3.4}$$

where the symbols are the same as above. In theory, the higher the order of the polynomial, the more closely the coefficients should model the geometric error in the original (unrectified) input image and place the pixels in their correct planimetric positions in the rectified output matrix. Higher-order polynomials often produce a more accurate fit for areas immediately surrounding GCPs (Wolf et al. 2004; McGlone et al. 2004).

A general rule of thumb is to select a higher-order polynomial (e.g., second or third order) to model the image geometric distortions, where different types of error are often found during imaging processing, such as aerial platforms where roll, pitch, and yaw by the aircraft introduce unsystematic, nonlinear distortions; lens distortions by radial distortion, decentering model, etc. These distortions can only be modeled using a higher-order polynomial (Buiten and Van Putten, 1997). In other words, a first-order affine rectification was not sufficient to rectify the significant geometric errors caused by the aircraft drifting off course during data collection (Wolf et al. 2004; McGlone et al. 2004).

The typical application of affine transformation, either linear or higher-order polynomial equation, is to perform an image-to-map orthorectification, which is widely used in many commercial software. This process makes the geometry of an image be planimetric, but with a low-accuracy of orthorectification (Wolf et al. 2004).

3.4 DIFFERENTIAL ORTHORECTIFICATION

The affine transformation, either linear or higher-order polynomial equation, cannot effectively model all of the geometric distortions of an image, resulting in an incomplete orthorecification. This is because

1. these coefficients (or called *parameters*) in the affine transformation equation cannot correctly model the different types of geometric distortions of an image, resulting in many errors, typically systemic errors remain;

2. the elevation information of Earth's surface and buildings is not used, i.e., relief displacement is not completely orthorectified;
3. the imaging mechanism of an image (also called sensor model) is not modeled, i.e., a sensor's inherent systemical error is not modeled;
4. the coefficients solved using an affine transformation model are only used for orthorectifying one image, i.e., different images usually need to establish individual models, resulting in seam errors of neighbor DOM when mosaicking them.

Therefore, the affine transformation method is only used for low-accuracy of orthorectification. A rigorous orthorectification method in photogrammetry is called *differential orthorectification,* which rigorously model sensor imaging process, thus many random errors and systematic errors can rightly be corrected.

3.4.1 Establishment of Differential Orthorectification Model

The traditional differential rectification model is based on photogrammetric collinearity, in which the interior and exterior orientation elements and digital elevation model (e.g., *X*-, *Y*-, and *Z*-coordinates of each grid) are known. Moreover, the model is for a single frame image, i.e., each image must have its own coordinate system and its own GCPs. The image coordinates and the associated projective center of a single image define a *spatial bundle of rays.* The advantage of the model is that the relief displacement is effectively corrected (Wolf et al. 2004). The input to the adjustment system includes the measured image coordinates of features that are identifiable in images, together with the coordinates of some of the GCPs. The system outputs the exterior orientation parameters of the images. If we consider the interior orientation parameters as unknown parameters, the model is further expanded into the so-called *self-calibration block bundle adjustment* (Wang 1991; Wolf et al. 2004). Figure 3.2 illustrates the principle of an adjustment model.

With Figure 3.2, for any point *P*, the collinearity equation is expressed as (Wolf et al. 2004)

$$x_p - x_0 = -f\frac{r_{11}(X_P - X_S) + r_{12}(Y_P - Y_S) + r_{13}(Z_P - Z_S)}{r_{31}(X_P - X_S) + r_{32}(Y_P - Y_S) + r_{33}(Z_P - Z_S)} = f_x \tag{3.5}$$

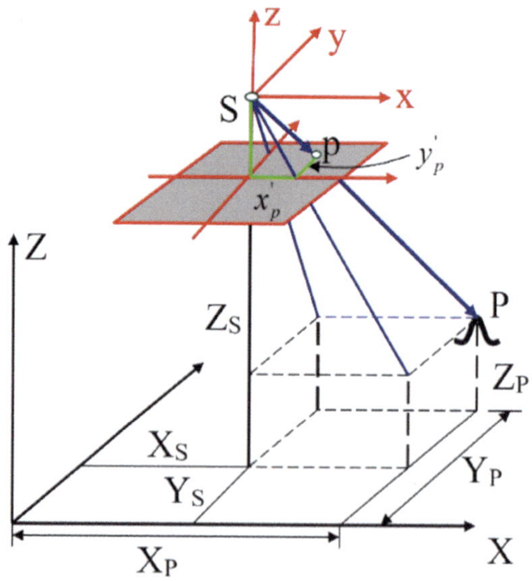

FIGURE 3.2 Collinearity condition of a ground point, corresponding image point, and perspective center.

$$y_p - y_0 = -f\frac{r_{21}(X_P - X_S) + r_{22}(Y_P - Y_S) + r_{23}(Z_P - Z_S)}{r_{31}(X_P - X_S) + r_{32}(Y_P - Y_S) + r_{33}(Z_P - Z_S)} = f_y \tag{3.6}$$

where (x_p, y_p) are image coordinates; (x_0, y_0, f) are interior orientation parameters; (X_P, Y_P, Z_P) are ground coordinates of a control point (GCP), which may be treated as knowns and unknowns differently in various situations depending upon the types of observations; $r_{i,j}$ (i = 1, 2, 3; j = 1, 2, 3) are components of the rotation matrix R, where

$$R = \begin{pmatrix} \cos\phi\cos\kappa & \cos\omega\sin\kappa + \sin\omega\sin\phi\cos\kappa & \sin\omega\sin\kappa - \cos\omega\sin\phi\cos\kappa \\ -\cos\phi\sin\kappa & \cos\omega\cos\kappa - \sin\omega\sin\phi\sin\kappa & \sin\omega\cos\kappa + \cos\omega\sin\phi\sin\kappa \\ \sin\phi & -\sin\omega\cos\phi & \cos\omega\cos\phi \end{pmatrix} \tag{3.7}$$

ϕ, ω, and κ are three rotation angles (see Wolf et al. 2004 or Figure 3.2).

As mentioned above, the film and atmospheric refraction can be corrected by models in advance. The lens distortions can be modeled by the following equations (Zhou, Uzi et al. 1998; Zhou, Uan et al. 1996)

$$\delta x = -x'_p\left(\kappa_1 r + \kappa_2 r^3\right)$$

$$\delta y = -y'_p\left(\kappa_1 r + \kappa_2 r^3\right) \tag{3.8}$$

where (x'_p, y'_p) are the distortion-free image coordinates; κ_1 and κ_2 are distortion parameters; r is the radial distance. Thus, the image coordinates to be inputted into the collinearity equations are

$$x_p = x'_p + \delta x - x_0$$

$$y_p = y'_p + \delta y - y_0 \tag{3.9}$$

Since the modified collinearity equation (3.5) is not linear, the equation must be linearized with respect to the parameters in order to build the observation equation for least-square adjustment. The linearized equation can be expressed as

$$v_1 = A_1X_1 + A_2X_2 - L \tag{3.10}$$

where X_1 represents exterior orientation parameters, X_2 denotes interior orientation and lens distortion parameters, A_1, and A_2 are their coefficients, and v_1 is the residual error. The partial derivatives with respect to the unknowns are (for details, see Appendix A)

$$A_1 = \begin{bmatrix} \frac{\partial f_x}{\partial X_P} & \frac{\partial f_x}{\partial Y_P} & \frac{\partial f_x}{\partial Z_P} & \frac{\partial f_x}{\partial \phi} & \frac{\partial f_x}{\partial \omega} & \frac{\partial f_x}{\partial \kappa} \\ \frac{\partial f_y}{\partial X_P} & \frac{\partial f_y}{\partial Y_P} & \frac{\partial f_y}{\partial Z_P} & \frac{\partial f_y}{\partial \phi} & \frac{\partial f_y}{\partial \phi} & \frac{\partial f_y}{\partial \kappa} \end{bmatrix} \tag{3.11}$$

$$A_2 = \begin{bmatrix} 1 & 0 & \dfrac{\partial f_x}{\partial f} & \dfrac{\partial f_x}{\partial \kappa_1} & \dfrac{\partial f_x}{\partial \kappa_2} \\ 0 & 1 & \dfrac{\partial f_y}{\partial f} & \dfrac{\partial f_y}{\partial \kappa_1} & \dfrac{\partial f_y}{\partial \kappa_2} \end{bmatrix} \tag{3.12}$$

$$A_3 = \begin{bmatrix} \dfrac{\partial f_x}{\partial X_P} & \dfrac{\partial f_x}{\partial Y_P} & \dfrac{\partial f_x}{\partial Z_P} \\ \dfrac{\partial f_y}{\partial X_P} & \dfrac{\partial f_y}{\partial Y_P} & \dfrac{\partial f_y}{\partial Z_P} \end{bmatrix} \tag{3.13}$$

$$X_1 = \begin{bmatrix} X_S & Y_S & Z_S & \phi & \omega & \kappa \end{bmatrix}$$

$$X_2 = \begin{bmatrix} x_0 & y_0 & f & \kappa_1 & \kappa_2 \end{bmatrix} \tag{3.14}$$

$$L = \begin{pmatrix} L_1 \\ L_2 \end{pmatrix} = \begin{bmatrix} x - x_0 + f\dfrac{r_{11}(X_P - X_S) + r_{12}(Y_P - Y_S) + r_{13}(Z_P - Z_S)}{r_{31}(X_P - X_S) + r_{32}(Y_P - Y_S) + r_{33}(Z_P - Z_S)} \\ y - y_0 + f\dfrac{r_{21}(X_P - X_S) + r_{22}(Y_P - Y_S) + r_{23}(Z_P - Z_S)}{r_{31}(X_P - X_S) + r_{32}(Y_P - Y_S) + r_{33}(Z_P - Z_S)} \end{bmatrix} \tag{3.15}$$

As observed from equations (3.5) to (3.10), there are 6 exterior orientation parameters for each image, plus 3 interior orientation and 2 lens distortion parameters, i.e., there are 11 unknown parameters in total. This means that at least six GCPs and their corresponding image points are needed for solution of equation (3.10).

3.4.2 Orthorectification Procedures

An orthorectification procedure for DOM generation involves the following steps listed below (Zhou et al. 2005), and the flowchart of orthorectifying an image is depicted in Figure 3.3:

1. Input source images and the corresponding exterior orientation parameters, which are solved above;
2. Input digital surface model (DSM), digital terrain model (DTM), or digital building model;
3. Determine the size of the output orthoimage;
4. Compute the geodetic coordinate from the column and row of pixels;
5. Estimate the *Z*-coordinate for the pixels;
6. Transform the geodetic coordinate into image coordinates for the pixels in the source image plane;
7. Compute the distance of the pixels to the imaging nadir in the source image;
8. Generate the ancillary data matrix from the distance and the coordinates;
9. Detect the occlusions using the data matrix;
10. Generate the occlusion index image;
11. Resample the output pixel value;
12. Generate a DOM;
13. Conduct necessary spectral smooth processing; and
14. Generate the final version of DOM.

Following up the flowchart in Figure 3.3, the major steps are detailed below.

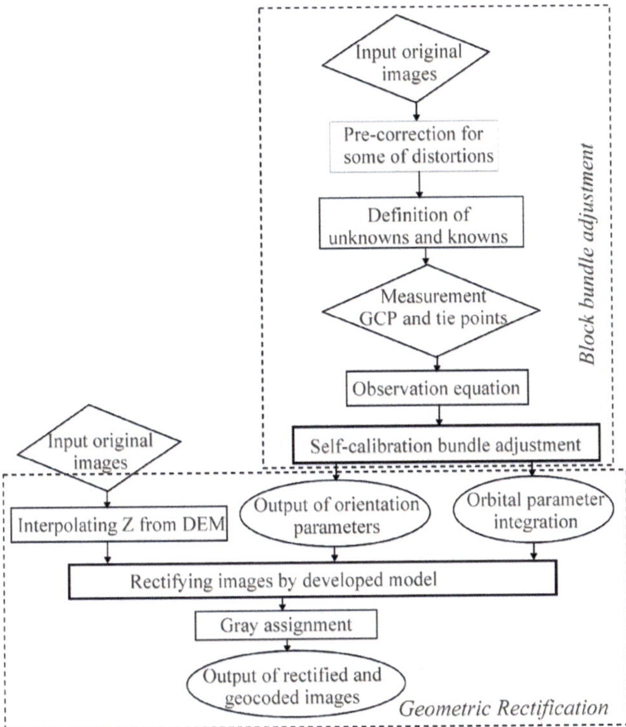

FIGURE 3.3 The flowchart of geometric rectification using the block bundle adjustment model. (Courtesy of Zhou et al. 2005.)

FIGURE 3.4 Three cases of overlap between the source and DSM images: (a) source fully covering the DSM, (b) source smaller than the DSM, and (c) partial overlap.

3.4.3 Determination of Size of the Orthoimage

The resulting orthoimage is generally expressed as a raster image, with pixels arranged in rows and columns. Since the resulting orthoimage is rectified from raster image input (called *source image*) using the DTM data (image), the size of the output orthoimage is a function of three factors: the size of the source image; the size of the DTM image; and the pixel resolution of the output orthoimage. As to the first two factors, there are generally three cases that describe the relationship of the source to the DTM (see Figure 3.4). Consequently, the size of the output image can be determined in three ways. When the size of the source image is larger than the size of the DTM image (Figure 3.4a), the size of output image will be the same as that of DTM image. In this case, the source image will not be fully presented in the output orthoimage. For the second case, when source image is smaller than the DTM image (Figure 3.4b), the size of the output image will be the same as the source image. The third case is that the source image partly overlays the DTM image (Figure 3.4c).

The output image then only covers the overlapping portions of the images. Coordinates of the output images for the three cases can be expressed as follows:

Case 1: $X_0 = \min\{X_D\}$ and $Y_0 = \min\{Y_D\}$
$X_1 = \max\{X_D\}$ and $Y_1 = \max\{Y_D\})$
Case 2: $X_0 = \min\{X_I\}$ and $Y_0 = \min\{Y_I\}$
$X_1 = \max\{X_I\}$ and $Y_1 = \max\{Y_I\})$
Case 3: $X_0 = \max\{\min\{X_I\}, \min\{X_D\}\}$ $Y_0 = \max\{\min\{Y_I\}, \min\{Y_D\}\}$
$X_1 = \min\{\max\{X_I\}, \max\{X_D\}\}$ $Y_1 = \min\{\max\{Y_I\}, \max\{Y_D\}\}$

where X_0 and Y_0 are the coordinates of the lower-left corner of the output image; X_1 and Y_1 are the coordinates of the upper-right corner of the output image; X_D and Y_D are the X and Y coordinates of the DSM image; X_I and Y_I are the X- and Y-coordinates of the source image; max and min denote the maximum and minimum of the elements in the blanket. All of the coordinates here refer to the geodetic coordinate system required in the resulting orthoimage.

3.4.4 Computation of Coordinate of Each Pixel

For a given pixel $P(K, L)$ in the output image, we need to convert its raster row and column into coordinates of the output image.

$$X = X_0 + P_x * K$$

$$Y = Y_0 + P_y * L \tag{3.16}$$

where K denotes the column of a pixel in the output image, L denotes the row of the pixel; X and Y are the coordinates of the pixel; P_x and P_y are the pixel size in the X and Y directions, generally $P_x = P_y$. Similarly, for the DSM image, we have

$$X_D = X_{0D} + P_D * K_D$$

$$Y_D = Y_{0D} + P_D * L_D \tag{3.17}$$

The subscript D represents the DSM image.

3.4.5 Computation of Z-Coordinates of Pixel

To perform orthorectification, we also need to know the Z-coordinates of the pixel $P\,(K, L)$ in output image. Generally, this is obtained from DSM image. First, we have to convert back to the pixel's geodetic coordinate into the column and row of the DSM image, using the following formula:

$$K_D = (X - X_{0D}) / P_D$$

$$L_D = (Y - Y_{0D}) / P_D \tag{3.18}$$

where P_D denotes the pixel size of the DSM image. X_{0D} and Y_{0D} are the lower-left corner coordinates of the DSM image. K_D and L_D will not generally be exact integers. Thus, an interpolation must be performed to determine Z. Usually, a bilinear interpolation method of the following form is employed (see Figure 3.5),

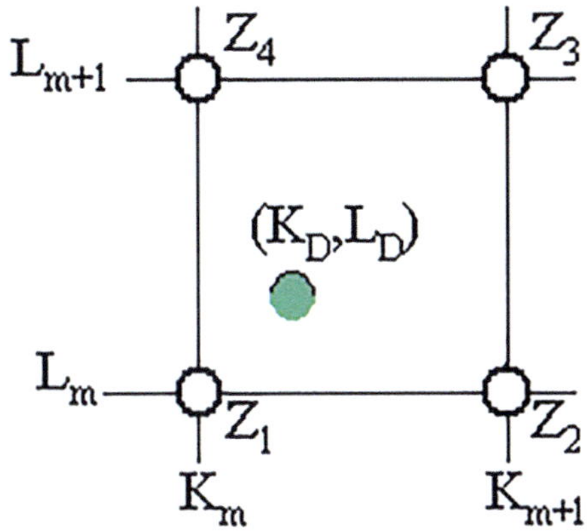

FIGURE 3.5 Estimating the altitude of the pixel from a DSM image.

$$Z = \left\{ \begin{matrix} \left[Z_1 dX + (1-dX) Z_2 \right] + \left[Z_4 dX + (1-dX) Z_3 \right] \\ + \left[Z_1 dY + (1-dY) Z_2 \right] + \left[Z_4 dY + (1-dY) Z_3 \right] \end{matrix} \right\} / 4 \tag{3.19}$$

where d$X = K_D - K_m$ and d$Y = L_D - L_m$, in which K_m is the K_D rounded to its maximal integer and L_m is L_D round to its maximal integer. After this estimation, we then know the coordinate (X, Y, Z) of the pixel.

3.4.6 Computation of the Corresponding Coordinate in Source Image

To orthorectify the source image, we have to find the corresponding coordinate of the source image pixel in the output image. This can be carried out by using a collinearity equation:

$$x_I = x_{0I} - f \frac{a_1 (X - X_s) + b_1 (Y - Y_s) + c_1 (Z - Z_s)}{a_3 (X - X_s) + b_3 (Y - Y_s) + c_3 (Z - Z_s)}$$

$$y_I = y_{0I} - f \frac{a_2 (X - X_s) + b_2 (Y - Y_s) + c_2 (Z - Z_s)}{a_3 (X - X_s) + b_3 (Y - Y_s) + c_3 (Z - Z_s)} \tag{3.20}$$

where x_I and y_I are the corresponding coordinates of the pixel $P(X, Y)$ in the source image; X_s, Y_s, and Z_s are the exposure station; f is the focal length; $a_i = \{a_1, a_2, a_3)$, $b_i = \{b_1, b_2, b_3\}$, and $c_i = \{c_1, c_2, c_3\}$ are the elements of the rotation matrix, which is a function of the three exterior orientation angles (ϕ, ω, κ). These elements have to be computed before the equation can be used for the coordinate transformation. These orientation parameters have been solved above.

3.4.7 Assignment of Gray Value to Each Pixel

Since the grid of pixels in the source image rarely matches the grid of the output orthoimage, resampling of the pixels has to be done in order to assign the gray value to the pixels in the output image. There are several ways of resampling, the most popular being the nearest neighbor, bilinear interpolation, or cubic convolution methods. We employ the nearest neighbor because it directly transfers the original data values without averaging them. To assign a gray value to the output pixel with coordinate $P(X, Y)$, we have to retransform it back to its adjusted coordinate in the source

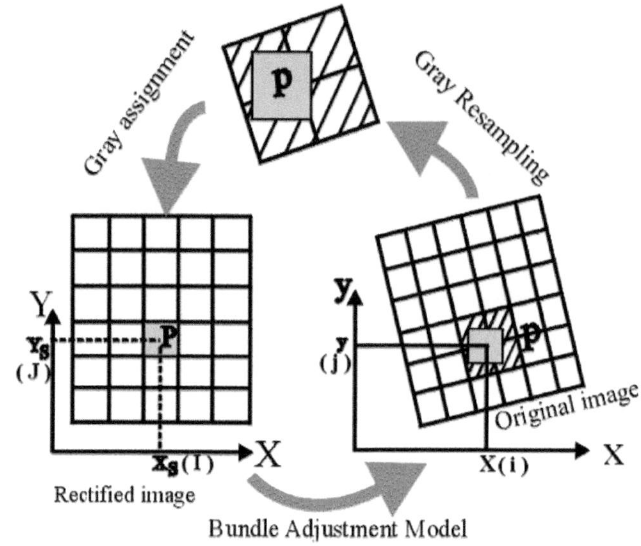

FIGURE 3.6 The procedure of geometric rectification.

image. Suppose the pixel $P(X, Y)$ in the output image has its corresponding coordinate $P(X_l, Y_l)$ in the source image. Once $P(X_l, Y_l)$ is computed, it is easy to find the nearest pixel from which to assign an output (gray) value for the pixel in the output image. The computational procedure is illustrated in Figure 3.6.

3.5 SELECTING THE GROUND CONTROL POINT (GCP) PAIRS

The optimum method for selecting the GCPs is to have both the rectified base year image (or reference map) and the image to be rectified on the screen at the same time. This dual display greatly simplifies GCP selection. Some image processing systems even allow the GCP selected to be reprojected onto the image to be corrected (with the appropriate transformation coefficients, to be discussed shortly) to determine the quality of the GCP point. Also, some systems allow the analyst to extract floating point row and column coordinates of GCPs (instead of just integer values) through the use of a chip extraction algorithm that zooms in and does subpixel sampling. GCP subpixel row and column coordinates often improve the precision of the image-to-map rectification or image-to-image registration. Some scientists have developed methods of automatically extracting GCPs common to two images, which can be used during image-to-image registration. However, most image-to-map rectification still relies heavily on human interaction.

REFERENCES

Baker, J. R.; Marks, G. W.; Mikhail, E. M., Analysis of digital multispectral scanner (MSS) data. *Bildmessung und Luftbildwesen*, 43, 1975, pp. 22–27.

Buiten, H. J.; Van Putten, B., Quality assessment of remote sensing registration-analysis and testing of control point residuals. *ISPRS Journal of Photogrammetry & Remote Sensing*, 52, 1997, pp. 57–73.

Craig, M. D.; Green, A. A., Registration of distorted images from airborne scanners. *The Australian Computer Journal*, 19, 1987, pp. 148–153.

De Leeuw, A. J.; Veugen, L. M. M.; Van Stokkom, H. T. C., Geometric correction of remotely sensed imagery using ground control points and orthogonal polynomials. *International Journal of Remote Sensing*, 9, 1988, pp. 1751–1759.

Devereux, A. J.; Fuller, R. M.; Carter, L.; Parsell, R. J., Geometric correction of airborne scanner imagery by matching Delaunay triangles. *International Journal of Remote Sensing*, 11, 1990, pp. 2237–2251.

Goshtasby, A., Registration of images with geometric distortions. *IEEE Transactions on Geoscience and Remote Sensing*, 26, 1988, pp. 60–64.

Leckie, D. G., Use of polynomial transformations for registration of airborne digital line scan images. In: *Proceedings of the 6th Canadian Symposium on Remote Sensing*, Halifax, Nova Scotia, 21–23 May, 1980, pp. 635–641.

McGlone, J. C.; Mikhail, E. M; Bethel, J. S; Mullen, R., *Manual of Photogrammetry*, 6th ed.; American Society for Photogrammetry and Remote Sensing: Bethesda, MD, 2004.

Novak, K., Rectification of digital imagery. *Photogrammetric Engineering & Remote Sensing*, 58, 1992, pp. 339–344.

Roy, D. P., Parametric geometric correction of airborne thematic mapper imagery. *International Journal of Remote Sensing*, 18, 1997, pp. 1865–1887.

Ton, J.; Jain, A. K., Registering landsat images by point matching. *IEEE Transactions on Geoscience and Remote Sensing*, 27, 1989, pp. 642–651.

Wang, Z., *Principles of Photogrammetry*, English ed.; Wuhan University Press: Wuhan, 1991.

Wolf, P.; Dewitt, B.; Wilkinson, B., *Elements of Photogrammetry with Application in GIS*, 4th ed.; McGraw Hill: New York, 2004.

Zheng, Q.; Klemas, V.; Yan, X.-H.; Wang, Z.; Kagleder, K., Digital orthorectification of space shuttle coastal ocean photographs. *International Journal of Remote Sensing*, 18, 1997, pp. 197–211.

Zhou, G.; Chen, W.; Kelmelis, J., A comprehensive study on urban true orthorectification. *IEEE Transactions on Geoscience and Remote Sensing*, 43, No. 9, 2005, pp. 2138–2147.

Zhou, G.; Uzi, E.; Feng, W; Yuan, B., Charge-coupled device camera calibration based on natural landmark. *Pattern Recognition*, 31, 1998, pp. 1715–1724.

Zhou, G.; Yuan, B.; Tang, X., The intrinsic and extrinsic factor of charge-coupled device calibration, *Acta Electronica Sinica*, 24, 1996, pp. 12–17.

4 Orthorectification above 60°N Latitude Area

4.1 INTRODUCTION

Early satellite photographs appearing at the beginning of the space age provided the research community with a unique opportunity to investigate historic phenomena. In order to use early 1960s declassified intelligence satellite photograph (DISP) imagery for ice sheet research in the glaciological community, the mosaic of the entire Greenland ice sheet needs to be created. To this end, individual photographs must be orthorectified with respect to a chosen map-based coordinate system. Conventionally available orthorectification algorithms, for example, Seto et al. (1991), Chen and Ran (1993), Devereux et al. (1990), Toutin et al. (1992), and Shin et al. (1997), cannot obtain satisfactory results due to the following complicating attributes (Zhou et al. 2002):

1. Not each of the DISP images contains sufficient GCPs due to featureless or homogeneous surfaces, ice sheet coverage, etc., and photogrammetric GCPs in Greenland are not available for the early 1960s.
2. Cloud-covered or partially cloud-covered photographic imagery reaches over 40% due to the nature of the study region and the climatic conditions present during these particular satellite orbits.
3. The Greenland ice sheet covers the area from latitude 55°N through 85°N. This high latitude causes additional distortions and creates challenges with respect to traditional ground coordinate systems, such as UTM.
4. Relief displacement caused by the difference in elevation in some areas, such as Jakobshen Isbræ (the difference in elevation exceeds 1000 m (Ekholm 1996)), is large.

A large number of orthorectification models have been presented, ranging from a simple affine transformation to projective transformations, and then differential rectification with relief displacement correction. Both parametric and nonparametric approaches require the collection of enough GCPs, which may in fact not be feasible in the area above 60°N altitude for the early 1960s satellite imagery. This chapter describes a satellite navigation parameter-assisted mathematical model for the orthorectification of DISP images over 60°N altitude. The model minimizes the request of GCPs and solves the sensor's exterior orientation parameters of all images and satellite navigation parameters simultaneously. A geocentric coordinate system is selected as a reference coordinate system to avoid the several-kilometer Earth's curvature distortion (Zhou et al. 2002).

4.2 MATHEMATICAL MODEL OF ORTHORECTIFICATION

4.2.1 Declassified Intelligence Satellite Photographs (DISPs)

Of the series of early reconnaissance satellite photographs, we chose 24 of the best films acquired during the ARGON 9034A missions in May 1962 to assemble the first mosaic and 36 films acquired during the ARGON 9058A and 9059A missions in October 1963 to assemble the second mosaic. The ARGON missions carried a frame camera with a focal length of 76.2 mm at a flying height of 322 km, providing high ground resolution (about 140 m) and extensive coverage (McDonald 1995). Some technical specification of the ARGON missions is listed in Table 4.1. All of the films

DOI: 10.1201/9781003632108-4

TABLE 4.1
Technical Specification of KH-5 (9034A) Mission

Attributes	Parameters	Attributes	Parameters	Attributes	Parameters
Film width	127 mm	Inclination	82.3°	Nominal photo scale	1:4,425,000
Camera type	Frame	Focal length	76.2 mm	Nominal orbit altitude	322 km
Imaging mode	Panchromatic	Ground resolution	140 m	Nominal ground coverage	556 × 556 km^2

were digitized at 7 μm resolution using the INTERGRAPH PhotoScan TD® Scanner, which is a high-resolution, radiometrically, and geometrically precise flatbed scanning system.

4.2.2 Bundle Adjustment Model

The early investigation on the orthorectification of the DISP imagery of Greenland demonstrated that the Earth's curvature distortion is significant. The maximum distortion caused by Earth's curvature on corner points of the photograph achieves 2.2 m, which is equivalent to 9 km in ground distance, inward from its actual position (Zhou et al. 2002) because of the extremely small ratio of flying height to focal length. The approaches usually taken for Earth's curvature distortion rectification are given below.

4.2.2.1 Model Rectification

This method models the Earth's curvature distortion using flying height, focal length of the camera, radial distance from the principal point to the image point, and the Earth's radius. Figure 4.1 illustrates the principle of this correction method. Point A is a ground point and A' is its corresponding position in a map plane, which is tangent to the earth at the ground nadir point. Points a and a' are the images of points A and A', respectively. From Figure 4.1, the distortion, dr, for earth curvature is (Wolf et al. 2004)

$$dr = \frac{Hr^3}{2Rf^2} \tag{4.1}$$

where H is the flying height above ground, r is the radial distance from the principal point to the image point, R is the Earth's radius, and f is the camera focal length. As seen from Figure 4.1 and equation (4.1), if the photograph covers a large area, such as each DISP image with 483 km by 483 km coverage, the Earth's radius, R, should no longer be a constant within a photograph as the Earth is ellipsoidal. Thus, the Earth curvature distortion in the margin of the photograph cannot be completely corrected when using this method.

4.2.2.2 Geocentric Coordinate System

This approach corrects the Earth's curvature distortion by establishing a geocentric coordinate system (the definition is given in Section 4.3.1). The adjustment is then carried out in the geocentric coordinate system.

4.2.2.3 Tangential Plane Coordinate System

The method translates the ground coordinate system to the tangential plane coordinate system, in which the nadir is chosen as the origin (see Kraus 1997, p. 161). This method therefore corrects the Earth's curvature distortion by means of the coordinate system transformation, which is able to reduce the projection error caused by the Earth's curvature (see Figure 4.2). This approach also cannot completely correct the distortion because there is still a projection error in the margin of the photograph when a photograph covers a large area.

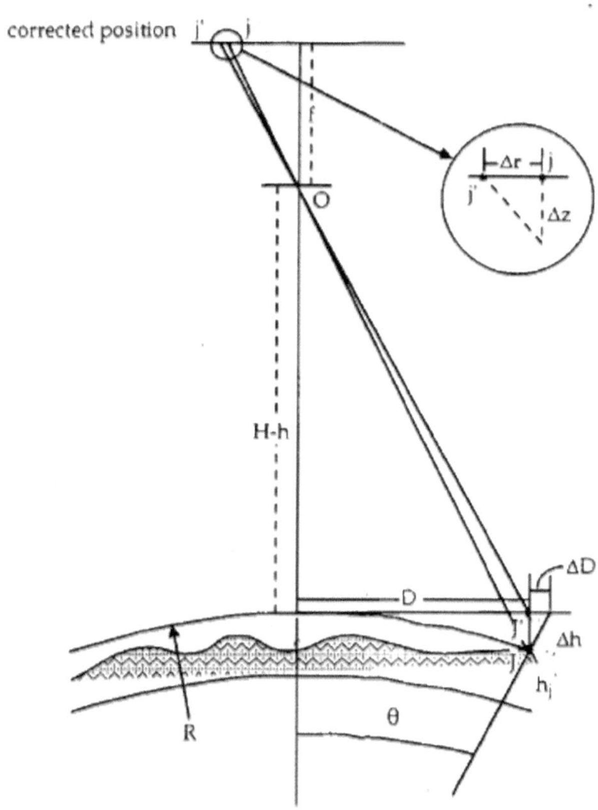

FIGURE 4.1 Earth curvature distortion correction. (Courtesy of Chapman 1995.)

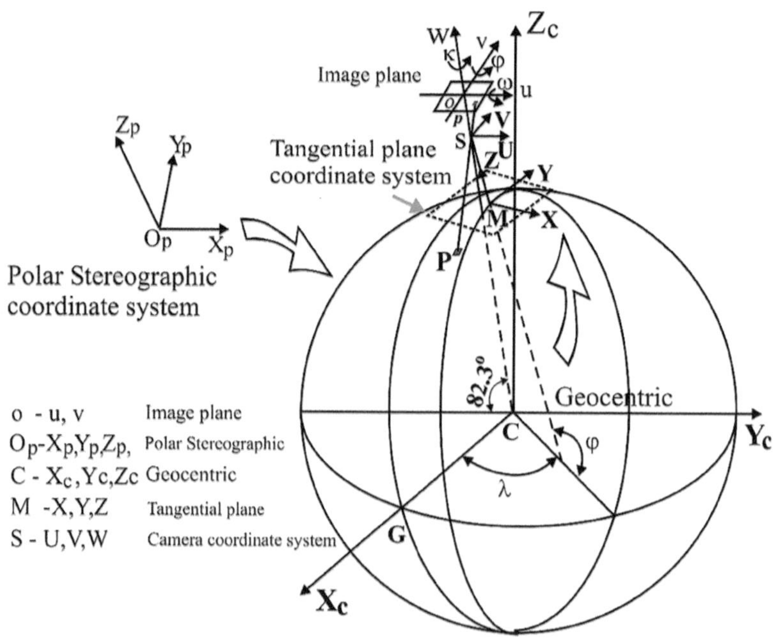

FIGURE 4.2 The configuration of DISP imaging geometry in ARGON 9034A mission. (Courtesy of Zhou and Jezek 2002, 2004.)

The traditional differential rectification model (nonparametric approach) is able to effectively rectify the relief displacement caused by the difference in elevation. The fundamental principle of the method is based on collinearity. However, this method requests that each image must have its own coordinate system and its own GCPs and only rectifies for a single frame image. The model can be extended for a block of images by tie points (Mikhail et al. 2001; Zhou et al. 2002). The objective of extending into a block of images is to tie overlapping images together while minimizing the need, even without the absolute need, for ground control points (GCPs) in each image. The input to the adjustment system includes the measured image coordinates of GCPs and features that are identifiable in as many images as possible, together with the 3D ground coordinates of the GCPs. The system outputs the exterior orientation parameters of all images and the 3D ground coordinates of the tie points.

Considering the significant Earth curvature distortion in Greenland, the geocentric coordinate system is established as the basis of bundle adjustment. The orthophoto is referred to as the Polar Stereographic projection with the reference ellipsoid of WGS-84 since some control datasets, such as both the digital ERS-1 SAR mosaic map developed by National Aeronautics and Space Administration (NASA) Jet Propulsion Laboratory (JPL) and the DEM of Greenland developed by Demark Kort-og Matrikelsyrelsen (KMS), have adopted this projection system. In this projection system, the latitude of the true scale is at 70°N and the longitude of the central meridian is at 315°. The false easting and false northing are 0.0. Therefore, the implementation of our model will have to conduct the following coordinate system transformations (see Figure 4.2).

4.2.2.4 Polar Stereographic Coordinate System

All control datasets to be used, orthorectification of DISP images, and the mosaicked products will take place in this coordinate system.

4.2.2.5 Geocentric Coordinate System

A number of different geocentric coordinate system have been defined and used within space physics. The definition of the axes varies from system to system, but they use the Earth center as origin and are all orthogonal, right-handed systems (Hapgood 1992). Usually, four geocentric coordinate systems, Geographic Coordinate System (GEO), Geomagnetic Coordinate System (MAG), Corrected Geomagnetic Coordinates (CGM), and Dipole Meridian (DM), which corotate with the Earth, have frequently been used. A transformation between these systems can thus be achieved by a maximum of two rotations. In this article, the GEO geocentric coordinate system, in which X axis is the intersection of the Greenwich meridian and the geographic equator and Z axis is the geographic North pole, is used.

4.2.2.6 Image Coordinate System (*o–uv*)

It is a right-hand coordinate system, in which the line connecting two opposite fiducial marks is taken as the u and v axes, respectively, with the direction of flight as positive. The intersection of u and v axes is the origin.

4.2.2.7 Camera Coordinate System (*O–UVW*)

The lens center of camera is set as the origin O, the line connecting the lens center of the camera and the principal point is taken W axis, and the U and V axes are parallel to the u and v axes in the image plane, respectively.

For any point, such as P, the collinearity equations based on the geocentric coordinate system are expressed as

$$x_p - x_0 = -f\frac{r_{11}(X_P - X_S) + r_{12}(Y_P - Y_S) + r_{13}(Z_P - Z_S)}{r_{31}(X_P - X_S) + r_{32}(Y_P - Y_S) + r_{33}(Z_P - Z_S)} = f_x$$

$$y_p - y_0 = -f\frac{r_{21}(X_P - X_S) + r_{22}(Y_P - Y_S) + r_{23}(Z_P - Z_S)}{r_{31}(X_P - X_S) + r_{32}(Y_P - Y_S) + r_{33}(Z_P - Z_S)} = f_y \tag{4.2}$$

where (x_p, y_p) are image coordinates; (x_0, y_0, f) are interior orientation parameters; (X_P, Y_P, Z_P) are ground coordinates of a control point or a tie point, which may be treated as known or/and unknown parameters differently in various situations depending upon the types of observations; $r_{i,j}$ $(i = 1, 2, 3; j = 1, 2, 3)$ are elements of the rotation matrix R, which is a function of three rotation angles, ω, ϕ, and κ with respect to the geocentric coordinate system, i.e.,

$$R = \begin{pmatrix} \cos\phi\cos\kappa & \cos\omega\sin\kappa + \sin\omega\sin\phi\cos\kappa & \sin\omega\sin\kappa - \cos\omega\sin\phi\cos\kappa \\ -\cos\phi\sin\kappa & \cos\omega\cos\kappa - \sin\omega\sin\phi\sin\kappa & \sin\omega\cos\kappa + \cos\omega\sin\phi\sin\kappa \\ \sin\phi & -\sin\omega\cos\phi & \cos\omega\cos\phi \end{pmatrix} \tag{4.3}$$

In this model, the interior orientation parameters, including the focal length of camera and lens distortions, are considered known parameters. They will not be resolved by equation (4.2).

4.2.3 Satellite Navigation Parameter-Assisted Orthorectification Model

The basic fact for the 1960s DISP image in Greenland is that it is almost impossible to find existing GCPs and/or to place artificial GCPs in the area above 60°N latitude prior to the 1960s. On the other hand, some images appeared homogeneous or featureless in central Greenland because they show nothing but the ice sheet. Each of images cannot therefore contain sufficient GCPs, which are required in traditional photogrammetric processing. It means that it is almost impossible to apply traditional 2D-to-3D photogrammetric techniques for single-image orthorectification. For this reason, an algorithm to obtain the elements of exterior orientation of the no-GCP imagery by interpolating or extrapolating the known navigation parameters (three position parameters and three orientation parameters) is proposed (see Figure 4.3). Previous research demonstrated that the position and attitude parameters of a camera in the same strip can be expressed by a third-order polynomial (Zhou et al. 2002)

$$X_S(t) = a_0 + a_1 t + a_2 t^2 + a_3 t^3$$

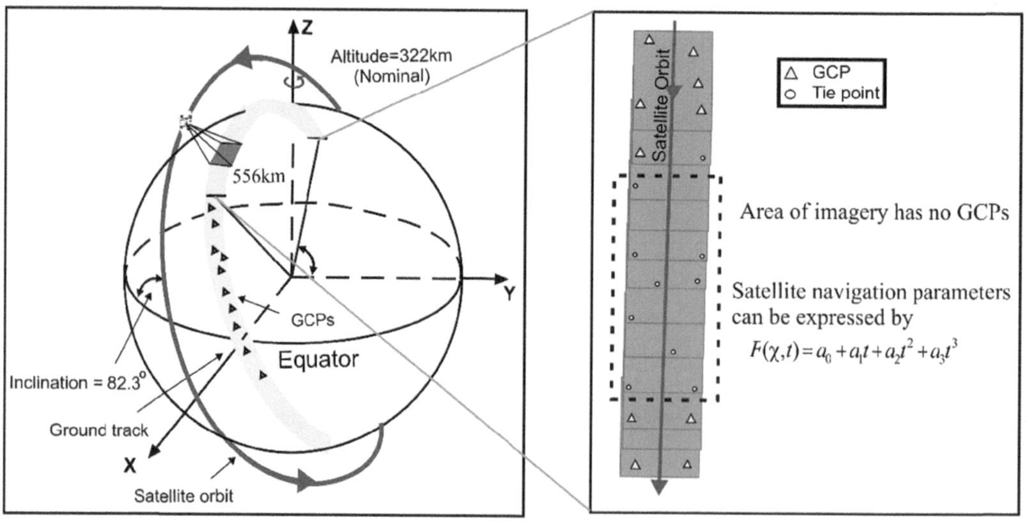

FIGURE 4.3 Satellite navigation parameter-assisted model. (Courtesy of Zhou and Jezek 2002, 2004.)

$$Y_S(t) = b_0 + b_1 t + b_2 t^2 + b_3 t^3$$

$$Z_S(t) = c_0 + c_1 t + c_2 t^2 + c_3 t^3$$

$$\varphi_S(t) = d_0 + d_1 t + d_2 t^2 + d_3 t^3$$

$$\omega_S(t) = e_0 + e_1 t + e_2 t^2 + e_3 t^3$$

$$\kappa_S(t) = f_0 + f_1 t + f_2 t^2 + f_3 t^3 \tag{4.4}$$

where a_i, b_i, c_i, d_i (i = 0, 1, 2, 3) are coefficients, they are unknown; t represents the time, which may be beginning time at a certain epoch from a certain orbit position. The unknown coefficients in equation (4.4) can be determined either independently or into equation (4.2). Here we recommend the second method, i.e. combining equation (4.4) into equation (4.2), to calculate the elements of exterior orientation of all images simultaneously. Thereby, when there are N_3 points, the observation equations for equation (4.4) are

$$\begin{pmatrix} v_S^{X_1} \\ v_S^{X_2} \\ \vdots \\ v_S^{X_{N3}} \end{pmatrix} = \begin{pmatrix} 1 & t_1 & t_1^2 & t_1^3 \\ 1 & t_2 & t_2^2 & t_2^3 \\ & & \vdots & \\ 1 & t_{N3} & t_{N3}^2 & t_{N3}^3 \end{pmatrix} \begin{pmatrix} da_0 \\ da_1 \\ da_2 \\ da_3 \end{pmatrix} + \begin{pmatrix} X_S^0(t_1) - X_S^1 \\ X_S^0(t_2) - X_S^2 \\ \vdots \\ X_S^0(t_{N_3}) - X_S^{N_3} \end{pmatrix}$$

$$\begin{pmatrix} v_S^{y_1} \\ v_S^{Y_2} \\ \vdots \\ v_S^{Y_{N3}} \end{pmatrix} = \begin{pmatrix} 1 & t_1 & t_1^2 & t_1^3 \\ 1 & t_2 & t_2^2 & t_2^3 \\ & & \vdots & \\ 1 & t_{N3} & t_{N_3}^2 & t_{N_3}^3 \end{pmatrix} \begin{pmatrix} db_0 \\ db_1 \\ db_2 \\ db_3 \end{pmatrix} + \begin{pmatrix} Y_S^0(t_1) - Y_S^1 \\ Y_S^0(t_2) - Y_S^2 \\ \vdots \\ Y_S^0(t_{N_3}) - Y_S^{N_3} \end{pmatrix}$$

$$\begin{pmatrix} v_S^{Z_1} \\ v_S^{Z_2} \\ \vdots \\ v_S^{Z_{N3}} \end{pmatrix} = \begin{pmatrix} 1 & t_1 & t_1^2 & t_1^3 \\ 1 & t_2 & t_2^2 & t_2^3 \\ & & \vdots & \\ 1 & t_{N3} & t_{N_3}^2 & t_{N_3}^3 \end{pmatrix} \begin{pmatrix} dc_0 \\ dc_1 \\ dc_2 \\ dc_3 \end{pmatrix} + \begin{pmatrix} Z_S^0(t_1) - Z_S^1 \\ Z_S^0(t_2) - Z_S^2 \\ \vdots \\ Z_S^0(t_{N_3}) - Z_S^{N_3} \end{pmatrix} \tag{4.5}$$

Also, the observations for rotation angles are

$$\begin{pmatrix} v_S^{\phi_1} \\ v_S^{\phi_2} \\ \vdots \\ v_S^{\phi_{N3}} \end{pmatrix} = \begin{pmatrix} 1 & t_1 & t_1^2 & t_1^3 \\ 1 & t_2 & t_2^2 & t_2^3 \\ & & \vdots & \\ 1 & t_{N_3} & t_{N_3}^2 & t_{N_3}^3 \end{pmatrix} \begin{pmatrix} dd_0 \\ dd_1 \\ dd_2 \\ dd_3 \end{pmatrix} + \begin{pmatrix} \phi_S^0(t_1) - \phi_S^1 \\ \phi_S^0(t_2) - \phi_S^2 \\ \vdots \\ \phi_S^0(t_{N3}) - \phi_S^{N_3} \end{pmatrix}$$

$$\begin{pmatrix} v_S^{\omega_1} \\ v_S^{\omega_2} \\ \vdots \\ v_S^{\omega_{N3}} \end{pmatrix} = \begin{pmatrix} 1 & t_1 & t_1^2 & t_1^3 \\ 1 & t_2 & t_2^2 & t_2^3 \\ & & \vdots & \\ 1 & t_{N3} & t_{N3}^2 & t_{N3}^3 \end{pmatrix} \begin{pmatrix} de_0 \\ de_1 \\ de_2 \\ de_3 \end{pmatrix} + \begin{pmatrix} \omega_S^0(t_1) - \omega_S^1 \\ \omega_S^0(t_2) - \omega_S^2 \\ \vdots \\ \omega_S^0(t_{N3}) - \omega_S^{N3} \end{pmatrix}$$

$$\begin{pmatrix} v_S^{\kappa_1} \\ v_S^{\kappa_2} \\ \vdots \\ v_S^{\kappa_{N3}} \end{pmatrix} = \begin{pmatrix} 1 & t_1 & t_1^2 & t_1^3 \\ 1 & t_2 & t_2^2 & t_2^3 \\ & & \vdots & \\ 1 & t_{N3} & t_{N3}^2 & t_{N3}^3 \end{pmatrix} \begin{pmatrix} df_0 \\ df_1 \\ df_2 \\ df_3 \end{pmatrix} + \begin{pmatrix} \kappa_S^0(t_1) - \kappa_S^1 \\ \kappa_S^0(t_2) - \kappa_S^2 \\ \vdots \\ \kappa_S^0(t_{N3}) - \kappa_S^{N3} \end{pmatrix} \quad (4.6)$$

For the same strip of satellite, we may be able to find a number of "GCPs", i.e., 2D-to-3D corresponding points from the existing data. These GCPs will not mean the traditional photogrammetric target points, i.e., these GCPs are probably from the identical nature features from the NASA/JPL ERS-1 SAR mosaic map of Greenland (the detailed description is given in Section 4.4). The partial derivatives of equation (4.2) can be referenced in Appendix A. So, the observation equations for a known point (GCP), e.g., Point P, are

$$\begin{aligned} v_P^x = {} & \varepsilon_{11}da_0 + \varepsilon_{12}db_0 + \varepsilon_{13}dc_0 + \varepsilon_{14}dd_0 + \varepsilon_{15}de_0 + \varepsilon_{16}df_0 \\ & + \varepsilon_{17}da_1 + \varepsilon_{18}db_1 + \varepsilon_{19}dc_1 + \varepsilon_{110}dd_1 + \varepsilon_{111}de_1 + \varepsilon_{112}df_1 \\ & + \varepsilon_{113}da_2 + \varepsilon_{114}db_2 + \varepsilon_{115}dc_2 + \varepsilon_{116}dd_2 + \varepsilon_{117}de_2 + \varepsilon_{118}df_2 \\ & + \varepsilon_{119}da_3 + \varepsilon_{120}db_3 + \varepsilon_{121}dc_3 + \varepsilon_{122}dd_3 + \varepsilon_{123}de_3 + \varepsilon_{124}df_3 + l_P^x \end{aligned}$$

$$\begin{aligned} v_P^y = {} & \varepsilon_{21}da_0 + \varepsilon_{22}db_0 + \varepsilon_{23}dc_0 + \varepsilon_{24}dd_0 + \varepsilon_{25}de_0 + \varepsilon_{26}df_0 \\ & + \varepsilon_{27}da_1 + \varepsilon_{28}db_1 + \varepsilon_{29}dc_1 + \varepsilon_{210}dd_1 + \varepsilon_{211}de_1 + \varepsilon_{212}df_1 \\ & + \varepsilon_{213}da_2 + \varepsilon_{214}db_2 + \varepsilon_{215}dc_2 + \varepsilon_{216}dd_2 + \varepsilon_{217}de_2 + \varepsilon_{218}df_2 \\ & + \varepsilon_{219}da_3 + \varepsilon_{220}db_3 + \varepsilon_{221}dc_3 + \varepsilon_{222}dd_3 + \varepsilon_{223}de_3 + \varepsilon_{224}df_3 + l_P^y \end{aligned} \quad (4.7)$$

where

$$l_P^x = -x_P + f\frac{r_{11}(X - X_S(t)) + r_{12}(Y - Y_S(t)) + r_{13}(Z - Z_S(t))}{r_{31}(X - X_S(t)) + r_{32}(Y - Y_S(t)) + r_{33}(Z - Z_S(t))}$$

$$l_P^y = -y_P + f\frac{r_{21}(X - X_S(t)) + r_{22}(Y - Y_S(t)) + r_{23}(Z - Z_S(t))}{r_{31}(X - X_S(t)) + r_{32}(Y - Y_S(t)) + r_{33}(Z - Z_S(t))} \quad (4.8)$$

The coefficients ε_{11} to ε_{224} are partial derivatives of equation (4.2) with respect to the polynomial parameters, and the details of coefficients ε_{11} to ε_{224} can be found in Zhou and Li (2000). Finally, equations (4.5) and (4.6) should be combined into equation (4.7) for solution of six navigation parameters (three position and three altitude parameters) of non-GCP images and GCP images simultaneously.

4.2.4 Orthorectification Process

With the orientation parameters of each image solved by the method above, the steps of orthorectifying the original image contain (1) the determination of the output image size, (2) coordinate transformation from the original image to the output image, (3) gray allocation for the output image through resampling the original image, and (4) registration of the output image into a map-based coordinate system.

4.2.4.1 Determination of Orthorectified Image Size

The orthorectified image (called *orthoimage*) is also usually expressed as a raster image, which is geocoded and registered to a chosen map-based coordinate system. Since an output orthoimage is rectified from an original raster image, called *source image*, using the DEM data (imagery), the size of the output orthoimage is therefore dependent upon the size of the source image, the size of the DEM imagery, and the resolution of the pixel, called *ground sample distance* (GSD). The GSD is usually specified by user requirement, and three cases between the source image and the DEM imagery usually happen (see Figure 4.4). Consequently, when the size of the source image is larger than the size of the DEM imagery (Figure 4.4a), the size of the output image will be the same as that of the DEM imagery, resulting in the source image not fully showing up in the output orthoimage. When the source image is smaller than the DEM imagery (Figure 4.4b), the size of the output image will be the same as the source image. The third case is when the source image partly overlays the DEM imagery (Figure 4.4c), and the output image only covers a part of their overlapping. The size of the output image can be determined by

Case 1: $X_0 = \min\{X_D\}$ and $Y_0 = \min\{Y_D\}$
$X_1 = \max\{X_D\}$ and $Y_1 = \max\{Y_D\}$
Case 2: $X_0 = \min\{X_I\}$ and $Y_0 = \min\{Y_I\}$
$X_1 = \max\{X_I\}$ and $Y_1 = \max\{Y_I\}$
Case 3: $X_0 = \max\{\min\{X_I\}, \min\{X_D\}\}$ $Y_0 = \max\{\min\{Y_I\}, \min\{Y_D\}\}$
$X_1 = \min\{\max\{X_I\}, \max\{X_D\}\}$ $Y_1 = \min\{\max\{Y_I\}, \max\{Y_D\}\}$

where X_0 and Y_0 are the coordinates of the lower-left corner of the output image; X_1 and Y_1 are the coordinates of the upper-right corner of the output image; X_D and Y_D are the X and Y coordinates of DEM imagery; X_I and Y_I are the X and Y coordinates of the source image; max and min denote the maximum and minimum of the elements in the blanket. All the coordinates here refer to the polar stereographic coordinate system.

4.2.4.2 Orthorectification

After the size of output orthoimage is determined and the elements of exterior orientation of all images are solved, the orthorectification can be carried out by (1) calculating the geographic coordinates of

FIGURE 4.4 Three cases for determining the relationship between source image and DEM: (a) source image fully covering DEM imagery, (b) source image not covering DEM imagery, and (c) partially covering DEM imagery.

individual pixels in output image, (2) resampling the original image, and (3) registering the image into a map-based coordinates system (Zhou et al. 2002; Zhou and Jezek 2002, 2004). The detailed steps are as follows:

- For any point, $P(I, J)$ in the output image, where I, J are row and column, but should be transformed into corresponding image coordinates in the image plane, compute the planimetric coordinates of the point P (X_P, Y_P) in the polar stereographic coordinate system by the given GSD;
- Point P (X_P, Y_P) is usually located within the grid of DEM, interpolate the vertical coordinates Z_P from the DEM using a bilinear interpolation algorithm, which requires the four neighbor grid points of DEM;
- Transform the coordinates (X_P, Y_P, Z_P) of the point P into the geographic coordinate system (ϕ, λ, H) and then transform the geographic coordinate (ϕ, λ, H) into the geocentric coordinates (X_C, Y_C, Z_C) by coordinate transformation algorithms described in Snyder (1997);
- Compute the photo coordinate (x_P, y_P) and the image coordinate (i_P, j_P) of the point P in the original image by equation (4.2), in which all of the parameters have been determined by the methods described in Section 4.3.3;
- The image coordinate (i_P, j_P) of the point P in the original image is usually not an integer. The gray value of g_{orig} needs to be calculated from neighbor pixels. A nearest neighbor resampling algorithm is employed for this purpose;
- The calculated gray value, g_{orig}, in Step (5) is assigned to the brightness, g_{result}, of the resulting (rectified) image pixel;
- The above procedure is then repeated for each pixel to be rectified until an entire image is orthorectified.

4.3 EXPERIMENTS AND ANALYSIS

4.3.1 Control Point Dataset

As described earlier, it was practically impossible to create photogrammetrically targeted GCPs prior to the 1960s in the study areas of Greenland, and in any event no such targets were prepared. The unique control datasets available for this study are a digital synthetic aperture radar (SAR) mosaic and a digital elevation model (DEM), both generated in the 1990s.

4.3.1.1 Digital SAR Mosaic Map of Greenland

The JPL and the Goddard Space Flight Center of the NASA assembled a digital SAR mosaic map of the Greenland ice sheet using several 1992 ERS-1 SAR image swaths. The mosaic image shows both the location of the ice edge and the distribution of melt-related "scatterers" on the surface (Figure 4.5) and was digitally mosaicked, geolocated, and resampled to a 100-m pixel size. This mosaic is distributed via CD and the internet from the U.S. National Snow Ice Data Center (NSIDC). Table 4.2 lists some parameters.

It should be mentioned that even though the ERS-1 SAR mosaic map of Greenland was constructed from microwave sensor, SAR, images, and the microwave radiation can penetrate 10 m or so into dry snow. The accuracy of ERS-1 SAR mosaic map is enough to be taken as planimetric control data because the pixel resolution of the DISP image is 140 m.

4.3.1.2 Digital Elevation Model (DEM) of Greenland

The Danish Geodetic and Cadastral Agency KMS in 1996 reported their product of a full-cover DEM of Greenland (Ekholm 1996). The DEM was constructed from satellite altimetry (Geosat, south of 72°N, ERS-1 north of 72°N), airborne radar altimetry (Greenland Aerogeophysics Project), and laser altimetry (Airborne Ice Mapping Project) over the ice sheet and from digitized maps of the coastal areas. The accuracy of the DEM is highly variable, ranging from a few meters at the

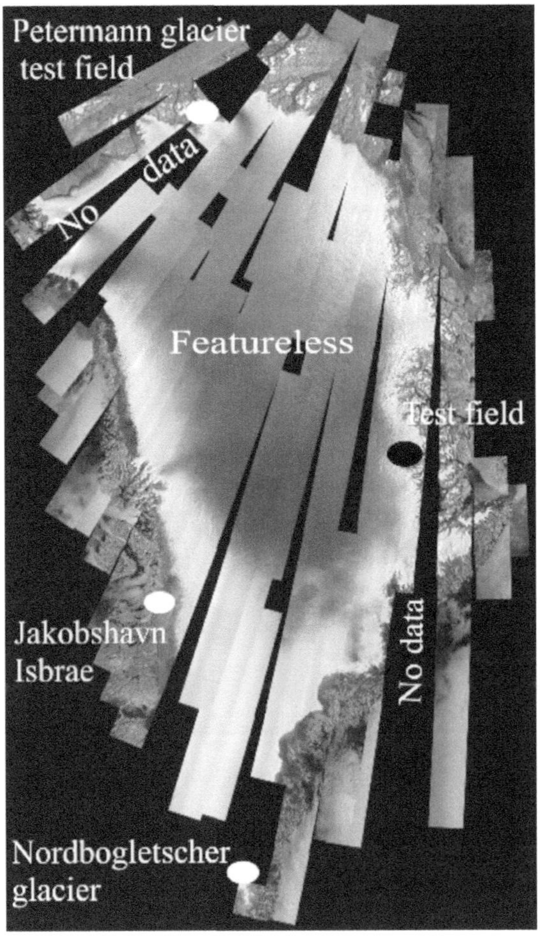

FIGURE 4.5 ERS-1 SAR mosaic assembled by JPL and GSFC of NASA, the blank area denotes no data, and the central area is featureless.

TABLE 4.2
Parameters of Digital SAR Mosaic Imagery of Greenland

Data Source		Pixel		Datum		Image Size
Sensor	Date	resolution	Projection Type	Planimetric	Vertical	(Pixel)
SAR by ERS-1	August 1992	100 m	Polar stereographic	WGS 84	WG S84	*X*: 15646 *Y*: 26266

center of the Greenland ice sheet to several hundred meters on the steepest portions on the coast of the Greenland. The cell size of the DEM, which is a grid form, is with an average of 1–2 km. The other parameters are listed in Table 4.3.

4.3.2 Experiments and Accuracy Analysis

Each DISP image covers approximately 483 × 483 km^2, while the area of Greenland's ice sheet is about 1600 × 2800 km^2. To assemble the mosaic image covering the entire Greenland ice sheet, we chose 24 of the best films (partially cloud-covered) from 8 orbits collected by the ARGON

TABLE 4.3
Parameters of the Digital Elevation Model of Greenland

Data Source	Elevation Accuracy (m)		Pixel Size	Projection	Datum		Image Size (Pixel)
					Planimetric	Vertical	
1. Satellite altimetry	Center	Steepest	1–2 km	Polar stereographic	WGS84	WGS84	*X*: 3631
2. Airborne radar altimetry	12–13	200–250					*Y*: 2850
3. Laser altimetry	25–35						

TABLE 4.4
Comparison of Accuracy in the First Test Field Using 144 Checkpoints

	Mean (m)		Point Position (m)	Comment
	μ_X	μ_Y	$\Delta_r = \sqrt{\mu_X^2 + \mu_Y^2}$	High latitude, low relief displacement
Second-order polynomial	4008	4883	6317	
Camera model	403	390	561	
Our method	106	90	139	

KH-5 (KeyHole) 9034A mission in October 1962, to assemble the first mosaics. In the same way, 36 images from 14 orbits acquired during the ARGON 9058A and 9059A missions in August and October 1963 are chosen to assemble the 1963 mosaic.

For the bundle adjustment of the 24 images of 1962, the 144 GCPs distributed in 21 images (GCPs were not identified in the remaining three images because of ice sheet coverage) and 362 tie points are measured. Observation equations for both GCPs and the tie points are established using equations (4.5)–(4.7), and then least square adjustment is employed for solving all unknown parameters through an iterative process. In each iteration, increments for all unknown parameters are obtained and used to update the approximations for the next iteration. Finally, the coefficients of the third-order polynomial (equation 4.4) and the elements of exterior orientation of 21 images are obtained. The elements of exterior orientation of the other three images are determined by equation (4.4). In the same way, the elements of exterior orientation of 31 images of 1963 are solved by bundle adjustment, for which 331 distributed GCPs and 800 tie points are measured in this group of computation. The elements of exterior orientation of the other five images are determined by equation (4.4). After the elements of exterior orientation of all images are determined, the orthorectification for the 1962 and 1963 images is carried out by procedures described in Section 4.3. Finally, the original images are orthorectified, resampled, and geocoded into the polar stereographic coordinate system. Three typical test fields are selected for the assessment of the accuracy of our method in comparison with the second-order polynomial method and the differential rectification method (called *camera model* in ERDAS/IMAGINE software, ERDAS Inc.).

The first test field is located at 81°N and 60°W on the northwestern flank of the Greenland ice sheet, which is centralized near the Petermann glacier. The terrain of this test field is relatively flat with the difference in elevation less than 300 m. In the process of the bundle adjustment, only seven GCPs, scattered over half of the image, are selected in this test field. The same GCPs (number and distribution) are employed for the camera model and the second-order polynomial method for evaluating the accuracy of orthorectification methods. A total of 144 checkpoints are selected for evaluation of the statistical accuracy relative to ERS-1 SAR. The accuracy for the three methods is shown in Table 4.4.

TABLE 4.5
Comparison of Accuracy at the Second Test Field Using 162 Checkpoints

	Mean (m)		Point position (m)	Comment
	μ_X	μ_Y	$\Delta_r = \sqrt{\mu_X^2 + \mu_Y^2}$	Mid-latitude, the biggest relief displacement
Second-order polynomial	3967	3733	5447	
Camera model	400	397	564	
Our method	103	97	141	

TABLE 4.6
Comparison of Accuracy at the Third Test Field Using 37 Checkpoints

	Mean (m)		Point position (m)	Comment
	μ_X	μ_Y	$\Delta_r = \sqrt{\mu_X^2 + \mu_Y^2}$	Lowest latitude, low relief displacement
Second-order polynomial	4800	4798	6788	
Camera model	466	493	678	
Our method	113	104	153	

The second test field is located at 69°10 N on the west coast of Greenland, which is centralized near Jakobshavns Isbrae. This is the world's fastest known glacier, moving at a speed of up to 7 km/year. The difference in elevation in Jakobshavns Isbrae reaches 1000 m (Zhou et al. 2002). It means that the relief displacement caused by the difference in elevation is significant. The eight GCPs, with a distribution that almost covers an entire image, are measured in the bundle adjustment model. The same number and distribution of the GCPs are employed for the camera model and the second-order polynomial method for evaluating the accuracy of orthorectification methods. The statistical accuracy of three methods relative to the ERS-1 SAR mosaic image is evaluated with 162 checkpoints, and the results are listed in Table 4.5.

The third test field is located at 61°28 N on the south of Greenland, which is centralized at the Nordbogletscher glacier. The five GCPs, clustered in the southeastern corner of the image, are measured in the bundle adjustment model, and the same GCPs are employed for the camera model and the second-order polynomial method for evaluating the orthorectification accuracy. The statistical accuracy evaluated with 37 checkpoints relative to the ERS-1 SAR mosaic image is shown in Table 4.6.

Comparing Table 4.4 with Table 4.5, we find that the relief displacement caused by the difference in elevation in the area of the Jakobshavns Isbrae is still less than 30 μm (about 120 m in the ground distance) (the computational formula is from p. 130 of Wolf (1983)), while the distortion caused by the Earth's curvature can reach 9 km in the ground distance (Zhou et al. 2002). Comparing Table 4.4 with Table 4.6, it is clear that the accuracy of the orthorectification in the area of the Nordbogletscher glacier is lower than that in the area of the Petermann glacier. This may be because the number of GCPs and their strength and distribution in the Nordbogletscher glacier are less optimal than those in the Petermann glacier. To better illustrate this, another test field, which is located in eastern Greenland, is selected to evaluate the influence of GCP distribution on the

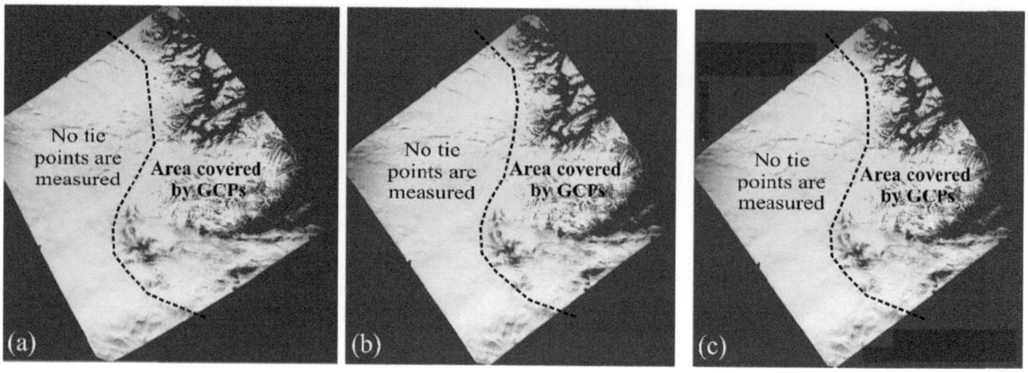

FIGURE 4.6 The comparison of accuracy for three images orthorectified by (a) second polynomial, (b) camera model, and (c) our model.

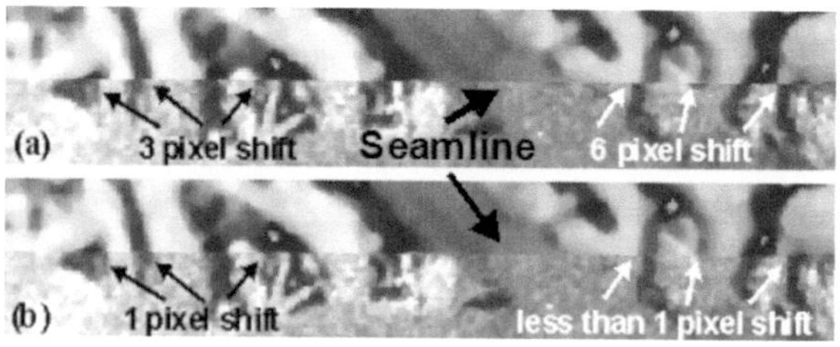

FIGURE 4.7 The accuracy comparison of seam lines in the first test field. Part (a) is produced from the camera model, while (b) is from our model.

accuracy of orthorectification (Figure 4.6). In this test field, seven GCPs with strong distribution are measured. From Figure 4.6, we conclude that the accuracy of orthorectification using polynomial largely depends on the distribution of GCPs, while the camera model seems to have similar accuracy within the area containing the GCPs, but has lower accuracy beyond the GCP region.

4.3.3 Comparison of Accuracy along Seam Line

During mosaicking of two neighboring images, it is very important for us to evaluate the accuracy of the seam line of the mosaicked imagery because the shift along the seam line is probably misunderstood as geophysical features if the mosaic is not seamless. We present the accuracy comparison along se eam line when mosaicking two adjacent images. The four seam lines from four glaciers: Petermann glacier, Jakobshavns Isbrae, Nordogletscher glacier, and East of Greenland, which are distributed around the Greenland ice sheet margin, are chosen for this purpose. As compared in Figures 4.7a and b through 4.10a and b, the size of the shift along the seam line in the camera model ranges from 3 to 7 pixels (approximately 420–980 m in the ground distance), which is largely dependent on the proximity to the GCPs. The error is not tolerable for glacier and ice sheet research in the glaciological community because the shifted "features" are probably misunderstood as geophysical features. Therefore, high-accuracy and seamless mosaics of neighbor images are essential. Our model demonstrates that the shift along seam lines of the two neighbor mosaicked images is about 1–1.3 pixels, even at the subpixel level (see Figures 4.7b through 4.10b). This accuracy ensures that a complete mosaic of the entire Greenland ice sheet is seamless. It also demonstrates that our bundle adjustment model is a useful technique for the construction of seamless mosaicked imagery covering a huge area. However, the mosaic of orthoimages generated by the camera model cannot guarantee that the seam line is in fact seamless, because the accuracy of orthorectification by this

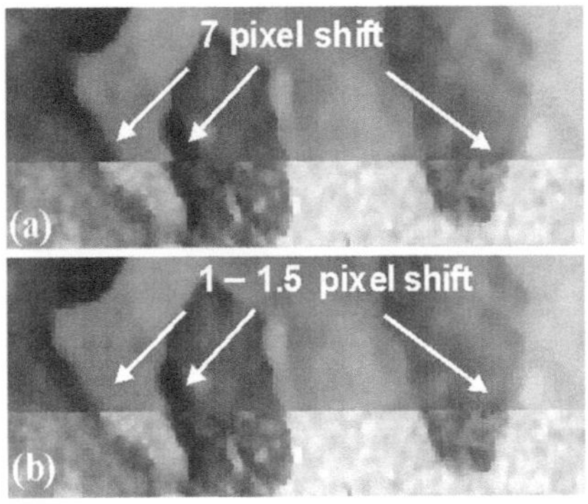

FIGURE 4.8 The accuracy comparison of seam lines in the second test field. Part (a) is produced from the camera model, while (b) is from our model.

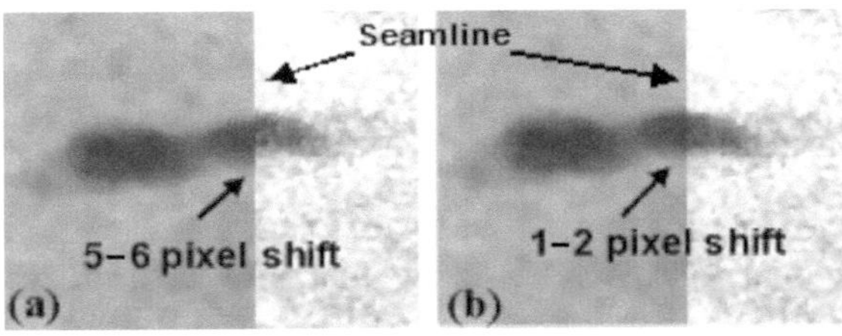

FIGURE 4.9 The accuracy comparison of seam lines in the third test field. Part (a) is produced from the camera model, while (b) is from our model.

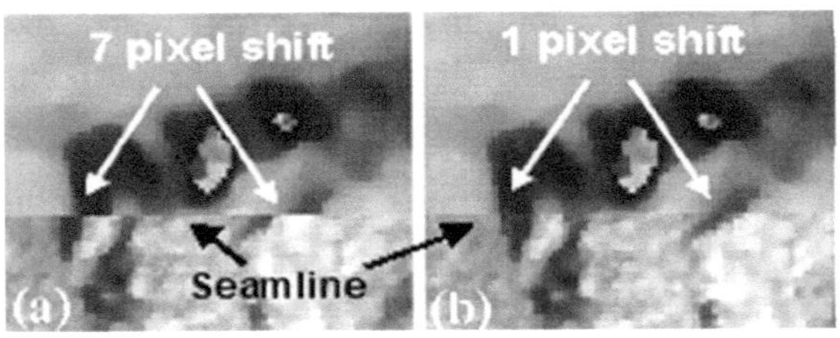

FIGURE 4.10 The accuracy comparison of seam lines in East of Greenland. Part (a) is produced from the camera model, while (b) is from our model.

method is largely dependent on the distribution of GCPs; our model can ensure virtually seamless transitions between images because a large number of tie points are employed. We also conclude that the polynomial method does not model all the sources of geometric error, such as the Earth's curvature, relief displacement, or film/lens distortion; thus, the polynomial method likely will not completely orthorectify all the distortions. The camera model method and our method are able to model all the sources of geometric errors and can orthorectify these distortions.

The polynomial method and camera model methods require a large number of GCPs with good distribution patterns, and the accuracy of rectification using the polynomial method and the camera model method is largely depends upon the number of GCPs and their optimal distribution patterns and strength.

The polynomial method and camera model methods cannot guarantee the seamless mosaic of two neighbor images, while our model can ensure the seamless mosaic because a large number of tie points, which link the overlap images together, are added in our model.

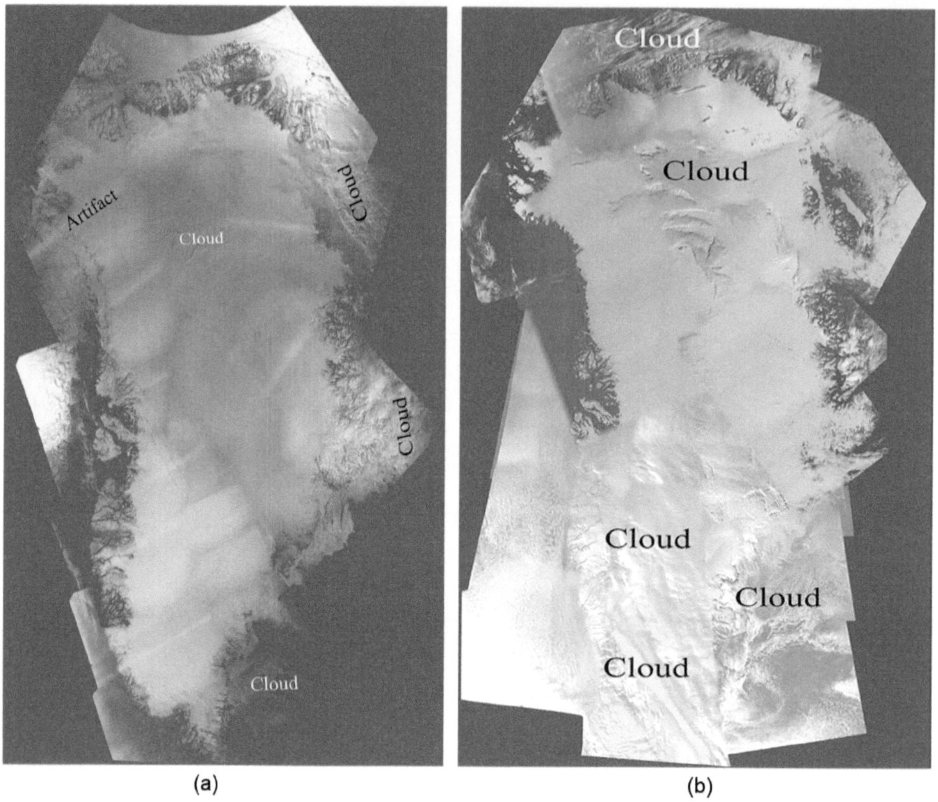

FIGURE 4.11 (a) 1962 mosaic of Greenland, (b) 1963 mosaic of Greenland. Much of southern Greenland is covered by clouds in the 1963 scene. (Courtesy of Zhou and Jezek 2002, 2004.)

4.3.4 Accuracy Evaluation of Final Products

With all of the individual image orthorectification for 1962 and 1963 images, two mosaic products covering the entire Greenland ice sheet are generated (Figure 4.11). Additional image processing, such as filtering and image enhancement, is undertaken. The accuracy of final products is evaluated by measuring 200 points in Greenland's topographically flat area and 261 points in a mountainous area. The statistical mean μ_x, μ_y along x and y directions are calculated relative to the NSIDC SAR mosaic map, and then the average position accuracy is computed by

$$\Delta_r = \sqrt{\mu_X^2 + \mu_Y^2} \tag{4.9}$$

The results demonstrate that the average position accuracy is approximately 168 m in a flat area and 184 m in a mountainous area. In addition, the absolute accuracy by error propagation (assuming a geolocational accuracy of the SAR mosaic map of 400 m (website in National Snow and Ice Data Center 1997)) is evaluated. The average absolute accuracy is estimated to be 430 m in a topographically flat area and 438 m in a mountainous area.

4.4 CONCLUSION

This chapter has presented a highly accurate satellite orbital parameter-assisted photogrammetric bundle adjustment technique for orthorectifying satellite images in the area above 60°N latitude. The model minimizes the need for GCPs in each photo. It is based on the fact that not every image of Greenland contains sufficient GCPs because of nature, such as ice sheet coverage, homologous, and featureless region in the study area. The early reconnaissance DISP satellite images appearing at the beginning of the space age are used for testing our method. The 24 images from 1962 and 36 images from 1963 are orthorectified and assembled into two mosaics of Greenland. The accuracy evaluation of orthorectification demonstrated that our method could reach the highest accuracy of the second-order polynomial method and the camera model method. The accuracy along the seam line achieves 1.0–1.5 pixels, even higher. Finally, two seamless mosaics covering the entire Greenland ice sheet are produced.

The two DISP mosaics are referenced to the WGS 84 datum and the Polar Stereographic projection coordinate system, with the latitude of true scale at 70°N and the longitude of the central meridian at 315°. The false easting and northing are 0.0. The pixel size is 100 m in the ground distance. The accuracy of two DISP mosaic images is approximately 168 m in a flat area and 184 m in a mountainous area. The two mosaic products have been distributed for the use of the research community via CD and the internet through the U.S. NSIDC.

REFERENCES

Chapman, M. A., Analytical Photogrammetry, Lecture Notes For Geomatics Engineering (ENGO 431), September 1995, The University of Calgary, Calgary, Alberta, Canada.

Chen, L. C.; Ran, J. Y., Geometric Correction of airborne scanner imagery using orthophotos and triangulated feature point matching. *International Journal of Remote Sensing*, 14, 1993, pp. 3041–3059.

Devereux, A. J.; Fuller, R. M.; Carter, L.; Parsell, R. J., Geometric correction of airborne scanner imagery by matching Delaunay triangles. *International Journal of Remote Sensing*, 11, 1990, pp. 2237–2251.

Ekholm, S. A full coverage, high-resolution, topographic model of Greenland computed from a variety of digital elevation data. *Journal of Geophysical Research*. 101, 1996, pp. 21961–21972.

Hapgood, M. A., Space physics coordinate transformations: A user guide. *Planetary and Space Science*, 40, 1992, pp. 711–717.

Kraus, K., *Photogrammetry: Fundamentals and Standard Processes*, Vol. 1. Duemmlers Verlag: Duemmler/Bonn, Ferd, Kaiserstrasse 31–37 (Duemmlerhaus), 5300 Bonl, 1997, p. 161.

McDonald, R. A., Opening the cold war sky to the public: Declassifying satellite reconnaissance imagery. *Photogrammetric Engineering and Remote Sensing*, 61, 1995, pp. 380–390.

Mikhail, E. M.; Bethel, J. S.; McGlone, J. C., *Introduction to Modern Photogrammetry*. John Wiley & Sons, Inc.: New York, 2001, p. 119.

National Snow and Ice Data Center, 1997. https://nsidc.org/data/nsidc-0052/versions/1

Seto, Y.; Homma, K.; Komura, F., Geometric correction algorithms for satellite imagery using a bi-directional scanning sensor. *IEEE Transactions on Geoscience and Remote Sensing*, 29, 1991, pp. 292–299.

Shin, D.; Pollard, J. K.; Muller, J. P., Accuracy geometric correction of ATSR images. *IEEE Transactions on Geoscience and Remote Sensing*, 3, 1997, pp. 997–1006.

Snyder, J. P., Map projections – A working manual. US Geological Survey (USGS), Code 1395, 1997, 383p.

Toutin, T.; Carbonneau, Y.; St-Laurent, L., An integrated method to rectify airborne radar imagery using DEM. *Photogrammetric Engineering and Remote Sensing*, 58, 1992, pp. 417–422.

Wolf, P.; Dewitt, B.; Wilkinson, B., *Elements of Photogrammetry with Application in GIS*, 4th ed.; McGraw Hill: New York, 2004..

Wolf, R. A., The quasi-static (slow-flow) region of the magnetosphere. *Solar-Terrestrial Physics: Principles and Theoretical Foundations Based Upon the Proceedings of the Theory Institute, Boston College, August 9–26, 1982*. Springer Netherlands: Dordrecht, 1983, pp. 303–368.

Zhou, G.; Jezek, K., Satellite photograph mosaics of Greenland from the 1960s era. *International Journal of Remote Sensing*, 23, 2002, pp. 1143–1160.

Zhou, G.; Jezek, K., Satellite navigation parameters-assisted orthorectification for over 60°N latitude satellite imagery. *Photogrammetric Engineering and Remote Sensing*, 70, 2004, pp. 1021–1029.

Zhou, G.; Jezek, K.; Wright, W.; Rand; Granger, J., Orthorectifying of 1960's satellite photographs Covering Greenland. *IEEE Transaction on Geoscience and Remote Sensing,* 40, 2002, pp. 1247–1259.

Zhou, G.; Li, R., Accuracy evaluation of ground points from high-resolution satellite imagery IKONOS. *Photogrammetric Engineering and Remote Sensing*, 66, 2000, pp. 1103–1112.

5 Orthorectification with Few GCPs

5.1 INTRODUCTION

The declassified intelligence satellite photograph (DISP) was collected by the first generation of United States photoreconnaissance satellites between 1960 and 1972 through the systems named CORONA, ARGON, and LANYARD. More than 860,000 images of the Earth's surface were declassified with the issuance of this executive order and were contracted to the United States Geological Survey (USGS) for sale. However, further processing and application of DISP has resulted in various problems:

1. The USGS does not provide all users with the parameters required to further process DISP. These parameters include satellite orbit parameters (e.g., inclination, flight height, and descent time) and the camera's interior orientation parameters (e.g., focal length, principal point coordinates, and fiducial marks). This implies that traditional bundle block adjustment based on the photogrammetric collinearity equation is not applicable (Zhou and Jezek 2002; Zhou et al. 2002).
2. It is very difficult to obtain sufficient ground control points (GCPs) in the historical DISP imagery due to the time intervals of several decades and cloudy coverage in Southern China. Thus, it is almost impossible to rectify each DISP image on a frame-by-frame basis.

For the two reasons above, this chapter describes a second-order polynomial equation-based rectification model (OPE-RM) for orthorectification of DISP images. The previous relevant studies on this topic are as follows. Kim et al. utilized a collinearity equation to rectify ARGON imagery from 1963 to study the seasonal variations of glaciers on the Queen Maud Land coast of Antarctica (Kim et al. 2001). Zhou and Jezek proposed a collinearity equation-based self-calibration block bundle adjustment method that integrates the bundle adjustment method and satellite orbital parameters, solving interior orientation (including lens distortion) and exterior orientation parameters simultaneously to rectify ARGON images from 1962 and 1963 (Zhou and Jezek 2002). The rectified ARGON imagery was employed to mosaic Greenland ice sheets from the 1960s, which were then quantitatively compared to the ice sheet extent over a 30-year interval (Zhou et al. 2002). Kim and Jezek applied the state-of-the-art digital imaging technology based on an extended block adjustment to rectify ARGON imagery from 1963 that covered Antarctica (Kim et al. 2007). They assembled all images into a quality mosaic of coastal Antarctica to study glaciers. In addition, due to the imaging model limitations of high-resolution satellites such as IKONOS, rational polynomial-based block adjustment, also called Rational Polynomial Coefficient (RPC), was proposed by multiple authors. For example, Tao et al. analyzed the accuracy of orthorectification of a Satellite Pour l'Observation de la Terre (SPOT) image and an aerial image using the RPC model (Tao and Hu 2001; Tao and Hu 2002). Yang suggested that the RPC model can replace the rigorous sensor model orthorectification of SPOT images (Yang 2000). Liu developed a stereotaxic method of IKONOS images based on the RPC model (Liu et al. 2004). Huang proposed a rational polynomial-based block adjustment for orthorectification of Synthetic Aperture Radar (SAR) images (Huang et al. 2008). Grodecki and Gene Dial rectified IKONOS satellite imagery using the RPC method. The RPC model incorporates prior constraints into the images described by RPC, and multiple independent images can be added in accordance with the needs of users (Grodecki and Dial 2003). However, the RPC model

DOI: 10.1201/9781003632108-5

requires several GCPs, and the computation is very time-consuming. Therefore, the RPC method is not applicable to DISP images that the USGS provides because the imaging model of DISP was not provided by the USGS. Additionally, a few GCPs are available in the study area. Thus, this chapter describes an effective and simple mathematical model for geometric rectification of DISP images, considerably improving the computational effectiveness.

5.2 THE SECOND-ORDER POLYNOMIAL EQUATION-BASED ORTHORECTIFICATION

5.2.1 Polynomial Equation-Based Block Adjustment Model

The objective of polynomial equation-based block adjustment is to tie overlapping images together without the absolute need for GCPs in each image and obtain coordinates of tie points (TPs) and conversion parameters for rectification. Because the study area is a karst landform with a large wavy terrain and large elevation differences, the relief displacement is large. For correction of relief displacement, relief displacement is introduced into the block adjustment model shown below.

Figure 5.1 shows the imaging geometry of DISP from the CORONA mission. $S-WVU(W'V'U')$ is a camera coordinate system, $o-xy$ is an image plane system, and $O-XYZ$ is a geographic coordinate system. There is relief displacement (Δh) in the imaging process; therefore, the relief displacement must be corrected in the rectification process. First, distortion caused by elevation differences should be corrected. Then, other distortions should be corrected by utilizing a polynomial model, and the reverse is true in the resampling process.

Because the relief displacement only occurs in the direction of scanning, CORONA images are panoramic camera images, and the panoramic projection scan direction is the x-direction. Therefore, as shown in the imaging equation above, there is no relief displacement in the y-direction. The relief displacement correction functions are as follows:

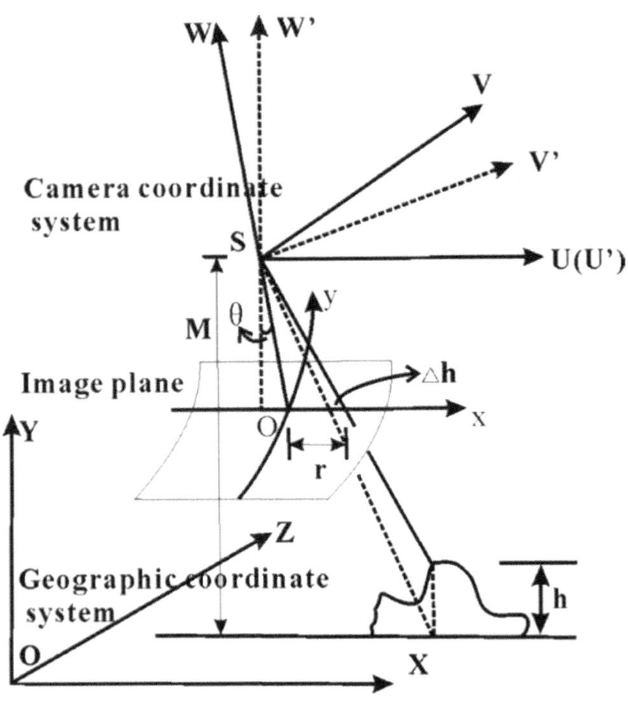

FIGURE 5.1 The imaging geometry of DISP from the CORONA mission.

$$\Delta h = Z \cdot h / M \tag{5.1}$$

$$\begin{cases} \Delta x = x \cdot Z / M \\ \Delta y = y \cdot Z / M = 0 \end{cases} \tag{5.2}$$

where x and y are image coordinates; Δx and Δy are image distortions in the x- and y-directions, respectively, caused by elevation differences; Z is elevation; h is the distance from the image point to the nadir point; and M is the satellite flight altitude. Because the relief displacement occurs in the direction of scanning and the KH-4A/B's images are panoramic camera images, the images can be rectified using the second-order polynomial equation-based model.

5.2.1.1 Traditional Second-Order Polynomial Equation

The traditional second-order polynomial model has been widely applied for image rectification. This chapter extends the traditional equation into a block situation by adding TPs, which tie overlapping images together. With the extended model, the 2D coordinates of TPs and the coefficients of second-order polynomial equations are solved. Furthermore, these parameters are used for orthorectification of DISP imagery without the absolute requirement of at least six GCPs in each DISP image.

The traditional second-order polynomial equations are expressed as follows (Valadan Zoej et al. 2002):

$$\begin{cases} x + \Delta x = a_0 + a_1 X + a_2 Y + a_3 XY + a_4 X^2 + a_5 Y^2 \\ y + \Delta y = b_0 + b_1 X + b_2 Y + b_3 XY + b_4 X^2 + b_5 Y^2 \end{cases} \tag{5.3}$$

where $\alpha = (a_0, a_1, a_2, a_3, a_4, a_5)^T$ and $\beta = (b_0, b_1, b_2, b_3, b_4, b_5)^T$ are coefficients; x and y are image coordinates; Δx and Δy are image distortions in the x- and y-directions, respectively; and X and Y are 2D coordinates in a given map coordinate system.

For a given GCP, equation (5.3) can be linearized using a Taylor series and is expressed as follows:

$$\begin{cases} v_x = \Delta a_0 + X\Delta a_1 + Y\Delta a_2 + XY\Delta a_3 + X^2\Delta a_4 + Y^2\Delta a_5 - l_x \\ v_y = \Delta b_0 + X\Delta b_1 + Y\Delta b_2 + XY\Delta b_3 + X^2\Delta b_4 + Y^2\Delta b_5 - l_y \end{cases} \tag{5.4}$$

where $\Delta a_i\,(i = 0,1,\ldots,5)$ and $\Delta b_i\,(i = 0,1,\ldots,5)$ are correction terms of coefficients; v_x, v_y are residuals; X and Y are 2D coordinates of GCPs; and l_x and l_y are constants expressed by equation (5.5).

$$\begin{cases} l_x = x - \left(a_0 + a_1 X + a_2 Y + a_3 XY + a_4 X^2 + a_5 Y^2\right) \\ l_y = y - \left(b_0 + b_1 X + b_2 Y + b_3 XY + b_4 X^2 + b_5 Y^2\right) \end{cases} \tag{5.5}$$

As shown in equation (5.4), one GCP only establishes two observations, but equation (5.4) has 12 unknown parameters. Therefore, six GCPs, which establish 12 observation equations, are needed to solve the 12 coefficients that are used to rectify a single image. Generally, more than six GCPs are observed in each image to establish more than 12 observation equations. The least-squares estimation is employed to calculate the 12 coefficients. Mathematically, the solution can be described as follows.

Assuming that N GCPs ($N \geq 6$) are observed, the observation equations are expressed in a matrix form as follows.

$$V = A \cdot \alpha - L \tag{5.6}$$

where

$$V = \begin{bmatrix} V_{x1} & V_{y1} & \cdots & \cdots & V_{xN} & V_{yN} \end{bmatrix}^T$$

$$A = \begin{pmatrix} 1 & X_1 & Y_1 & X_1Y_1 & X_1^2 & Y_1^2 & 0 & 0 & 0 & 0 & 0 & 0 & 0 \\ 0 & 0 & 0 & 0 & 0 & 0 & 0 & 1 & X_1 & Y_1 & X_1Y_1 & X_1^2 & Y_1^2 \\ \vdots & \vdots & \vdots & \vdots & \vdots & \vdots & \vdots & \vdots & \vdots & \vdots & \vdots & \vdots & \vdots \\ \vdots & \vdots & \vdots & \vdots & \vdots & \vdots & \vdots & \vdots & \vdots & \vdots & \vdots & \vdots & \vdots \\ 1 & X_N & Y_N & X_NY_N & X_N^2 & Y_N^2 & 0 & 0 & 0 & 0 & 0 & 0 & 0 \\ 0 & 0 & 0 & 0 & 0 & 0 & 0 & 1 & X_N & Y_N & X_NY_N & X_N^2 & Y_N^2 \end{pmatrix}$$

$$\alpha = \begin{pmatrix} a_0 & a_1 & a_2 & a_3 & a_4 & a_5 & b_0 & b_1 & b_2 & b_3 & b_4 & b_5 \end{pmatrix}^T$$

and

$$L = \begin{bmatrix} l_{x1} & l_{y1} & \cdots & \cdots & l_{xN} & l_{yN} \end{bmatrix}$$

The least-squares estimation, i.e., $V^T PV = \min$, gives the solutions of the coefficients of the second-order polynomial equation below.

$$\alpha = (A^T A)^{-1} A^T L \tag{5.7}$$

We can further obtain the following expressions from equation (5.7):

$$a_i = a_i^0 + \sum_{j=1}^{N^{ite}} \Delta a_i^j \quad \left(i = 1,\ldots,5;\ j = 1,\ldots,N^{ite}\right) \tag{5.8a}$$

$$b_i = b_i^0 + \sum_{j=1}^{N^{ite}} \Delta b_i^j \quad \left(i = 1,\ldots,5;\ j = 1,\ldots,N^{ite}\right) \tag{5.8b}$$

where a_i^0, b_i^0 are initial values; $\Delta a_i^j, \Delta b_i^j$ are increased during each iteration; and N^{ite} is the number of iterations.

5.2.1.2 The Second-Order Polynomial Equation-Based Rectification Model

As mentioned above, due to the shortage of GCPs in each of the DISP images, the TPs must be identified to tie images with the same overlapping areas. Under this condition, the TPs whose

XY-coordinates are unknown are introduced into the traditional second-order polynomial equation. This extended model is called the *second-order polynomial equation-based rectification model (2OPE-RM)* in this chapter (see Figure 5.1). Equation (5.3) is extended by considering TPs as unknown parameters and linearized into the following form:

$$\begin{cases} v_x = \Delta a_0 + X\Delta a_1 + Y\Delta a_2 + XY\Delta a_3 + X^2\Delta a_4 + Y^2\Delta a_5 \\ \quad +(a_1 + a_3Y + 2a_4X)\Delta X + (a_2 + a_3X + 2a_5Y)\Delta Y - l_x \\ v_y = \Delta b_0 + X\Delta b_1 + Y\Delta b_2 + XY\Delta b_3 + X^2\Delta b_4 + Y^2\Delta b_5 \\ \quad +(b_1 + b_3Y + 2b_4X)\Delta X + (b_2 + b_3X + 2b_5Y)\Delta Y - l_y \end{cases} \tag{5.9}$$

Then, equation (5.9) can be rewritten as follows:

$$\begin{cases} v_x = \Delta a_0 + X\Delta a_1 + Y\Delta a_2 + XY\Delta a_3 + X^2\Delta a_4 + Y^2\Delta a_5 + f_1\Delta X + f_2\Delta Y - l_x \\ v_y = \Delta b_0 + X\Delta b_1 + Y\Delta b_2 + XY\Delta b_3 + X^2\Delta b_4 + Y^2\Delta b_5 + g_1\Delta X + g_2\Delta Y - l_y \end{cases} \tag{5.10}$$

where

$$f_1 = a_1 + a_3Y + 2a_4X$$

$$f_2 = a_2 + a_3X + 2a_5Y$$

$$g_1 = b_1 + b_3Y + 2b_4X$$

$$g_2 = b_2 + b_3X + 2b_5Y$$

The symbols above are the same as those in equation (5.9).

Additionally, assuming that there are N GCPs ($N \geq 6$), M TPs in t images are collected at the GCPs. Similarly, equation (5.9) can be expressed in a matrix form as follows:

$$V = A \cdot \alpha + B \cdot \beta - L \tag{5.11}$$

where

$$V = \left[\begin{array}{cccccc|cccccc} v_{x1}^{GCP} & v_{y1}^{GCP} & \cdots & \cdots & v_{xn}^{GCP} & v_{xn}^{GCP} & v_{x1}^{TP} & v_{y1}^{TP} & \cdots & \cdots & v_{xM}^{TP} & v_{xM}^{TP} \end{array} \right]^T$$

$$\alpha = \left(\Delta\alpha_0^1, \Delta\alpha_1^1, \Delta\alpha_2^1, \Delta\alpha_3^1, \Delta\alpha_4^1, \Delta\alpha_5^1, \ldots \Delta\alpha_0^t, \Delta\alpha_1^t, \Delta\alpha_2^t, \Delta\alpha_3^t, \Delta\alpha_4^t, \Delta\alpha_5^t\right)$$

$$\beta = \left(\Delta X_1^1, \Delta Y_1^1, \ldots, \ldots, \Delta X_M^t, \Delta Y_M^t\right)$$

$$
A=\begin{pmatrix}
\left.\begin{matrix}
1 & X_1^1 & Y_1^1 & X_1^1Y_1^1 & X_1^{12} & Y_1^{12} & 0 & 0 & 0 & 0 & 0 & 0\\
\vdots & \vdots & \vdots & \vdots & \vdots & \vdots & \vdots & \vdots & \vdots & \vdots & \vdots & \vdots\\
0 & 0 & 0 & 0 & 0 & 0 & 1 & X_{N_1}^i & Y_{N_1}^i & X_{N_1}^iY_{N_1}^i & X_{N_1}^{1\,2} & Y_{N_1}^{1\,2}
\end{matrix}\right\}\\
\text{The first image with } N_1 \text{ GCPs}\\
\left.\begin{matrix}
1 & X_1^i & Y_1^i & X_1^iY_1^i & X_1^{i2} & Y_1^{i2} & 0 & 0 & 0 & 0 & 0 & 0\\
\vdots & \vdots & \vdots & \vdots & \vdots & \vdots & \vdots & \vdots & \vdots & \vdots & \vdots & \vdots\\
0 & 0 & 0 & 0 & 0 & 0 & 1 & X_{N_i}^i & Y_{N_i}^i & X_{N_i}^iY_{N_i}^i & X_{N_i}^{i\,2} & Y_{N_i}^{i\,2}
\end{matrix}\right\}\\
\text{The } i\text{th image with } N_i \text{ GCPs}\\
\left.\begin{matrix}
1 & X_1^N & Y_1^N & X_1^NY_1^N & X_1^{N2} & Y_1^{N2} & 0 & 0 & 0 & 0 & 0 & 0\\
\vdots & \vdots & \vdots & \vdots & \vdots & \vdots & \vdots & \vdots & \vdots & \vdots & \vdots & \vdots\\
0 & 0 & 0 & 0 & 0 & 0 & 1 & X_{N_n}^N & Y_{N_n}^N & X_{N_n}^NY_{N_n}^N & X_{N_n}^{N\,2} & Y_{N_n}^{N\,2}
\end{matrix}\right\}\\
\text{The } N\text{th image with } N_n \text{ GCPs}
\end{pmatrix}
$$

$$
B=\begin{pmatrix}
\left.\begin{matrix} f_1^1 & f_2^1\\ \vdots & \vdots\\ f_2^{M_1} & f_2^{M_1}\end{matrix}\right\}\text{The first image with } M_1 \text{ TPs}\\
\left.\begin{matrix} f_1^i & f_2^i\\ \vdots & \vdots\\ f_1^{M_i} & f_2^{M_i}\end{matrix}\right\}\text{The } i\text{th image with } M_i \text{ TPs}\\
\left.\begin{matrix} f_1^M & f_2^M\\ \vdots & \vdots\\ f_1^{M_n} & f_2^{M_n}\end{matrix}\right\}\text{The } M\text{th image with } M_n \text{ TPs}
\end{pmatrix}.
$$

Equation (5.11) is the 2OPE-RM model derived in this chapter. Relative to the traditional model in equation (5.6), this model introduces TPs as unknown parameters.

Equation (5.11) is usually solved using least-squares estimation, which is expressed as follows:

$$\Phi = V^T V = \min \tag{5.12}$$

With least-squares estimation, the normal equation matrix can be written as follows:

$$\begin{pmatrix} A^T A & A^T B \\ B^T A & B^T B \end{pmatrix}\begin{pmatrix} \delta\alpha \\ \delta\beta \end{pmatrix}=\begin{pmatrix} A^T L \\ B^T A \end{pmatrix} \tag{5.13}$$

Thus, the solution of the unknown parameters is given by equation (5.14):

$$\begin{cases} \Delta a = -\left(Q_{\alpha\alpha} A^T L + Q_{\alpha\beta} B^T L\right) \\ \Delta b = -\left(Q_{\beta\alpha} A^T L + Q_{\beta\beta} B^T L\right) \end{cases} \tag{5.14}$$

where Q_{ij} $(i, j = 1, 2)$ gives the components of the covariance matrix, which is the inverse of the normal matrix, as shown in equation (5.15).

$$Q_{ij} = \begin{pmatrix} A^T A & A^T B \\ B^T A & B^T B \end{pmatrix}^{-1} = \begin{pmatrix} Q_{\alpha\alpha} & Q_{\alpha\beta} \\ Q_{\beta\alpha} & Q_{\beta\beta} \end{pmatrix} (i, j = 1, 2) \tag{5.15}$$

The coefficients of the 2OPE-RM in each image and the 2D coordinates (XY) of each TP are as follows:

$$a_i = a_i^0 + \sum_{j=1}^{N^{ite}} \Delta a_i^j \quad \left(i = 1, \ldots, 5;\ j = 1, \ldots, N^{ite}\right) \tag{5.16a}$$

$$b_i = b_i^0 + \sum_{j=1}^{N^{ite}} \Delta b_i^j \quad \left(i = 1, \ldots, 5;\ j = 1, \ldots, N^{ite}\right) \tag{5.16b}$$

$$X_i^{t_i} = X_i^0 + \sum_{j=1}^{N^{ite}} \Delta X_i^j \quad \left(i = 1, \ldots, 5;\ j = 1, \ldots, N^{ite};\ t_i = 1, \ldots, t\right) \tag{5.17a}$$

$$Y_i^{t_i} = Y_i^0 + \sum_{j=1}^{N^{ite}} \Delta Y_i^j \quad \left(i = 1, \ldots, 5;\ j = 1, \ldots, N^{ite};\ t_i = 1, \ldots, t\right) \tag{5.17b}$$

where X_i, Y_i are coordinates of the ith TP in image t_i; $\Delta X_i, \Delta Y_i$ are increased in X_i and Y_i; a_i^0 and b_i^0 are initial values; and Δa_i and Δb_i are increased in the coefficients in each iteration.

As shown in equation (5.11), each image has 12 unknown parameters $\left(a_i, b_i; i = 0, 1, \ldots, 5\right)$, and each TP has two unknown parameters (*XY* coordinates). Two equations can be established for each GCP or TP. Moreover, the TPs and/or GCPs should be well distributed in each image. For example, there are 4 images, 12 GCPs, and 9 TPs in Figure 5.2. The 4 images imply that there are 48 unknown parameters. The 12 GCPs can be used to establish 42 observation equations (i.e., seven GCPs in Image #1 can be used to establish 14 observations, three GCPs in Image #2 can be used to establish six observations, six GCPs in Image #3 can be used to establish 12 observations, and five GCPs in Image #4 can be used to establish 10 observations). The nine TPs can be used to establish 34 observation equations (i.e., three TPs in Image #1 can be used to establish six observations, three TPs in Image #2 can be used to establish six observations, six TPs in Image #3 can be used to establish 12 observations, and 5 GCPs in Image #4 can be used to establish 10 observations). With this model, we have 76 (76 = 42 + 34) observations and 66 (66 = 48 + 18) unknown parameters. Thus, 2OPE-RM does not require each DISP image to have more than six GCPs.

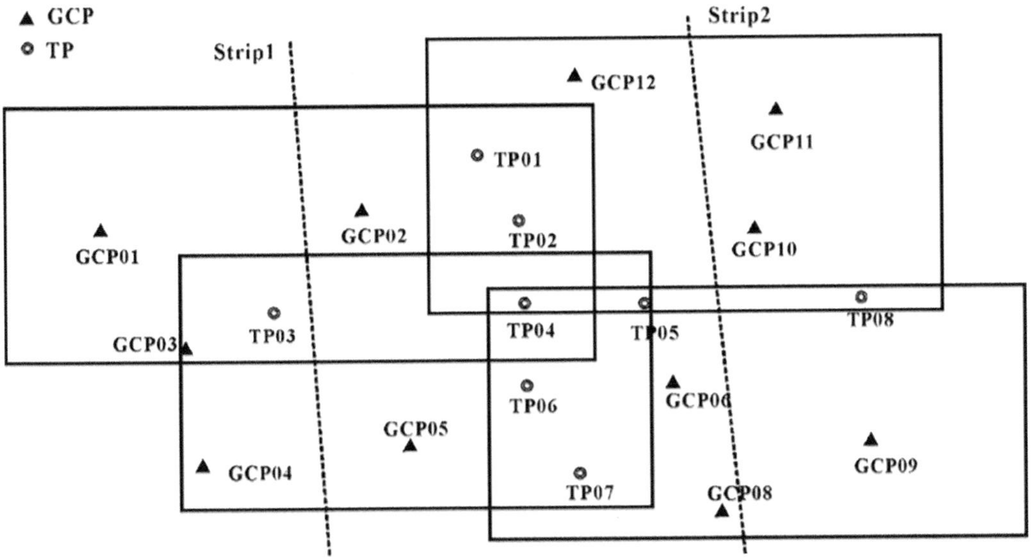

FIGURE 5.2 Illustration of the second-order polynomial equation-based rectification model (2OPE-RM).

The accuracy of the adjustment computation is evaluated using equation (5.18):

$$\delta_o = \sqrt{\frac{V^T V}{r}} \tag{5.18}$$

where δ_o is the standard deviation of the unit weight, V is the matrix of residuals, and r is the number of redundant observations. Thus, the standard deviations of individual unknown parameters can be calculated as follows.

$$\delta_{X_i} = \delta_o \sqrt{Q_{X_i}} \tag{5.19}$$

To evaluate the accuracies of TPs, assuming that there are n TPs, the average of δ_{X_i} is as follows:

$$\mu_X = \frac{1}{n} \sum \delta_{X_i} \tag{5.20}$$

where n is the number of TPs.

5.2.2 Orthorectification of DISP Images

With the established model and the coefficients determined in Section 5.2.1, each original DISP image can be orthorectified. The steps are as follows.

1. *Step 1*: Determination of the rectified image size

   To properly establish the storage space of the orthorectified image, the size of the resulting image (upper left, lower left, upper right, and lower right) must be determined in advance. This procedure is proposed as follows.

- *Determination of the four corner coordinates*: The four corner coordinates of the original image are projected into the UTM coordinate system. Then, eight coordinates are obtained:

$$(X_{ul}, Y_{ul}),\ (X_{ll}, Y_{ll}),\ (X_{ur}, Y_{ur}),\ (X_{lr}, Y_{lr})$$

The maximum and minimum values of X and Y ($X_{\min}$, $X_{\max}$, $Y_{\min}$, and $Y_{\max}$) are calculated from the eight coordinates above to constitute four coordinate pairs. These pairs are the map coordinates of the four boundaries of the resulting image's scope.

$$\begin{aligned} X_{\min} &= \min(X_{ul}, X_{ll}, X_{ur}, X_{lr}), \quad X_{\max} = \min(X_{ul}, X_{ll}, X_{ur}, X_{lr}) \\ Y_{\min} &= \min(Y_{ul}, Y_{ll}, Y_{ur}, Y_{lr}), \quad Y_{\max} = \min(Y_{ul}, Y_{ll}, Y_{ur}, Y_{lr}) \end{aligned} \tag{5.21}$$

- *Determination of the resulting image's size*: The size of the resulting image can be determined by M and N as follows:

$$M = \frac{Y_{\max}}{Y_{GSD} \dfrac{X_{\min}}{X_{GSD}}} \tag{5.22}$$

where $M = row$, $N = col$, and Y_{GSD}, X_{GSD} are the ground-sampled distances (GSDs) in the resulting image.

2. *Step 2*: Coordinate transformation

   Because the orthorectification model only expresses the relationship between the original coordinates (x_{ori}, y_{ori}) and ground coordinates (X_{gro}, Y_{gro}), the ground coordinates should be transformed into the coordinates of the resulting image (x_{re}, y_{re}) as follows:

$$x_{re} = \frac{Y_{gro\max}}{Y_{GSDre} \dfrac{X_{gro} - X_{\min}}{X_{GSD}}} \tag{5.23}$$

   where Y_{gro}, X_{gro} are the ground coordinates of the pixel after rectification.

3. *Step 3*: Orthorectification

   The calculation of the geographic coordinates of individual pixels, resampling of the original image, and registration of the chosen map coordinate system are carried out as follows.

   - The process can be applied to any point $P(I, J)$ in the resulting image with image coordinates (I, J).
   - In accordance with image coordinates (I, J) and GSD, calculate the geographic coordinates (X, Y).
   - Compute the image coordinates (i, j) of point P in the original image using equation (5.5).
   - Calculate the gray value g_{ori} via bilinear resampling interpolation.
   - Assign the gray value g_{ori} to point P as g_{res} in the resulting (rectified) image/pixel.

The above procedure is then repeated for each pixel that must be rectified until the entire image is completely rectified.

5.2.3 Study Area and Dataset

5.2.3.1 Study Area

The study area is located in Guangxi, China, spanning from 20.54°N to 26.24°N latitudes and 104.26°E to 112.04°E longitudes (Figure 5.3) and encompassing 23,790.8 km². The study area is in the south-central subtropics of China.

5.2.3.2 Dataset

DISP imagery: In total, 444 DISP images from five orbits of different missions, including the CORONA 1035-1 Mission (24 images) on September 25, 1966, the CORONA 1102-2 Mission (48 images) on December 18, 1967, and the 1106-1/2 Mission (39 images) on February 7, 1969, were purchased from the USGS (Figure 5.4).

Aerial photos: Five aerial photos with film formats of $18 \times 18\ cm^2$ from 1961 were acquired at a photographic scale of 1:14,000. Each photo covers approximately 6.35 km². Five aerial photos were purchased from the Guangxi Bureau of Geospatial Information, China.

Coordinate data of GCPs: The coordinate data associated with GCPs in the KRD area were collected from Google Earth.

5.3 RESULTS AND ACCURACY ANALYSIS

5.3.1 Image Preprocessing

The DISP film was scanned into digital images, producing film-grain noise and resulting in image quality degradation. Many noise filters have been used in the public domain. However, most of these approaches are either time-consuming because of complex modeling or they erroneously remove geophysical features because of noise in the overall image. The filter algorithm developed by Zhou et al. was used to remove noise in this study (Zhou et al. 2002). One of the advantages of the algorithm is that it avoids the problems noted above because this approach performs statistical calculations within variable-size and variable-shape sub-windows (see Figure 5.5) that are determined individually for every pixel in the image, rather than modeling the noise in the overall image. The algorithm is briefly described as follows.

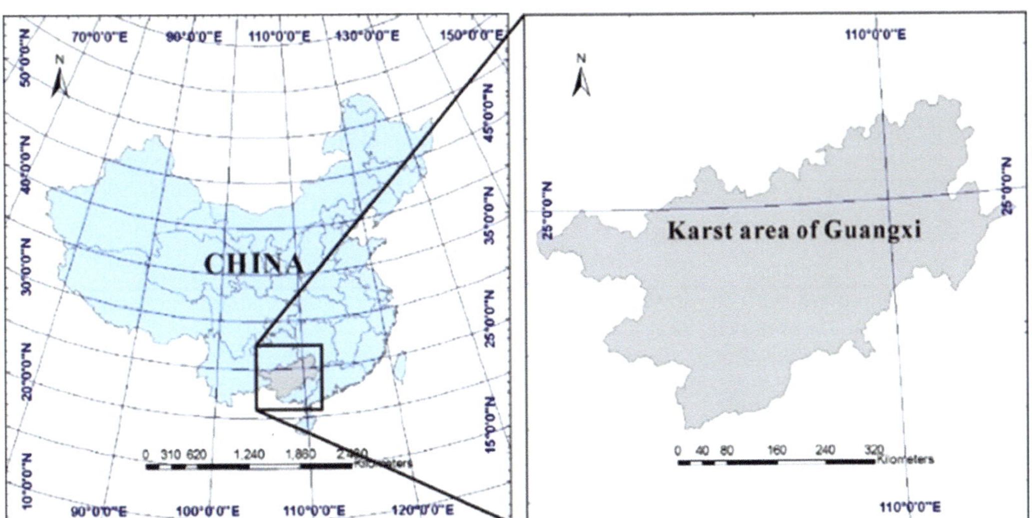

FIGURE 5.3 Study area.

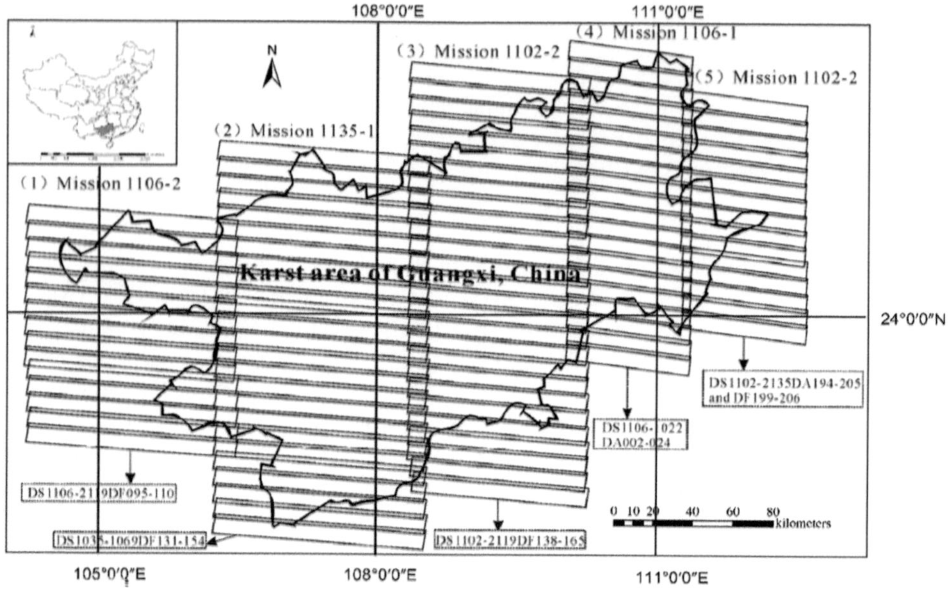

FIGURE 5.4 DISP image dataset.

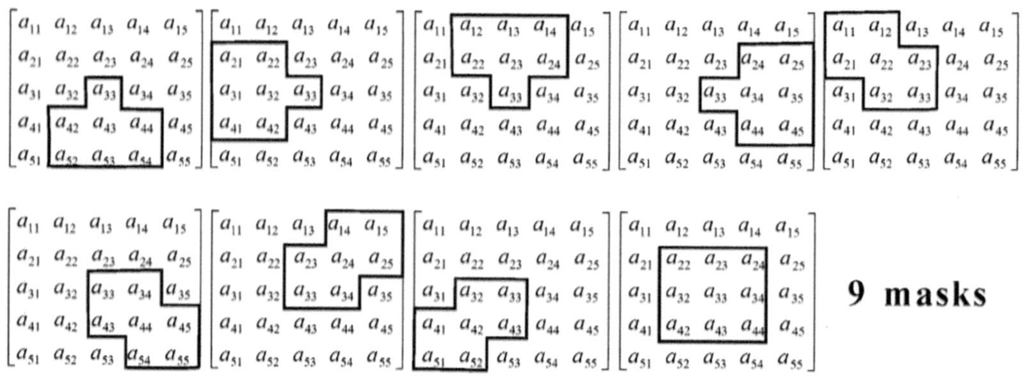

FIGURE 5.5 The adaptive filter algorithm.

1. Select a window of 5×5 pixels.
2. Calculate the mean $n_i \, (i = 1,2,\ldots,9)$ and variance $\alpha_i \, (i = 1,2,\ldots,9)$ of nine masks.
3. Select one mask with the lowest variance α_k and mean n_i, and calculate the weights of every pixel within the kth mask using the following equation:

$$\omega_i = e^{|\Delta_i|}, \quad \Delta_i = gray_i - n_k \tag{5.24}$$

4. Calculate the output using equation (5.25):

$$gray_{output} = \frac{\sum_{i=1}^{M} W_i \cdot gray_i}{\sum_{i=1}^{M} \omega_i} \tag{5.25}$$

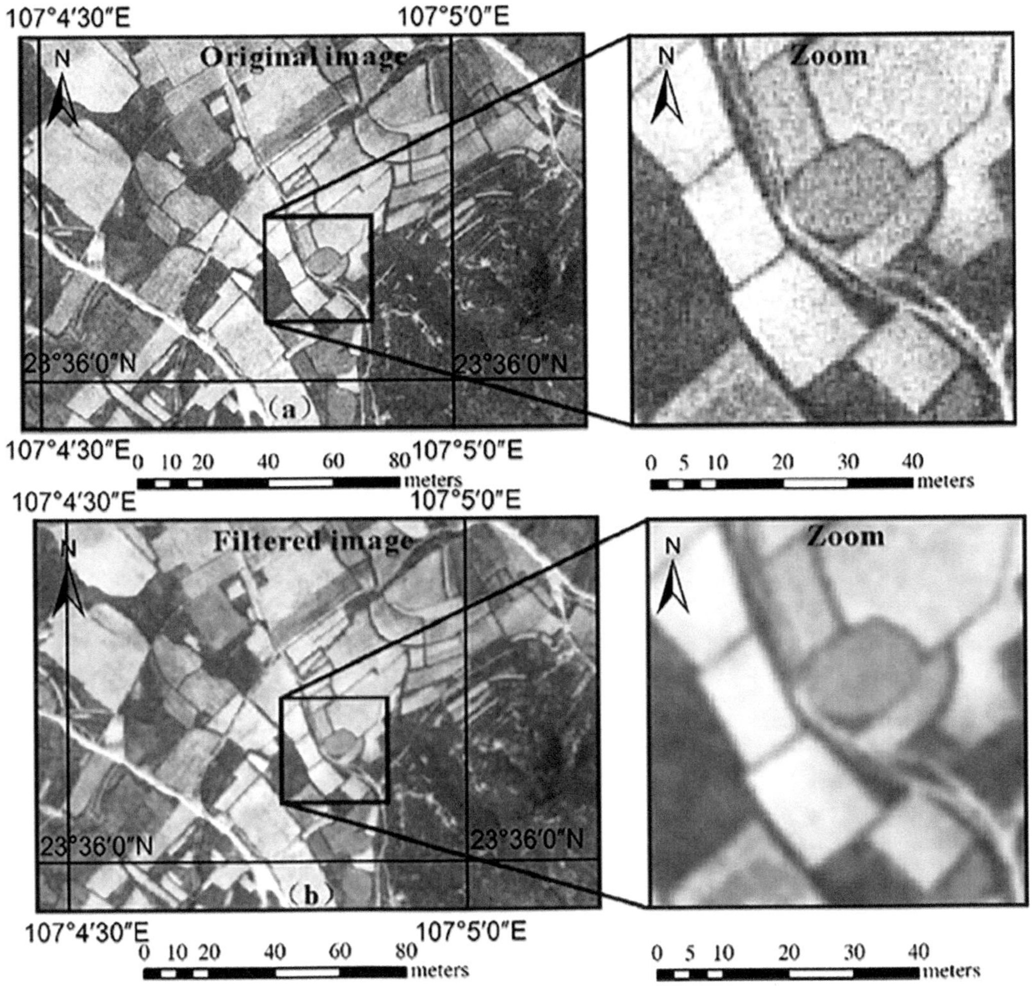

FIGURE 5.6 The results of noise removing (a) the original image and (b) the filtered image.

where M is the number of pixels in the kth mask and $gray_i \left(i = 1, 2, \ldots, 9\right)$ is the intensity.

With the filter algorithm above, the results of removing the DISP image noise are depicted in Figure 5.6, which demonstrates the effectiveness of the proposed approach.

5.3.2 DISP Image Orthorectification and Accuracy Analysis

5.3.2.1 DISP Image Orthorectification

Because sufficient numbers of GCPs are not observed in each DISP image, TPs are identified to tie overlapping images together and solve for the coefficients of the 2OPE-RM. The study area consists of 355 DISP images (there are 444 DISP images in total, but we only employed 355 high-quality images). Thus, it is impractical to construct a block in the entire study area and then solve for the orthorectification parameters of all DISP images simultaneously because such a huge block will produce a significantly large number of observation equations, resulting in a huge computational burden during matrix inversion. Therefore, this chapter divides the study area into 24 blocks consisting of various DISP images (see Figure 5.7a). Each block was rectified independently. For example, Block 1 consists of nine images in Figure 5.7b, in which 20 GCPs and 29 TPs were identified

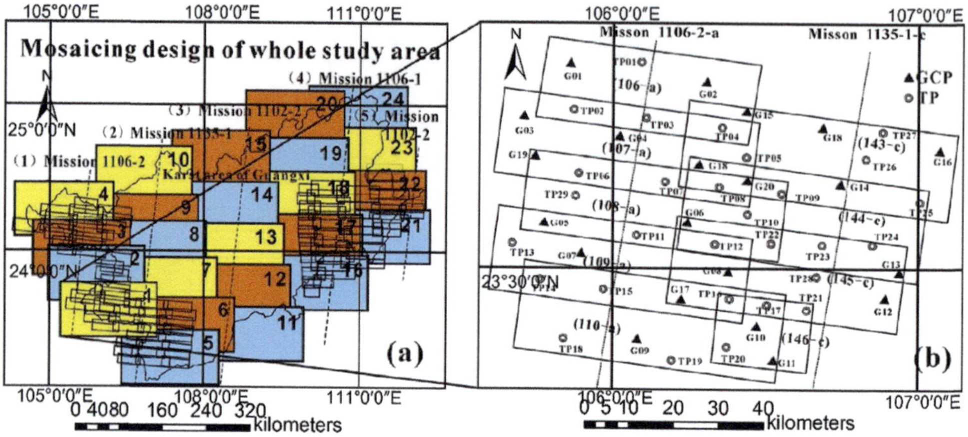

FIGURE 5.7 The polynomial block adjustment: (a) design of polynomial block adjustment in the entire area, which is divided into 24 blocks. (b) One block, which is used to explain the establishment of the observation equations.

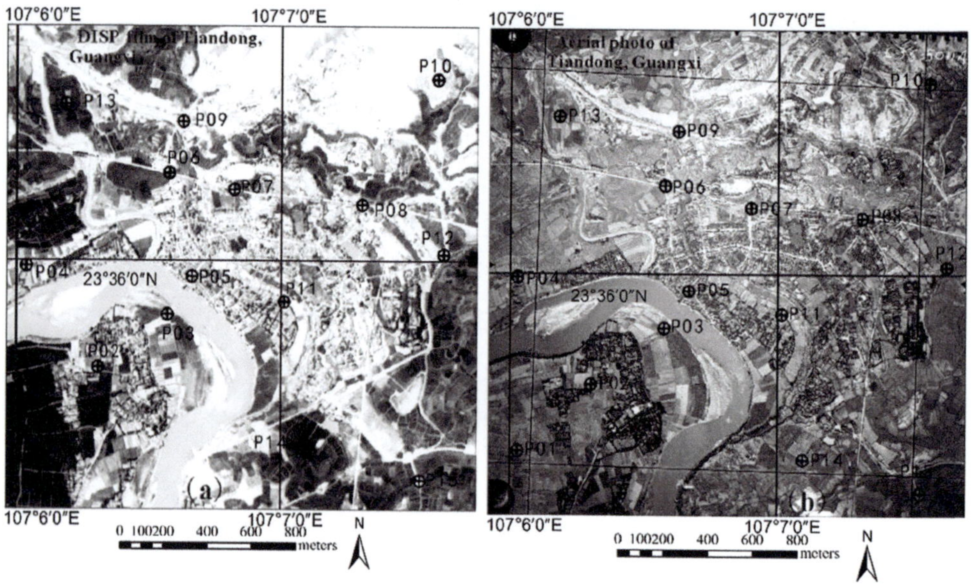

FIGURE 5.8 Accuracy verification using orthorectified aerial photos from 1961: (a) DISP image orthorectified using the proposed method and (b) orthorectified aerial photo from 1961.

and measured. The 20 GCPs are employed 36 times to establish 72 observation equations. The 29 TPs are employed 60 times to establish 120 observation equations. Thus, 192 observation equations (72 + 120 = 192) are established in Block 1. There are 166 unknown parameters (9 × 12 + 29 × 2 = 166). There are 26 redundant observations (i.e., 192 – 166 = 26), meeting the requirements of least-squares adjustment.

With the 192 observation equations established using equation (5.11), the parameters used to rectify the nine DISP images are solved simultaneously using equation (5.14). The 2D coordinates of TPs are obtained using equation (5.17). With the solved coefficients and TP coordinates in each image, orthorectification is performed for each DISP image at a GSD of 2.0 m. Figure 5.8a is part of one orthorectified DISP image.

The computational accuracies of TPs using the 2OPE-RM are evaluated by equation (5.20). The standard deviations of TPs (μ_X and μ_Y) are, on average, 0.34 and 0.23 m, respectively. In addition, the "absolute" accuracy of the orthorectified aerial photo created in 1961 is calculated using the following equations:

$$\Delta X_{RMSE} = \sqrt{\frac{\sum_{k=1}^{n} (X_k - x_k)^2}{n-1}} \tag{5.26}$$

$$\Delta Y_{RMSE} = \sqrt{\frac{\sum_{k=1}^{n} (Y_k - y_k)^2}{n-1}} \tag{5.27}$$

where X_k and Y_k are XY-coordinates of TPs in the orthorectified DISP image, x_k and y_k are XY-coordinates in the orthorectified aerial photo created in 1961, and n is the total number of TPs. Using equations (5.26) and (5.27), ΔX_{RMSE} and ΔY_{RMSE} are 2.0 and 1.6 m, respectively. These values are equivalent to approximately 2.0 pixels in the orthorectified DISP imagery.

5.3.2.2 Accuracy Comparison Analysis

Accuracy comparison between the DISP images orthorectified using the traditional second-order polynomial model and the 2OPE-RM was conducted. Two test fields, which are located in mountainous and flat areas, were selected for the accuracy comparison.

1. The Bameng field is a mountainous area located in Bameng County to the west of the city of Baise, Guangxi, China, at 23.671°N–24.135°N and 106.941°W–107.698°W. This test area covers the entire DS1106–2119DF107a image. The maximum and minimum elevations are 1128 and 790 m, respectively, above mean sea level. Therefore, the relief displacement is significant. There are 12 GCPs and seven TPs scattered throughout the test field. The 12 GCPs are used for second-order polynomial equations to solve for the 12 rectification coefficients, and the 12 GCPs and seven TPs are used in the 2OPE-RM to calculate the coefficients. Twenty-three checkpoints were chosen to evaluate the achievable accuracy. The orthorectified aerial photo provided by the Bureau of Guangxi Geomatics and Geographic Information is considered to represent the "true" values for validation. The results are listed in Table 5.1.
2. The Longzhou field is a flat area located in Longzhou County to the west of the city of Chongzuo, Guangxi, China, at 22.105°N–22.469°N and 106.593°W–106.878°W. This test field completely covers the entire DS1106–2119DF110a image. In this test field, 11 GCPs and seven TPs are scattered throughout the DISP image. The same GCPs are employed in the traditional second-order polynomial model and the 2OPE-RM. Twenty-three checkpoints were chosen to evaluate the accuracy. The planimetric accuracies of the two models relative to the orthorectified aerial image are shown in Table 5.1.

TABLE 5.1
Accuracy Comparison of Mountainous and Flat Areas

Test Area / Models	Bameng Field (Located in a Mountainous Area)		Longzhou Field (Located in a Flat Area)	
	ΔX_{RMSE} (m)	ΔY_{RMSE} (m)	ΔX_{RMSE} (m)	ΔY_{RMSE} (m)
Second-order polynomial model	1.96	1.84	1.85	1.57
Our model	1.85	1.69	1.67	1.49

5.3.3 Image Mosaicking

Based on the individual image orthorectification above, the next work is to mosaic the individual orthorectified DISP images into an image map. First, the characteristics of the study area and DISP images must be understood.

1. The study area covers 23,790.8 km^2 (between latitudes 20.54° and 26.24°N and longitudes 104.26° and 112.04°E), which consists of 355 DISP images of total 100 GB. A good mosaicking scheme may save computational time and computer storage.
2. The study area is located in a karst landscape, where mountainous and hilly terrain areas account for two-thirds of the total area.
3. The overlap between neighboring images must be less than 30%.
4. The study area is covered by five strips of DISP images from four missions (Figure 5.9).

To minimize the influence of error propagation and avoid repeatedly sampling images, based on the characteristics above, the mosaicking is designed as follows (see Figure 5.9).

1. The 16 DISP images from Mission 1106 were first mosaicked, covering the western portion of the study area. The mosaicked map is depicted in Figure 5.10a. Twenty DISP images from Mission 1102-2 were mosaicked, and the mosaicked map is depicted in Figure 5.10b. Twenty-eight DISP images from Mission 1106 were mosaicked, and the mosaicked map is depicted in Figure 5.10c. Twenty-three DISP images from Mission 1106 were mosaicked, as the mosaicked map is depicted in Figure 5.10d. Finally, 18 DISP images from Mission 1106 were mosaicked, and the mosaicked map is depicted in Figure 5.10d, covering the eastern portion of the area.
2. With the five mosaicked maps above, a map image of the entire study area was assembled by merging the five mosaicked images. The order of mosaicking is from the east and west to the middle of the study area (see Figure 5.10).

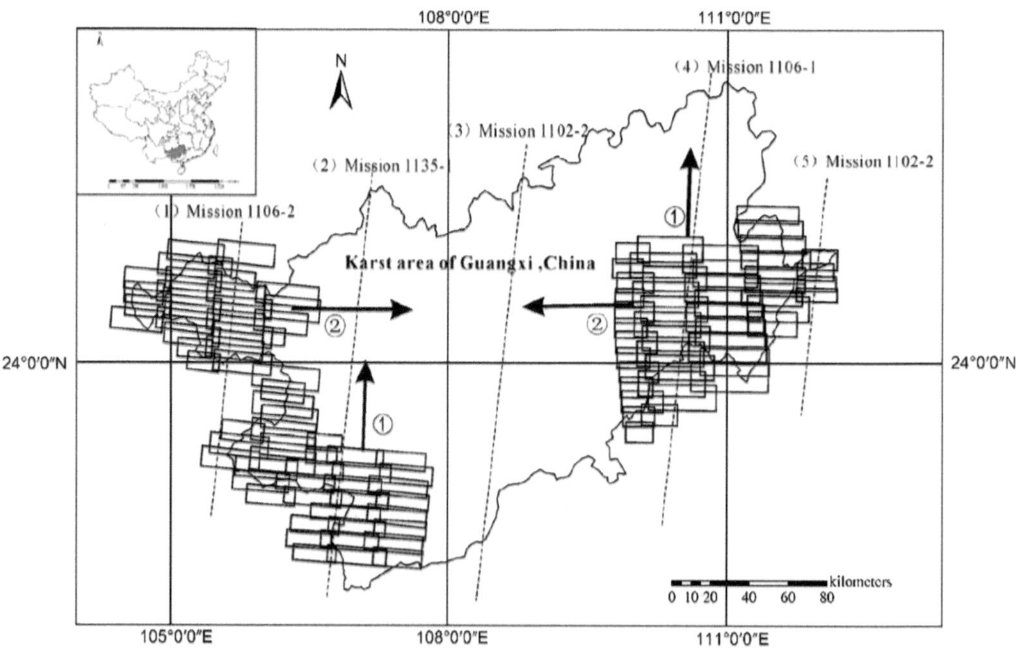

FIGURE 5.9 Mosaicking process from east/west to the middle.

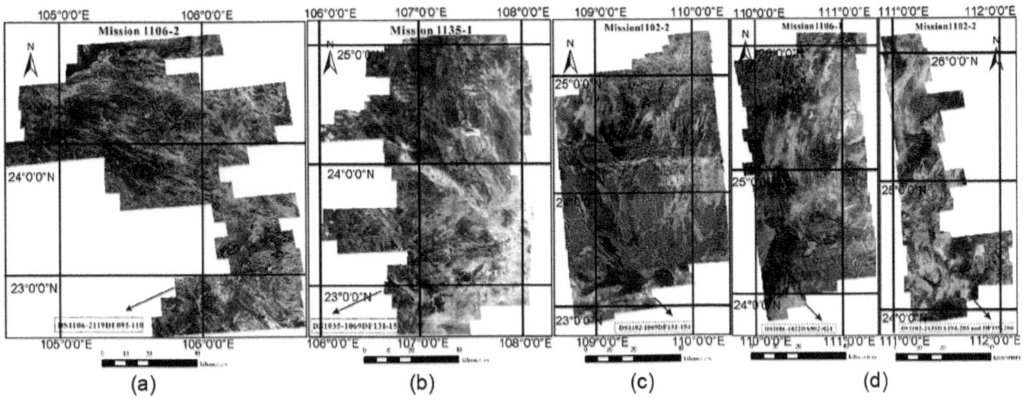

FIGURE 5.10 The mosaicked images of the various missions: (a) for Mission 1106-2, (b) for Mission 1135-1, (c) for Mission 1102-2, (d) for Mission 1106-1, and (e) for Mission 1102-2.

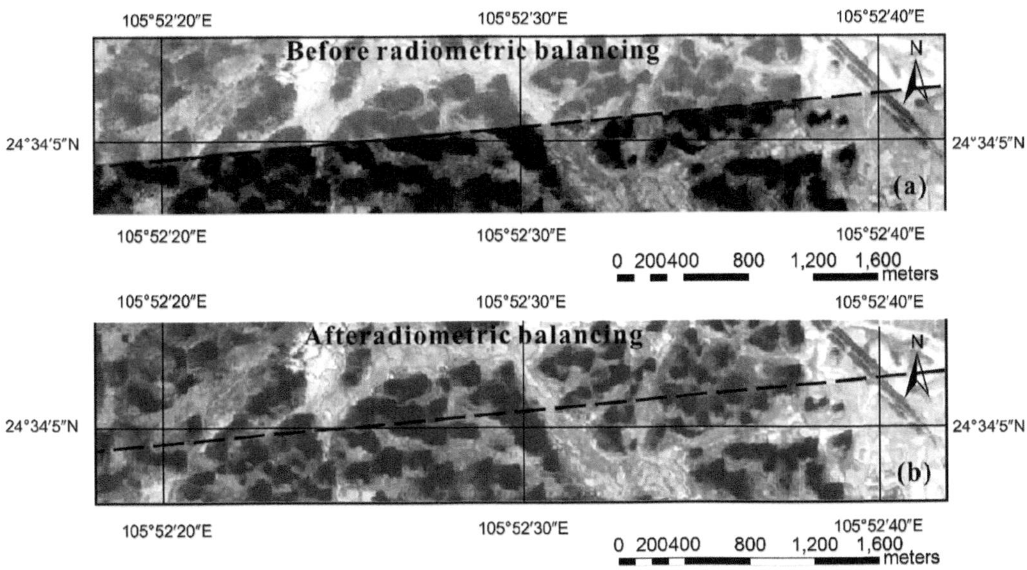

FIGURE 5.11 Radiometric balance of neighboring strips: (a) the image before radiometric balancing and (b) the image after radiometric balancing.

5.3.4 Radiance Balance

Due to the differences in the imaging date/time and different imaging conditions during different missions, brightness differences between neighboring strips are unavoidable. In addition, patchwork lines are also unavoidable. To produce a seamless mosaic of the entire study area, this chapter used a histogram equalization method to adjust the brightness of two neighboring strips. The boundary line was chosen along the center image, and overlapping areas were feathered. Figure 5.11 shows the result of the radiance balance.

5.3.5 Mosaicked Result and Accuracy Evaluation

The entire study area has been mosaicked by 355 orthorectified DISP images (Figure 5.12a). A mountainous area located in Du'an County (Figure 5.12b) and a flat area located in Xingbin

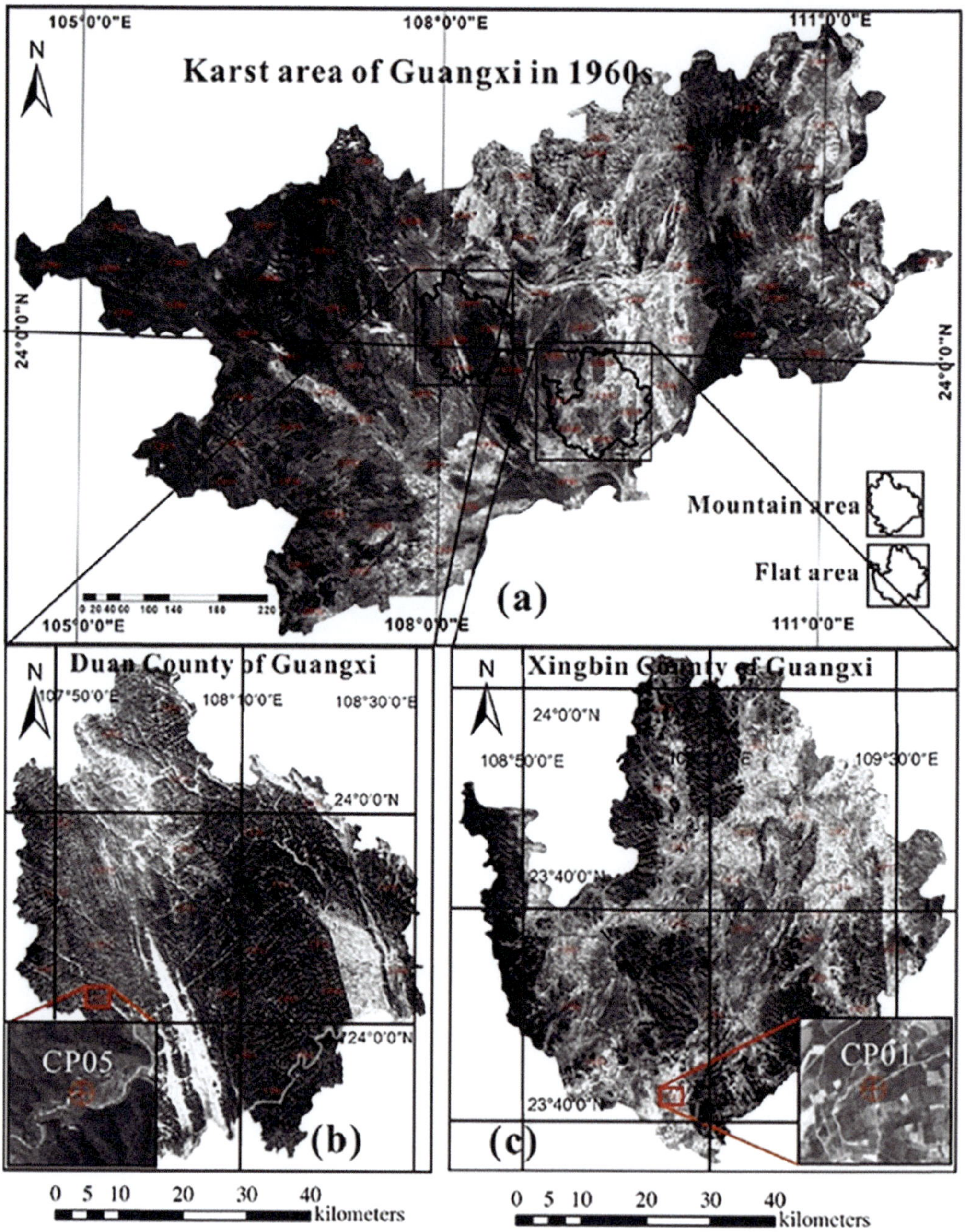

FIGURE 5.12 Mosaic results: (a) a completely assembled 1960s mosaic of the Guangxi karst area, (b) accuracy validation in a mountainous area, and (c) accuracy validation in a flat area.

County (Figure 5.12c) are selected as samples for accurate validation. Seventy-eight GCPs, which were measured by RTK GPS measurements, are uniformly distributed in other countries throughout the entire study area. These include 25 checkpoints (CPs) scattered throughout the two test fields. ΔX_{RMSE} and ΔY_{RMSE} in equations (5.26) and (5.27) are used to measure the accuracy. The results are listed in Table 5.2. As shown in Table 5.2, the accuracy in flat areas is better than that in mountainous areas, and the overall accuracies of the entire study area are 2.11 and 1.74 m.

TABLE 5.2
Final Accuracies of the Assembled DISP Image Map in the Study Area

Area	ΔX_{RMSE} (m)	ΔY_{RMSE} ⟨B⟩ (m)⟨/B⟩
Mountainous area	2.07	1.60
Flat area	1.86	1.79
The entire study area	2.11	1.74

5.4 DISCUSSION

From the accuracy comparison between the DISP images orthorectified by the 2OPE-RM and the DISP images orthorectified by the traditional second-order polynomial model (Valadan Zoej et al. 2002) in Table 5.1, it is demonstrated that the RMSEs of *XY*-coordinates in the images orthorectified by the 2OPE-RM are smaller than those orthorectified by the traditional second-order polynomial model in both the mountainous area and the flat area. This experiment result demonstrated that (1) the proposed 2OPE-RM can successfully solve the problem of the lack of sufficient GCPs in the historical DISP imagery and (2) the proposed 2OPE-RM can obtain a better result than the traditional second-order polynomial method does. In the 2OPE-RM, TPs are the key parameters of the model. The accuracy of TPs, which can be evaluated by equation (5.20), will affect the accuracy of the orthorectification result. The proposed method will be suitable for the condition that the historical images lack sufficient GCPs.

From Table 5.2, it can be seen that the mosaicked images of the entire KRD area in Guangxi still keep a high accuracy, and the accuracy in flat areas is higher than that in mountainous areas.

5.5 CONCLUSIONS

This chapter describes a highly effective, simple, practical mathematical model for the orthorectification of CORONA DISP images from the 1960s, whose interior and exterior parameters are unknown and in which GCPs are lacking. The model is called *the second-order polynomial equation-based block rectification model* (2OPE-RM). With the proposed model, all images can be orthorectified at an accuracy level of 2.0 pixels, corresponding to approximately 2.0–4.0 m with respect to the WGS 84 datum. All the images covering the entire karst area of Guangxi, China, are assembled into a high-quality image map. The sampled distance of the assembled mosaicking map is 2.0 m. The proposed model can solve the problems associated with the traditional second-order polynomial model, such as the lack of GCPs, yielding acceptable and improved accuracy. The assembled image map of the entire rock desertification area in Guangxi, China, will be delivered to the Guangxi Geological Library for use by the research community.

REFERENCES

Grodecki, J.; Dial, G., Block adjustment of high-resolution satellite images described by rational polynomials, *Photogrammetric Engineering and Remote Sensing*, 69, 2003, pp. 59–68.

Huang, M.; Yue, X. J; Zhao, Z.; Fan, H. D., Block adjustment with airborne SAR images based on polynomial ortho-rectification. *Geomatics and Information Science of Wuhan University*, 33, 2008, pp. 569–572.

Kim, K.; Jezek, K. C.; Liu, H., Orthorectified image mosaic of Antarctica from 1963 Argon satellite photography: Image processing and glaciological applications. *International Journal of Remote Sensing*, 28, 2007, pp. 5357–5373.

Kim, K.; Jezek, K. C.; Sohn, H., Ice shelf advance and retreat rates along the coast of Queen Maud Land, Antarctica. *Journal of Geophysical Research*, 106, 2001, pp. 7097–7106.

Liu, J.; W, D. H.; Mao, G. M., High accuracy stereo positioning of IKONOS satellite image based on RPC model. *Bulletin of Surveying and Mapping*, 0911, 2004, pp. 1–4.

Tao, C. V.; Hu, Y., 3D reconstruction methods on the rational function model. *Photogrammetric Engineering& Remote Sensing*, 68(7), 2002, pp. 705–714.

Tao, C. V.; Hu, Y., A comprehensive study of the rational function model for photogrammetric processing. *Photogrammetric Engineering & Remote Sensing*, 67, 2001, pp. 1347–1357.

Valadan Zoej, M.; Mansourian, A.; Mojaradi, B.; Sadeghian, S., 2D geometric correction of IKONOS imagery using genetic algorithm. In: *International Archives of Photogrammetry and Remote Sensing and Spatial Information Sciences*, Ottawa, Ontario, July, 2002, pp. 9–12.

Yang, X. H., Accuracy of rational function approximation in photogrammetry. In: *Proceedings of ASPRS Annual Convention XXXIII (B3)*, Amsterdam, Netherlands, July, 2000, pp. 16–22.

Zhou, G; Jezek, K. C., Satellite photograph mosaics of Greenland from the 1960s era. *International Journal of Remote Sensing*, 23, 2002, pp. 1143–1159.

Zhou, G.; Jezek, K. C.; Wright, W.; Granger, J., Orthorectification of 1960s satellite photographs covering Greenland. *IEEE Transactions on Geoscience and Remote Sensing*, 40, 2002, pp. 1247–1259.

6 Near Real-Time Video Orthorectification

6.1 INTRODUCTION

Applications of small, low-cost, moderately functional, varying in size, and long-endurance unmanned aerial vehicle (UAV) systems for private sector businesses and nonmilitary government agencies to meet geospatial needs focusing on small areas of interest are largely attracting many researchers. One of the most important image data processing in the UAV system is real-time orthorectification and mosaic of video images so that the geo-referenced UAV video image can be merged with geospatial data for fast response to time-critical events (Zhou et al. 2006). Many approaches of image orthorectification have been presented by researchers in the past decades. The previous methods have handled different operation platforms, such as spaceborne (e.g., Sheng and Alsdorf 2005) and airborne (e.g., Bielski et al. 2007; Shin et al. 1997; Zhou et al. 2005), and different sensors, such as radar (Simard et al. 1999), visible and multispectral images (e.g., Li et al. 2003; Zhou et al. 2002; Du et al. 2001). The previously mathematical models, ranging from a simple affine transformation (which utilizes higher-order polynomials), to projective transformations, have also been discussed (Zhou et al. 2002).

This chapter describes a method of real-time orthorectification of UAV-based video streams in order to meet the need of UAV data processing for fast response to time-critical events.

6.2 MATHEMATICAL MODEL OF ORTHORECTIFICATION

For our UAV system (see Zhou et al. 2006), the geometric configuration between the two navigation sensors and the video camera is depicted in Figure 6.1. The mathematical model can be expressed by

$$r_G^M = r_{GPS}^M(t) + R_{Att}^M(t) \cdot \left[s_G \cdot R_C^{Att} \cdot r_g^C(t) + r_{GPS}^C \right] \tag{6.1}$$

where r_G^M is a vector to be computed for any ground point G in the given mapping frame; $r_{GPS}^M(t)$ is a vector of the GPS antenna phase center in the given mapping frame, which is determined by the onboard GPS at a certain epoch (t); s_G is a scale factor between the camera frame and the mapping frame; $r_g^C(t)$ is a vector observed in the image frame for point g, which is captured and synchronized with GPS epoch (t); R_C^{Att} is the so-called *boresight matrix* (orientation offset) between the camera frame and the attitude sensor body frame; and r_{GPS}^C is the vector of position offset between the GPS antenna geometric center and the camera lens center, which is usually determined by terrestrial measurements as part of the calibration process. $R_{Att}^M(t)$ is a rotation matrix from the UAV attitude sensor body frame to the given mapping frame and is a function of the three attitude angles, i.e.,

$$R_{Att}^M = \begin{pmatrix} \cos\psi\cos\zeta & \cos\xi\sin k + \sin\xi\sin\psi\cos\zeta & \sin\xi\sin\zeta - \cos\xi\sin\psi\cos\zeta \\ -\cos\psi\sin\zeta & \cos\xi\cos k - \sin\xi\sin\psi\sin\zeta & \sin\xi\cos\zeta + \cos\xi\sin\psi\sin\zeta \\ \sin\psi & -\sin\xi\cos\psi & \cos\xi\cos\psi \end{pmatrix} \tag{6.2}$$

where ξ, ψ, and ζ represent roll, pitch, and yaw, respectively. So, the relationship between the two sensors is, in fact, to mathematically determine a matrix R_C^{Att} through equation (6.1). The determination

DOI: 10.1201/9781003632108-6

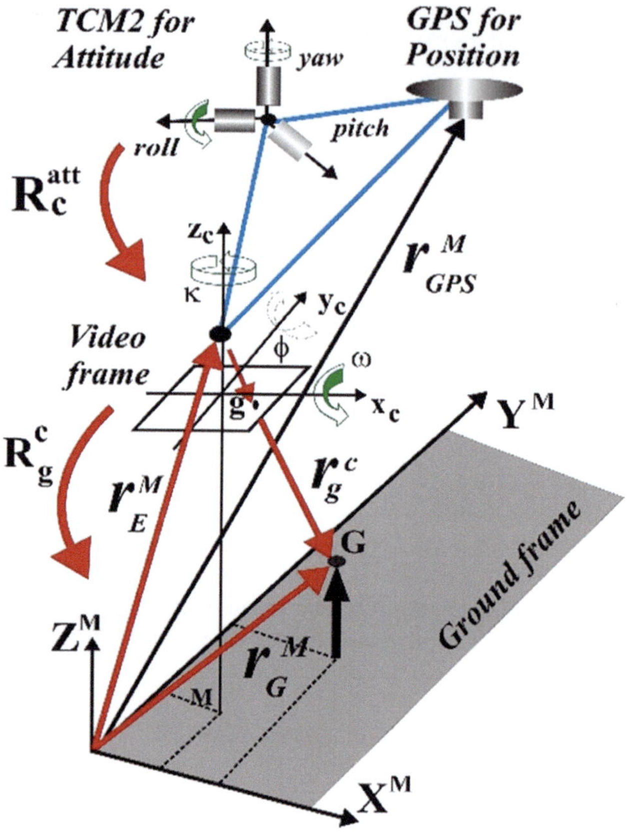

FIGURE 6.1 The geometric configuration for UAV-based multisensors, including a video camera, GPS, and attitude sensor. (Courtesy of Zhou et al. 2006.)

of R_C^{Att} is usually solved by a least-squares adjustment on the basis of a number of well-distributed ground control points (GCPs). Once this matrix is determined, its value is assumed to be a constant over the entire flight time in the traditional airborne mapping system (Grejner-Brzezinska 1999; Yastikli and Jacobsen 2005; Cramer and Stallmann 2002; Mostafa et al. 1998; Schwarz et al. 1993). The basic procedures of UAV-based orthorectification and mosaic are as follows.

6.2.1 Calibration of Video Camera

Video camera would calibrate the parameters such as focal length, principal point coordinates, and lens distortion, which are called interior orientation parameters (IOPs). We used the direct linear transformation (DLT) method, which was originally presented by Abdel-Aziz and Karara (1971). This method requires a set of GCPs whose object space and image coordinates are already known. In this step, the calibration process only considers the focal length and principal point coordinates because the solved IOPs and EOPs (exterior orientation parameters) will be employed as initial values in the later bundle adjustment model. The DLT model is

$$x_{g1} - x_0 + \rho_1\left(x_{g1} - x_0\right)r_1^2 = \frac{L_1X_G + L_2Y_G + L_3Z_G + L_4}{L_9X_G + L_{10}Y_G + L_{11}Z_G} = f_x^1 \quad (6.3a)$$

$$y_{g1} - y_0 + \rho_1\left(y_{g1} - y_0\right)r_1^2 = \frac{L_5X_G + L_6Y_G + L_7Z_G + L_8}{L_9X_G + L_{10}Y_G + L_{11}Z_G} = f_y^1 \quad (6.3b)$$

where $r_{(i)}^2 = (x_{g(i)} - x_0)^2 + (y_{g(i)} - y_0)^2$ $(i = 1,2)$; (x_{g1}, y_{g1}) and (x_{g2}, y_{g2}) are the coordinates of the image points g_1 and g_2 in the first and second image frames, respectively; (X_G, Y_G, Z_G) are the coordinates of the ground point, G; (x_0, y_0, f, ρ_1) are the IOPs; and $L_i\,(i = 1, \ldots, 9)$ are unknown parameters.

Equation (6.3) is a nonlinear equation and must be linearized using a Taylor series. The linearized equation is

$$-\left[X_G L_1 + Y_G L_2 + Z_G L_3 + L_4 + x_{g1} X_G L_9 + x_{g1} Y_G L_{10} + x_{g1} Z_G L_{11}\right]/A \\ +\left(x_{g1} - x_0\right) r_1^2 \rho_1 + x_{g1}/A = v_x \tag{6.4a}$$

$$-\left[X_G L_5 + Y_G L_6 + Z_G L_7 + L_8 + y_{g1} X_G L_9 + y_{g1} Y_G L_{10} + y_{g1} Z_G L_{11}\right]/A \\ +\left(y_{g1} - y_0\right) r_1^2 \rho_1 + y_{g1}/A = v_y \tag{6.4b}$$

The matrix form of equation (6.4) is

$$V = C\Delta + L \tag{6.5}$$

where

$$C = -\frac{1}{A}\begin{bmatrix} X_G & Y_G & Z_G & 1 & 0 & 0 & 0 & 0 & x_{g1}X_G & x_{g1}Y_G & x_{g1}Z_G & \left(x_{g1} - x_0\right)r_1^2 \\ 0 & 0 & 0 & 0 & X_G & Y_G & Z_G & 1 & y_{g1}X_G & y_{g1}Y_G & y_{g1}Z_G & \left(y_{g1} - y_0\right)r_1^2 \end{bmatrix}$$

$$\Delta = \begin{bmatrix} L_1 & L_1 & L_1 & L_1 & L_1 & L_1 & L_1 & L_1 & L_1 & L_1 & L_1 & \rho_1 \end{bmatrix}^T$$

$$V = \begin{bmatrix} v_x \\ v_y \end{bmatrix}$$

$$L = -\frac{1}{A}\begin{bmatrix} x \\ y \end{bmatrix}$$

With the iteration computation, the 11 parameters can be solved. With the solved 11 parameters, the IOPs can be calculated by

$$x_0 = -\left(L_1 L_9 + L_2 L_{10} + L_3 L_{11}\right)/\left(L_9^2 + L_{10}^2 + L_{11}^2\right) \tag{6.6}$$

$$y_0 = -\left(L_5 L_9 + L_6 L_{10} + L_7 L_{11}\right)/\left(L_9^2 + L_{10}^2 + L_{11}^2\right) \tag{6.7}$$

$$f_x^2 = -x_0^2 + \left(L_1^2 + L_2^2 + L_3^2\right)/\left(L_9^2 + L_{10}^2 + L_{11}^2\right) \tag{6.8a}$$

$$f_y^2 = -y_0^2 + \left(L_5^2 + L_6^2 + L_7^2\right) / \left(L_9^2 + L_{10}^2 + L_{11}^2\right) \tag{6.8b}$$

$$f = \frac{f_x + f_y}{2} \tag{6.9}$$

The EOPs can be calculated by

$$a_3 = L_9 / \sqrt{\left(L_9^2 + L_{10}^2 + L_{11}^2\right)}$$

$$b_3 = L_{10} / \sqrt{\left(L_9^2 + L_{10}^2 + L_{11}^2\right)}$$

$$c_3 = L_{11} / \sqrt{\left(L_9^2 + L_{10}^2 + L_{11}^2\right)}$$

$$a_1 = \frac{1}{f_x}\left[L_1 / \sqrt{\left(L_9^2 + L_{10}^2 + L_{11}^2\right)} + a_3 x_0\right]$$

$$b_1 = \frac{1}{f_x}\left[L_2 / \sqrt{\left(L_9^2 + L_{10}^2 + L_{11}^2\right)} + b_3 x_0\right]$$

$$c_1 = \frac{1}{f_x}\left[L_3 / \sqrt{\left(L_9^2 + L_{10}^2 + L_{11}^2\right)} + c_3 x_0\right]$$

$$a_2 = \frac{1}{f_y}\left[L_5 / \sqrt{\left(L_9^2 + L_{10}^2 + L_{11}^2\right)} + a_3 y_0\right]$$

$$b_2 = \frac{1}{f_y}\left[L_6 / \sqrt{\left(L_9^2 + L_{10}^2 + L_{11}^2\right)} + b_3 y_0\right]$$

$$c_2 = \frac{1}{f_y}\left[L_7 / \sqrt{\left(L_9^2 + L_{10}^2 + L_{11}^2\right)} + c_3 y_0\right]$$

The rotation matrix can be expressed by

$$R_M^C = \begin{pmatrix} a_1 & a_2 & a_3 \\ b_1 & b_2 & b_3 \\ c_1 & c_2 & c_3 \end{pmatrix} \tag{6.11}$$

The exposure center coordinates (X_S, Y_S, Z_S) can be calculated by solving the following equations:

$$a_3 X_S + b_3 Y_S + c_3 Z_S + L' = 0 \tag{6.12a}$$

$$x_0 + f_x\left(a_1 X_S + b_1 Y_S + c_1 Z_S\right) / L' + L_4 = 0 \tag{6.12b}$$

$$y_0 + f_y\left(a_2 X_S + b_2 Y_S + c_2 Z_S\right) / L' + L_8 = 0 \tag{6.12c}$$

6.2.2 Determination of the Offset between GPS Antenna and Camera

The GPS antenna geometric center and the camera lens center cannot occupy an identical center. The offset $\left(r_{GPS}^{M}\right)$ between the two centers must precisely be measured so that the correction can be carried out in equation (6.1). The precise measurement of the offset was conducted using a Topcon® GTS-2B Total Station, whose specification can be referenced at https://vm.civeng.unsw.edu.au/surveying/f_pall/html/e14.html. The process is as follows: (1) set up the Total Station far from 5 to 10 m away from the UAV aircraft, (2) take a shot to the GPS antenna and read the horizontal and vehicle distance and angles from the Total Station, (3) take a shot to the lens of camera, during which the vertical wire of telescope of the Total Station is aligned with the telescope axis and the horizontal wire of telescope of the Total station is aligned with the shut, (4) revise the telescope of total station and repeat the operations of Steps 2 and 3, (5) repeat the operation of Steps 2, 3, and 4 for three times, (6) suppose that the origin of a presumed local coordinate is at Total Station, calculate coordinates of the GPS antenna $\left(X_{GPS}, Y_{GPS}, Z_{GPS}\right)$ and the camera lens' $\left(X_{lens}, Y_{lens}, Z_{lens}\right)$, and (7) calculate the offset between two centers by $D_{offset} = \sqrt{\left(X_{GPS} - X_{lens}\right)^2 + \left(Y_{GPS} - Y_{lens}\right)^2 + \left(Z_{GPS} - Z_{lens}\right)^2}$. The measurement accuracy reached the millimeter level, since the Total Station has a measurement capability of the millimeter level.

6.2.3 Solution of Kinematic GPS Errors

For kinematic GPS errors, we limit the baseline length to ground reference stations for the onboard differential GPS (DGPS) survey. It has been demonstrated that a GPS receiver onboard an UAV can achieve an accuracy of a few centimeters using this limitation (Zhou et al. 2006). The other errors will be orthorectified in this chapter mathematically. Basically, the traditional differential rectification model is based on photogrammetric collinearity, in which the interior and exterior orientation elements and DEM (*X*, *Y*, and *Z* coordinates) are known.

6.2.4 Estimation of Boresight Matrix

With the solved EOPs, estimation of an initial boresight matrix, R_C^{Att}, can be realized through multiplication of the attitude sensor orientation data derived from the onboard TCM2 sensor with the three angular elements of the EOPs solved by DLT. The formula is expressed by

$$R_C^{Att}(t) = \left[R_M^C(t) \cdot R_{Att}^M(t)\right]^T \tag{6.13}$$

where R_C^{Att} and R_{Att}^M are the same as in equation (6.1); R_M^C is a rotation matrix, which is a function of three rotation angles (ω, ϕ, and κ) of a video frame, and is expressed by

$$R_M^C = \begin{pmatrix} a_1 & a_2 & a_3 \\ b_1 & b_2 & b_3 \\ c_1 & c_2 & c_3 \end{pmatrix} = \begin{pmatrix} \cos\phi\cos\kappa & \cos\omega\sin\kappa + \sin\omega\sin\phi\cos\kappa & \sin\omega\sin\kappa - \cos\omega\sin\phi\cos\kappa \\ -\cos\phi\sin\kappa & \cos\omega\cos\kappa - \sin\omega\sin\phi\sin\kappa & \sin\omega\cos\kappa + \cos\omega\sin\phi\sin\kappa \\ \sin\phi & -\sin\omega\cos\phi & \cos\omega\cos\phi \end{pmatrix} \tag{6.14}$$

With the initial values computed above, a rigorous mathematical model was established to simultaneously solve the camera's IOPs and EOPs of each video frame. In addition, because the stereo camera calibration method can increase the reliability and accuracy of the calibrated parameters due to coplanar constraints (Bethea et al. 1997), a stereo pair of images constructed by the first and the second video frames is selected. The mathematical model, for any ground point, *G*, can be expressed as

For the first video frame:

$$f_x^{g1} = -f \frac{r_{11}^1 \left(X_G - X_S^1\right) + r_{12}^1 \left(Y_G - Y_S^1\right) + r_{13}^1 \left(Z_G - Z_S^1\right)}{r_{31}^1 \left(X_G - X_S^1\right) + r_{32}^1 \left(Y_G - Y_S^1\right) + r_{33}^1 \left(Z_G - Z_S^1\right)} \tag{6.15a}$$

$$f_y^{g1} = -f \frac{r_{21}^1 \left(X_G - X_S^1\right) + r_{22}^1 \left(Y_G - Y_S^1\right) + r_{23}^1 \left(Z_G - Z_S^1\right)}{r_{31}^1 \left(X_G - X_S^1\right) + r_{32}^1 \left(Y_G - Y_S^1\right) + r_{33}^1 \left(Z_G - Z_S^1\right)} \tag{6.15b}$$

For the second video frame:

$$f_x^{g2} = -f \frac{r_{11}^2 \left(X_G - X_S^2\right) + r_{12}^2 \left(Y_G - Y_S^2\right) + r_{13}^2 \left(Z_G - Z_S^2\right)}{r_{31}^2 \left(X_G - X_S^2\right) + r_{32}^2 \left(Y_G - Y_S^2\right) + r_{33}^2 \left(Z_G - Z_S^2\right)} \tag{6.16a}$$

$$f_y^{g2} = -f \frac{r_{21}^2 \left(X_G - X_S^2\right) + r_{22}^2 \left(Y_G - Y_S^2\right) + r_{23}^2 \left(Z_G - Z_S^2\right)}{r_{31}^2 \left(X_G - X_S^2\right) + r_{32}^2 \left(Y_G - Y_S^2\right) + r_{33}^2 \left(Z_G - Z_S^2\right)} \tag{6.16b}$$

where $r_{(i)}{}^2 = (x_{g(i)} - x_0)^2 + (y_{g(i)} - y_0)^2$ $\left(i = 1, 2\right)$; $\left(x_{g1}, y_{g1}\right)$ and $\left(x_{g2}, y_{g2}\right)$ are the coordinates of the image points g_1 and g_2 in the first and second video frames, respectively; $\left(X_G, Y_G, Z_G\right)$ are the coordinates of the ground point, G; $\left(x_0, y_0, f, \rho_1\right)$ are the IOPs; $r_{i,j}^m \left(i = 1, 2, 3;\ j = 1, 2, 3\right)$ are elements of the rotation matrix R for the first video frame (when $m = 1$) and the second video frame (when $m = 2$), which are a function of three rotation angles $(\omega_1, \phi_1, \kappa_1)$ and $(\omega_2, \phi_2, \kappa_2)$. The expression is described in equation (6.14). In this model, the unknown parameters contain the camera's IOPs, $\left(x_0, y_0, f, \rho_1\right)$ and the EOPs of the first and second video frames $(X_S^1, Y_S^1, Z_S^1, \omega_1, \phi_1, \kappa_1)$ and $(X_S^2, Y_S^2, Z_S^2, \omega_2, \phi_2, \kappa_2)$, respectively. To solve these unknown parameters, equations (6.15) and (6.16) must be linearized by using a Taylor series expansion including only the first-order terms. The vector form of the linearized equation is expressed by

$$\boldsymbol{v}_1 = \boldsymbol{A}_1 \boldsymbol{X}_1 + \boldsymbol{A}_2 \boldsymbol{X}_2 - \boldsymbol{L} \tag{6.17}$$

where $\boldsymbol{X}_1$ represents a vector of the EOPs of two video frames; $\boldsymbol{X}_2$ denotes the vector of the camera IOPs; $\boldsymbol{A}_1$, and $\boldsymbol{A}_2$ are their coefficients; and $\boldsymbol{v}_1$ is a vector containing the residual error. Their components can be referenced to Zhou et al. (2006).

6.3 GEORECTIFICATION OF VIDEO STREAM

After the orientation parameters of the individual video frame are determined by the model described in Section 6.2, each original video frame can be orthorectified. The procedures include: (1) the determination of the size of the orthorectified image; (2) the transformation of pixel locations from the original image to the resulting (rectified) image using equation (6.1); and (3) resampling the original image pixels into the rectified image for assignment of gray values. The flowchart is illustrated in Figure 6.2.

6.3.1 Determination of Orthorectified Image Size

The orthorectification process registers the original image into a chosen map-based coordinate system, and invariably the size of the original image is changed. To properly set up the storage space

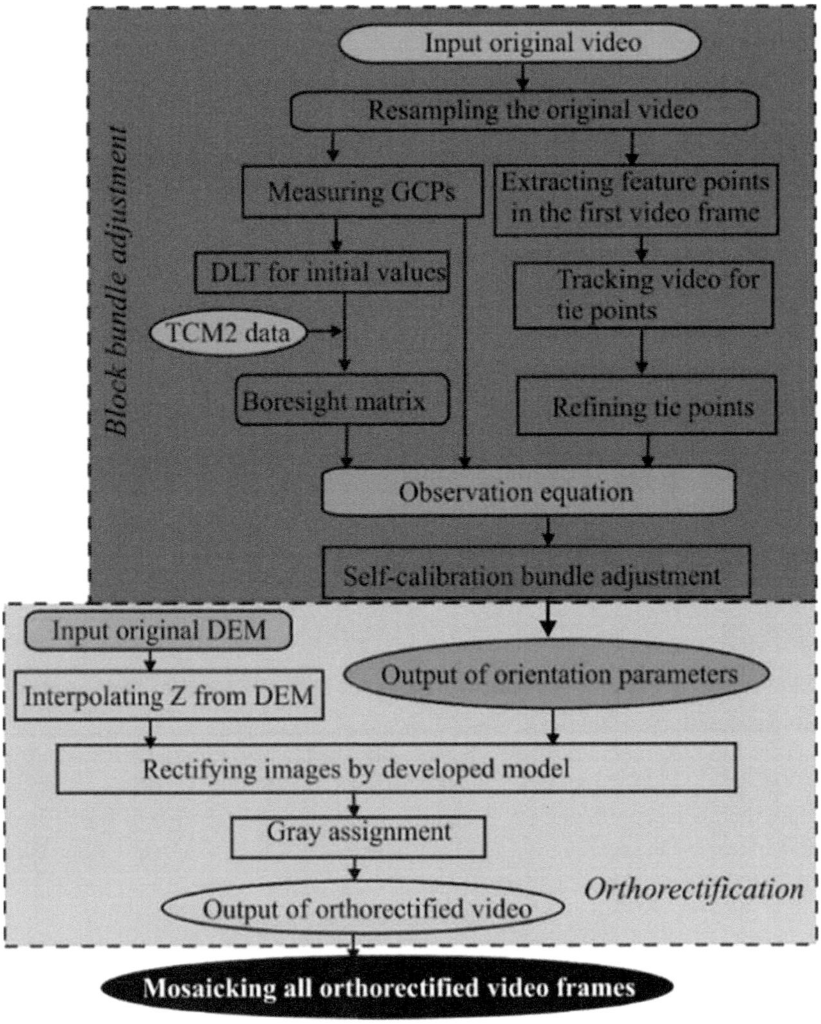

FIGURE 6.2 The flowchart of geometric rectification using the block bundle adjustment model.

requirements when programming, the size of the resulting image footprint (upper left, lower left, upper right, and lower right) has to be determined in advance. These procedures are as follows:

- *The determination of four corner coordinates*: For a given ground resolution of $\Delta_{Xsample}$ and $\Delta_{Ysample}$ along x and y directions in the original image, assume that the planimetric coordinates of any GCP are $\left(X_{GCP}, Y_{GCP}\right)$, whose corresponding location in the original image plane is $\left(row_{GCP}, col_{GCP}\right)$. The coordinates of four corner points can then be determined routinely. For example, for Corner 1, which can be calculated by

$$X_1 = X_{GCP} - col_{GCP} \cdot \Delta_{Xsample}$$

$$Y_1 = Y_{GCP} - row_{GCP} \cdot \Delta_{Ysample}$$

The other corners can also be calculated accordingly.

- *The determination of minimum and maximum coordinates from the above four corners*: For example, for the minimum x coordinate, which can be calculated by

$$X_{\min} = \min\left(X_1, X_3\right)$$

The maximum x ($X_{\max}$) and minimum and maximum y ($Y_{\min}$, $Y_{\max}$) can be calculated accordingly.

- *The determination of the size of the resulting image* is given by

$$N = Col = \frac{X_{\max} - X_{\min}}{\Delta X}, \quad M = Row = \frac{Y_{\max} - Y_{\min}}{\Delta Y}$$

where ΔX and ΔY are ground-sampled distances in the resulting image.

6.3.2 Orthorectification

The basic procedures of orthorectification are as follows:

- For any point P (I, J) in the resulting image, (I, J) are its image coordinates in the image plane;
- Compute the planimetric coordinates of the point $P\left(X_S, Y_S\right)$ with respect to the geodetic coordinate system by using the given cell size;
- Interpolate the vertical coordinates, Z_S from the given DEM using a bilinear interpolation algorithm;
- Compute the photo coordinate (x, y) and the image coordinate (i, j) of the point P in the original image by using equation (6.1), in which all of the parameters have been determined by the methods described in Section 6.2;
- Calculate the gray value g_{orig} by the nearest neighbor resampling algorithm;
- Assign the gray value g_{orig} as the brightness g_{result} of the resulting (rectified) image pixel.

The above procedure is then repeated for each pixel to be rectified. The details of the overall process of the orthorectification can be referenced to Zhou et al. (2002).

6.4 EXPERIMENTS AND ANALYSIS

6.4.1 Experimental Field Establishment

An experimental field, located in Picayune, Mississippi, approximately 15 minutes north of the NASA John C. Stennis Space Center, was established. This test field covers about 4 miles long along N.W. and 3.0 miles wide along S.W. In this field, 21 nontraditional GCPs using DGPS were collected. These "GCPs" are located in the corners of sidewalks, parking lots, crossroad, and curb ends (see Figure 6.2). Each point is observed for at least 30 minutes, and at least four GPS satellites are locked simultaneously. The height angle cutoff is 15°. The planimetric and vertical accuracy of the "GCPs" is decimeter level. This accuracy is enough for taking as GCPs in the late processing of UAV-based geo-referencing and 2D planimetric mapping because the accuracy evaluation of this system is carried out relative to the USGS DOQ (U.S. Geological Survey, digital orthophoto quadrangle), whose cell size is 1 m. In addition to the 21 nontraditional GCPs, 1-m USGS DOQ imagery (Figure 6.3) covering the control field was also downloaded from the USGS website for the accuracy evaluation of UAV-based real-time video data geo-referencing and 2D planimetric mapping.

FIGURE 6.3 USGS DOQ and the distribution of the measured 21 nontraditional GCPs.

TABLE 6.1
Specifications of a Low-Cost Civilian UAV Platform

Power plant	2 stroke, 1½ hp	Endurance	45 minutes at cruise speed	Fuel capacity	0.46 kg
Length × height	1.53 m × 1.53 m	Cruise speed	56 km/h	Wingspan	2.44 m
Gross weight	10 kg	Max speed	89 km/h	Payload	2.3 kg
Operating altitudes	152–619 m	Operating range	1.6–2.5 km		

6.4.2 UAV System

A cheap, small UAV system was developed by Zhou et al. (2006). The specifications of the UAV are listed in Table 6.1. This UAV system is specifically designed as an economical, moderately functional, small airborne platform intended to meet the requirement for fast response to time-critical events in many small private sectors or government agencies for small areas of interest. The cheap materials such as sturdy plywood, balsa wood, and fiberglass are employed to feature a proven versatile hi-wing design, tail dragger landing gear with excellent ground clearance that allows operation from semi-improved surfaces. Generous flaps enable short rolling take-offs and slow flight. The 1½ hp, two-stroke engine burns a commercial glow fuel mixed with gas (Figure 6.4).

In addition, the UAV is constructed to break down into a few easy-to-handle components that quickly pack into a small size van and are easily deployed, operated, and maintained by a crew of three. This UAV system including hardware and software is housed in a lightly converted

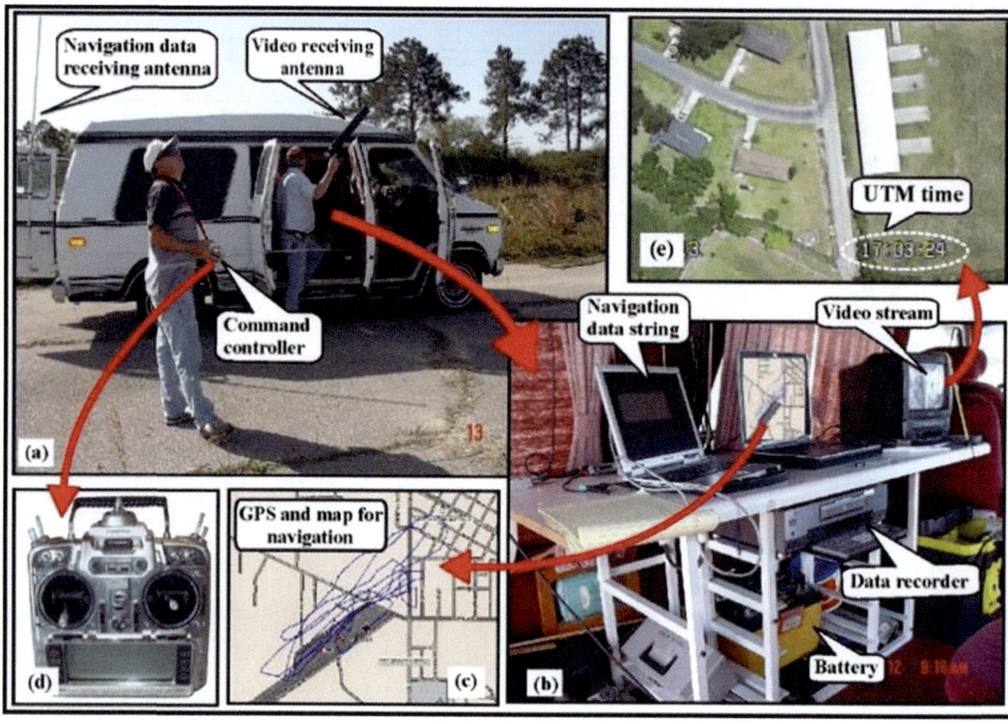

FIGURE 6.4 UAV ground control station and field data collection. (a) Field data collection, (b) Real-time visualization for field mobile platform, (c) 3D visualization of GPS waypoints, (d) UAV remote-control operation, and (e) Real-time image-video stream display (Courtesy of Zhou et al. 2006.).

(rear seat removed and bench top installed) van (Figure 6.4); on a mobile vehicle for providing command, control, and data recording to and from the UAV platform; and real-time data processing in order to meet the requirement of fast response to time-critical events. The field control station houses the data stream monitoring and UAV position interface computer, radio downlinks, antenna array, and video terminal. All data (GPS data, UAV position and attitude data, and video data) are transmitted to the ground receiver station via wireless communication, with real-time data processing in the field for fast response to rapidly evolving events. In this project, three onboard sensors, GPS, attitude sensor (TCM2), and video camera are integrated into a compact unit. GPS is a GARMIN eTrex Vista Personal Navigator Handheld GPS Receiver with 12 parallel channels, which continuously tracks and uses up to 12 satellites to compute and update the position. The eTrex Vista combines a basemap of North and South America, with a barometric altimeter and electronic compass. The compass provides bearing information and the altimeter determines the UAV precise altitude. An attitude navigation sensor (TCM2–20) is selected to provide the real-time UAV's attitude information. This sensor integrates a three-axis magneto-inductive magnetometer and a high-performance two-axis tilt sensor (inclinometer) in a package and provides tilt-compensated compass headings (azimuth, yaw, or bearing angle) and precise tilt angles relative to Earth's gravity (pitch and roll angle) for precise three-axis orientation. The electronic gimbaling eliminates moving parts and provides information about the environment of pitch and roll angles and three-dimensional magnetic field measurement. Data is output on a standard RS-232 serial interface with a simple text protocol that includes checksums. A video camera, *Topica Color TP 6001A CCD*, was used to acquire the video stream at a nominal focal length of 8.5 mm with auto and preset manual focus and program and manual exposure. The camera was installed in the UAV payload bay at a nadir-looking direction. The video stream is recorded with a size of 720 (h) × 480 (v) pixel^2 and delivered in an MPEG-I format.

6.4.3 Data Collection

The data was collected over the established test field on April 3, 2005. The UAV and all the other hardware including computers, monitors, antennas, and the periphery equipment (e.g., cable), and the software developed in this project are housed in and transported to the field test field via the field control station (see Figure 6.4). After the UAV was assembled, all the instruments, such as antennas, computers, video recorder, and battery, were set up, and the software system is tested. An autopilot avionics system was employed in this UAV system for command, control, autopilot telemetry, DGPS corrections uplink and the pilot in the loop (manual flight) modes. The Piccolo autopilot data link is built on a MHz 910/2400 radio modem (http://www.cloudcaptech.com/). The data link has up to 40k band throughput and is used. The data architecture allows multiple aircraft to be controlled by a single operator from a single ground control station. Data from the payload can be downlinked over the main data link. The autopilot has pressure ports for total and static pressure. Both the dynamic and static pressures are used in the autopilot primary control loops.

Video data stream was collected for approximately 60 minutes and is transmitted (downlinked) to the field control station in real time using a 2.4 GHz, S-band transmitter with a 3 dB transmit antenna. The data collection process demonstrated that the received video is much clear (Figure 6.4e). Moreover, the UTC time taken from the onboard GPS was overlaid onto the video in the lower right-hand corner (Figure 6.4e). Meanwhile, the video is recorded on digital tape using the DHR-1000NP SONY DV Edit Recorder. The video was then converted from tape to the MPEG I format and delivered through CD or the Internet.

6.4.4 Bundle Adjustment of Video

With the measurement of several high-quality nontraditional GCPs described in Section 6.4.1, all unknown parameters in equation (6.1) can be solved. In this model, 11 GCPs are employed and their imaged coordinates in the first and second images are also measured. The initial values of unknown parameters including $\left(x_0, y_0, f, \rho_1\right)$, $(X_S^1, Y_S^1, Z_S^1, \omega_1, \phi_1, \kappa_1)$, and $(X_S^2, Y_S^2, Z_S^2, \omega_2, \phi_2, \kappa_2)$ are provided by the above computation. With the initial values, an iterative computation with updating the initial values is carried out, and the final solved results for the first video frame are listed in Table 6.2.

The above computational processing can be extended into an entire strip, in which the interesting distinct points must be extracted and tracked. The final tracked distinct points in the video flow will be used as tie points to tie all overlapping images together in the bundle adjustment model (i.e., equation 6.17). The solution of equation (6.17), the EOPs of each video frame can be obtained. A statistical analysis of EOPs for the video flow (corresponding to 18,200 video frames) is listed in the last column of Table 6.2. From our experimental results, the standard deviation (σ_0) of the six unknown parameters can reach 0.42 pixels. In addition, the maximum, minimum, and average standard deviations of six EOPs are listed in Table 6.3. As seen, the average standard deviations of linear elements of EOPs are less than 1.5 m, and the average standard deviations of nonlinear elements of EOPs are less than 22 seconds.

TABLE 6.2
Results of the Three Methods (σ_0 is the Standard Deviation) for the First Video Frame

	Roll (ω) (°)	Pitch (ϕ) (°)	Yaw (κ) (°)	x_0 (Pixel)	y_0 (Pixel)	f (Pixel)	ρ_1	σ_0 (Pixel)
Onboard TCM2	0.07032	0.00245	1.08561	—	—	—	—	
DLT	−0.01039	0.00002	−1.06379	362.20	241.32	790.54	—	1.27
Our method	−0.01873	0.00032	−1.02943	361.15	239.96	804.09	−1.02E−7	0.42

TABLE 6.3
The Accuracy Statistics of Results of the Proposed Methods (σ_0 is Standard Deviation)

	X_S (m)	Y_S (m)	Z_S (m)	ω (s)	ϕ (s)	κ (s)
Minimum σ_0	0.17	0.09	1.33	10.5	8.4	17.1
Maximum σ_0	2.20	1.94	1.21	30.8	24.4	13.3
Average σ_0	1.54	1.11	1.25	21.2	17.5	15.8

FIGURE 6.5 The mosaicked orthovideo and the accuracy estimation of ground coordinates and seam lines in different sections, in which 11, 9, and 15 GCPs are selected in Sections 6.1–6.3, respectively.

6.4.5 Orthorectification and Accuracy Analysis

With the above-solved EOPs for each video frame, the generation of geo-referencing video can be implemented using the proposed method described in Section 6.3. More details of this method can be referenced to Zhou et al. (2002). With the method, we individually orthorectify each video frame and mosaic them together to create a 2D planimetric mapping covering the test area (Figure 6.5).

TABLE 6.4
Accuracy Evaluation of the 2D Planimetric Mapping Derived Using Three Orientation Parameters, and $\delta X = \sqrt{(X - X')^2 / n}$ and $\delta Y = \sqrt{(Y - Y')^2 / n}$, Where (X, Y) and (X', Y') Are Coordinates in the 2D Planimetric Mapping and the USGS DOQ, Respectively

Accuracy Relative to USGS DOQ	From Self-Calibration Bundle Adjustment	From Boresight Alignment	From GPS/TCM2
δX (m)	0.17	10.46	44.04
δY (m)	0.25	10.33	56.26

To quantitatively evaluate the accuracy (absolute accuracy) achieved by our method, we measured 55 checkpoints in both the mosaicked orthovideo and the USGS DOQ. The results are listed in Table 6.4. As seen from Table 6.4, the average accuracy can achieve 1.5–2.0 m (i.e., 1–2 pixels) relative to USGS DOQ. Meanwhile, we found that the lowest accuracy occurred in the middle area (Section 6.2), due to the paucity and poor distribution of GCPs used in the bundle adjustment model. Sections 6.1 and 6.3 in Figure 6.5 have a relatively higher accuracy due to more and better-distributed GCPs. So, the experimental results demonstrated that the algorithms developed and the proposed method in this chapter can rapidly and correctly rectify a video image within acceptable accuracy limits.

In addition, we also measured the accuracy of seam lines of two overlapping mosaicked images. The sub-windows of the magnified seam lines for the three sections are shown in Figure 6.5. The results showed that the accuracy of seam lines in the three sections can achieve less than 1.2 pixels.

6.5 CONCLUSIONS

This chapter presented a mathematical model for real-time orthorectification and mosaic of video flow acquired by a small and low-cost UAV. The developed model is based on the photogrammetry bundle model, in which the DLT algorithm is used for calculating the initial values of unknown parameters. The developed model can simultaneously solve the video camera's IOPs and the EOPs of each video frame. A test field, located in Picayune, Mississippi, has been established and 60-minute video data were acquired by the UAV platform to test our method. The experimental results demonstrated that the accuracy of the mosaicked video images (i.e., 2D planimetric map) is approximately 1–2 pixels, i.e., 1–2 meters, when compared with 55 checkpoints, which were measured by DGPS surveying. The accuracy of seam lines of two neighbor images is less than 1.2 pixels. The developed software and system is operated in the field ground station, which is housed in a van. Through the field test, the processing speed and achieved accuracy can meet the requirements of UAV-based real-time response to time-critical events. The suggested method is specifically proper for an economical, moderately functional, small UAV platform intended to meet the requirement for fast response to time-critical events in many small private sectors or government agencies for the small areas of interest, e.g., forest fire.

REFERENCES

Abdel-Aziz, Y. I.; Karara, H. M., Direct linear transformation from comparator coordinates into object space coordinates in close-range photogrammetry. In: *Proceedings of the Symposium on Close-Range Photogrammetry*, 1971, pp. 1–18. American Society of Photogrammetry: Falls Church, VA.

Bethea, M. D.; Lock, J. A.; Merat, F.; Crouser, P., Three-dimensional camera calibration technique for stereo imaging velocimetry experiments. *Optical Engineering*, 36(12), 1997, pp. 3445–3454.

Bielski, C.; Grazzini, J.; Soille, P., Automated morphological image composition for mosaicing large image data sets. In: *IEEE International Geoscience and Remote Sensing Symposium, IGARSS 2007*, Barcelona, Spain, 2007, pp. 4068–4071.

Cramer, M.; Stallmann, D., System calibration for direct georeferencing. In: *Proceedings of International Society for Photogrammetry and Remote Sensing*, Graz, September, Com. III, Part A, vol. 34, 2002, pp. 79–84.

Du, Y.; Cihlar, J.; Beaubien, J.; Latifovic, R., Radiometric normalization, compositing, and quality control for satellite high resolution image mosaics over large areas. *IEEE Transactions on Geoscience and Remote Sensing*, 39(3), 2001, pp. 623–634.

Grejner-Brzezinska, A. G., Direct exterior orientation of airborne imagery with GPS/INS system-performance analysis. *Journal of the Institute of Navigation*, 46(4), 1999, pp. 261–270.

Li, M.; Liew, S. C.; Kwoh, L. K., Producing cloud free and cloud-shadow free mosaic from cloudy IKONOS images. *IEEE International Geoscience and Remote Sensing Symposium. Proceedings*, 6, 2003, pp. 3946–3948.

Mostafa, M. R., Schwarz, K. P.; Chapman, M. A.; Development and testing of an airborne remote sensing multi-sensor system. In: *ISPRS Commission II, Symposium on Data Integration: Systems and Techniques*, Cambridge, 1998.

Schwarz, K. P., Chapman, M. E.; Cannon, E.; Gong, P., An integrated INS/GPS approach to the georeferencing of remotely sensed data. *Photogrammetric Engineering & Remote Sensing*, 59, 1993, pp. 1667–1674.

Sheng, Y.; Alsdorf, D. E., Automated georeferencing and orthorectification of Amazon basin-wide SAR mosaics using SRTM DEM data. *IEEE Transactions on Geoscience and Remote Sensing*, 43(8), 2005, pp. 1929–40.

Shin, D., Pollard, J. K.; Muller, J. P., Accuracy geometric correction of ATSR images. *IEEE Transactions on Geoscience and Remote Sensing*, 35(4), 1997, pp. 997–1006.

Simard, M.; Saatchi, S.; DeGrandi, G., Classification of the Gabon SAR mosaic using a wavelet based rule classifier. In: *IEEE* 1999 *International Geoscience and Remote Sensing Symposium.* IGARSS'99, Hamburg, Germany, vol. 5, 1999, pp. 2768–2770.

Yastikli, N.; Jacobsen, K., Influence of system calibration on direct sensor orientation. *Photogrammetric Engineering & Remote Sensing*, 71(5), 2005, p. 629.

Zhou, G.; Chen, W.; Kelmelis, J.; Zhang, D., A comprehensive study on urban true orthorectification. *IEEE Transactions on Geoscience and Remote Sensing*, 43(9), 2005, pp. 2138–2147.

Zhou, G.; Jezek, K.; Wright, W., Orthorectifying 1960's disclassified intelligence satellite photography (DISP) of Greenland. *IEEE Transactions on Geoscience and Remote Sensing*, 40(6), 2002, pp. 1247–1259.

Zhou, G., Wu, J.; Wright, S.; Gao, J., *High-resolution UAV video data processing for forest fire surveillance.* Technical Report to National Science Foundation, Old Dominion University: Norfolk, VA, August, 2006, 82p.

7 True Orthophotomap (TOM) Generation

7.1 INTRODUCTION

The USA National Digital Orthophoto Program (NDOP) was first proposed in 1990. The primary goal of this program is to ensure the public domain availability of data for the nation (USGS, Digital Orthophoto Program, 1998). The NDOP was based on quarter-quadrangle-centered aerial photographs (3.75 minutes of longitude and latitude in geographic extent) obtained at a nominal flying height of 20,000 feet above mean terrain with a 6-inch focal-length camera (photo scale = 1:40,000). Additionally, the early procedures and algorithms for digital orthophoto generation were based on earlier United States Geological Survey (USGS) mapping operations, such as field control, aerotriangulation (using photogrammetric equations derived in the early 1920s), and 2.5D digital elevation models.

The procedures and algorithms used in the 1990s are not appropriate for large-scale city orthorectification, because when generating large-scale urban orthophotos using the existing algorithms and methods, many problems, such as incomplete orthorectification, occlusion, ghost image, and shadow, have been disclosed (Zhou et al. 2005). The impacts of these problems have significantly influenced the usefulness of the digital orthophoto in the industry because the error of these incompletely rectified large-scale city orthoimage maps can no longer be tolerated when used for updating and planning urban tasks. Thus, the generation of the so-called *true orthophotomap* (*TOM*) has become increasingly interesting at the beginning of the 21st century, due to the need for a complete and geometrically correct image, which can only be fulfilled by TOM. This chapter describes orthorectification for TOM generation in urban areas, including algorithms and data processing procedures.

7.2 PROBLEM OF TRADITIONAL ORTHORECTIFICATION

7.2.1 DATASET

The experimental field is located in Downtown Denver, CO, where the highest buildings reach 125 m, and many buildings are around 100 m. The six original aerial images from two flight strips were acquired using an RC30 aerial camera at a focal length of 153.022 mm on April 17, 2000. The flying height was 1650 m above the mean ground elevation of the imaged area. The aerial photos were originally recorded in the film and later scanned into a digital format at a pixel resolution of 25 μm. Figure 7.1 illustrates the configuration of the photogrammetric flight mission and the available digital surface model (DSM). Two of the six aerial images (DV1119 and DV1120) covered the downtown area (see Figure 7.2). The endlap of the images is about 65%, and the sidelap is approximately 30%. A DSM in the central part of Downtown Denver is available (Figure 7.1). The accuracies of planimetric coordinate and vertical coordinate in the DSM are approximately 0.1 and 0. 2 m, respectively. The horizontal datum is GRS 1980, and the vertical datum is NAD83. The visualization of the DSM is depicted in Figure 7.3.

The projection information of ground control points (GCPs) is

Reference map projection: Lambert conformal conic
Spheroid: GRS 1980
Datum: NAD83
Map unit: feet

DOI: 10.1201/9781003632108-7

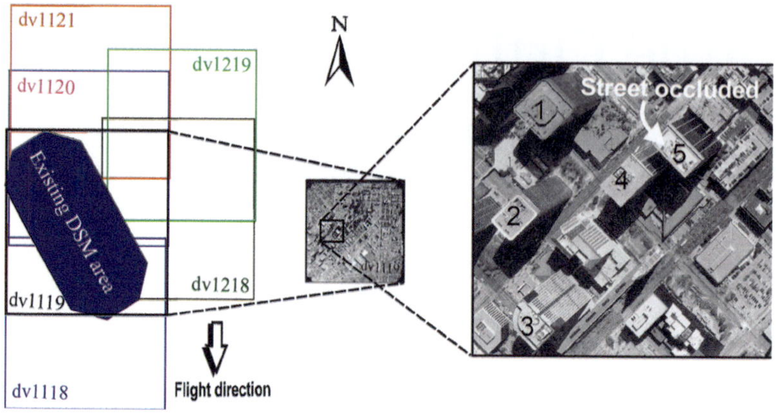

FIGURE 7.1 Six aerial images from two strips covering Downtown Denver, CO, USA. The heights of the five numbered buildings are 100, 103, 125, 100, and 107 m, respectively. (Courtesy of Zhou et al. 2005.)

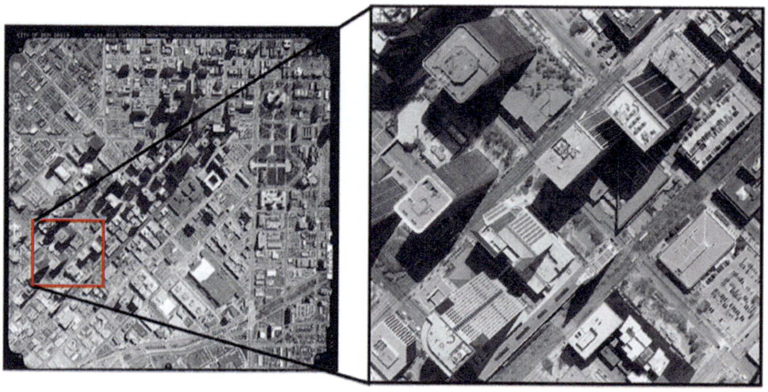

FIGURE 7.2 Source image (DV1119) of Downtown Denver, CO, USA.

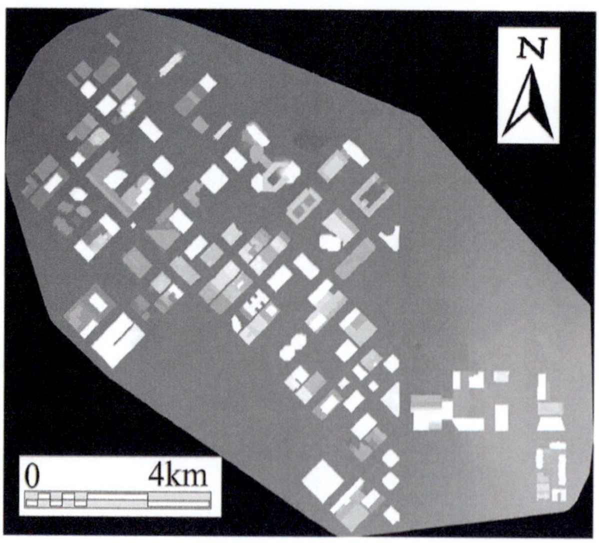

FIGURE 7.3 The digital surface model (DSM) in Downtown Denver, CO, used for the experiment. The brightness in the DSM represents the height of the surface (ground or buildings). (Courtesy of Zhou et al. 2005.)

We then collected 333 GCPs, which were visible in both the source images and the DSM. These GCPs cover most of the buildings in the available DSM. Generally, we selected pixels located at the building root and roof points as the GCPs. A very large number of GCPs allows for increased computation accuracy.

With these date sets, we used the camera model in the ERDAS/IMAGINE to orthorectify one image, DV1119, for which we input principal point coordinates (0.002, −0.004 mm) and a focal length (153.022 mm). We collected 333 well-distributed GCPs, which were visible in both the original aerial images, DV1119, and the DSM imagery. These GCPs do not mean the traditional photogrammetric target points. Instead, they are corners of building roofs. The measurement accuracy of the GCPs can reach the sub-pixel level. The purpose of collecting so many GCPs is to allow for a sufficient computational accuracy during the orthorectification process.

7.2.2 Problems and Analysis of Traditional Orthorectification

With these parameters and ERDAS/IMAGINE software, we produced the orthoimage of DV1119. Figure 7.3 shows the central part of the orthorectified image of DV1119. From this orthoimage, the following observations have been made.

7.2.2.1 Ghost Images

As observed from Figure 7.4, although the high buildings can be correctly orthorectified into their upright positions, the roofs of buildings are superimposed onto the bottom of the buildings and the roofs of buildings appear twice. This phenomenon is called "ghost image" (Rau et al. 2002). The reason for ghost images is due to the fact that the higher buildings occlude other objects, and the existing differential rectification model cannot compensate for the occlusion, leading to the result that the occluding objects remain, and the occluded objects cannot be detected and filled. During the orthorectification of images, the brightness of any pixel in the resulting orthoimage is assigned from the corresponding pixel's brightness in the original image using the collinearity equation. Both the resulting orthoimage and the original image share the identical DSM. When we orthorectify the relief displacement caused by buildings, the roofs of buildings are orthorectified in their upright positions; when we orthorectify the relief displacement caused by occluded objects, the occluding objects are again employed to fill the occluded objects. This kind of processing leads to the occluding pixels being doubly employed: the first is for the roof and the second is for the occluded objects. Thus, double roofs of the identical building are seen in the resulting orthoimage.

7.2.2.2 Occlusion and Building Lean

Again as observed in Figure 7.4, perspective displacements and occlusions still remain in the resulting orthoimage. The orthorectification of the image is partly geometrically inaccurate and/

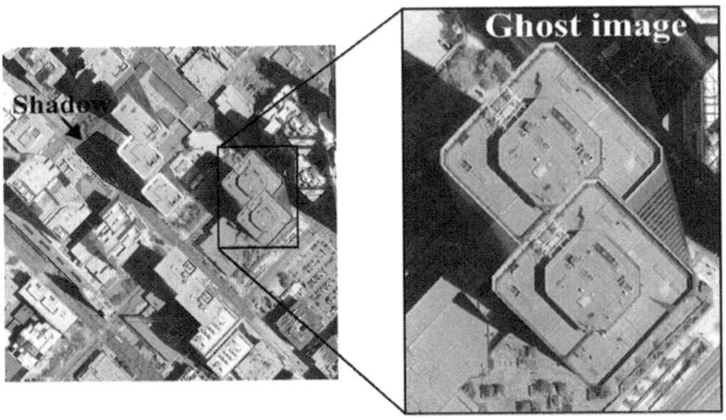

FIGURE 7.4 Part of the orthorectified image of Downtown Denver generated by the differential model.

or incomplete; the image features are distorted from their true locations. This distortion appears in the form of leaning buildings. As illustrated in Figure 7.3, a tall building leans over a street and occludes features like streets, manholes, fire hydrants, and utility poles. The existing differential rectification model is not able to identify occlusions and compensate for occlusions. Thus, the problem is how we can identify and compensate for the occlusion in urban areas, considering the occlusion is an inherent property of perspective projection.

7.2.2.3 Shadow of the Buildings

Shadow of the buildings is another important concern in urban areas. Figure 7.3 also demonstrates that the shadow cannot be removed using the differential rectification model. As observed in Figure 7.4, the shadow of the tall building deteriorates the quality of orthoimages; as a result, it is hard to identify the shadowed objects like buildings, bridges, and fire hydrants. To produce high-quality orthoimages, shadows must be removed. We have manually detected the shadowed areas by a human–computer interactive operation and then automatically conducted radiometric balance inside and outside the shadowed areas using the gray variance algorithm (Zhou et al. 2005). This operation is significantly time-consuming and labor-intensive. An operation for automatic detection and radiometric balance for shadowed areas must be done in true orthorectification.

7.2.2.4 Inaccuracy of DSM

The differential rectification method is based on the DSM, which is used to represent and model the buildings. As observed in Figure 7.5, we found that (1) small objects on top of the building were not completely orthorectified due to the lack of a model; (2) the wall of the building is incompletely orthorectified due to the inaccuracy of the DSM; and (3) the boundary of the building is inaccurately orthorectified due to the inaccuracy of the DSM. Thus, the accuracy of the DSM will greatly affect the quality of orthorectification. In fact, it has been demonstrated that the traditional DSM to represent the urban surface objects cannot obtain high-quality orthoimage since the DSM cannot accurately model the details of buildings, i.e., edges of buildings.

7.2.2.5 Incomplete Refilling to Occlusion

It is unavoidable to have occlusion in urban large-scale images. One of the tasks is therefore to identify the occluded areas from one orthoimage, called *master orthoimage*, and refill them from neighboring orthoimages, called *slave orthoimage*. The DV1119 and DV1120 images are orthorectified

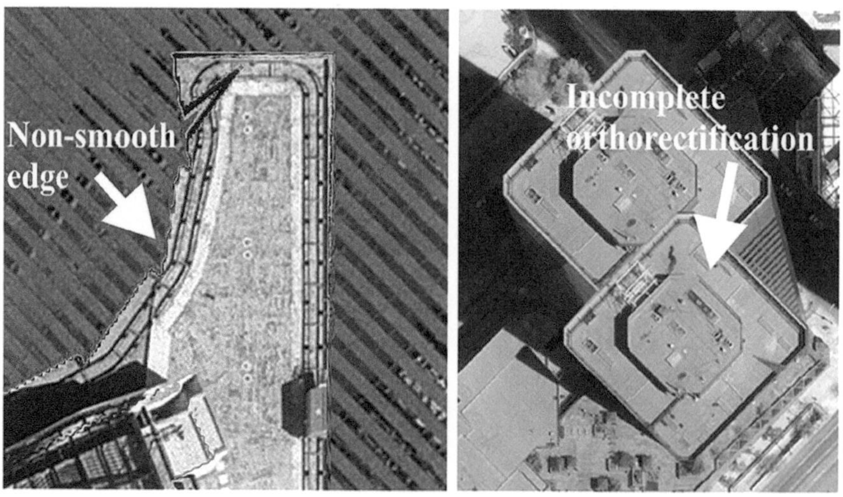

FIGURE 7.5 Incomplete orthorectification of buildings due to the inaccuracy of DSM and the lack of detailed representation.

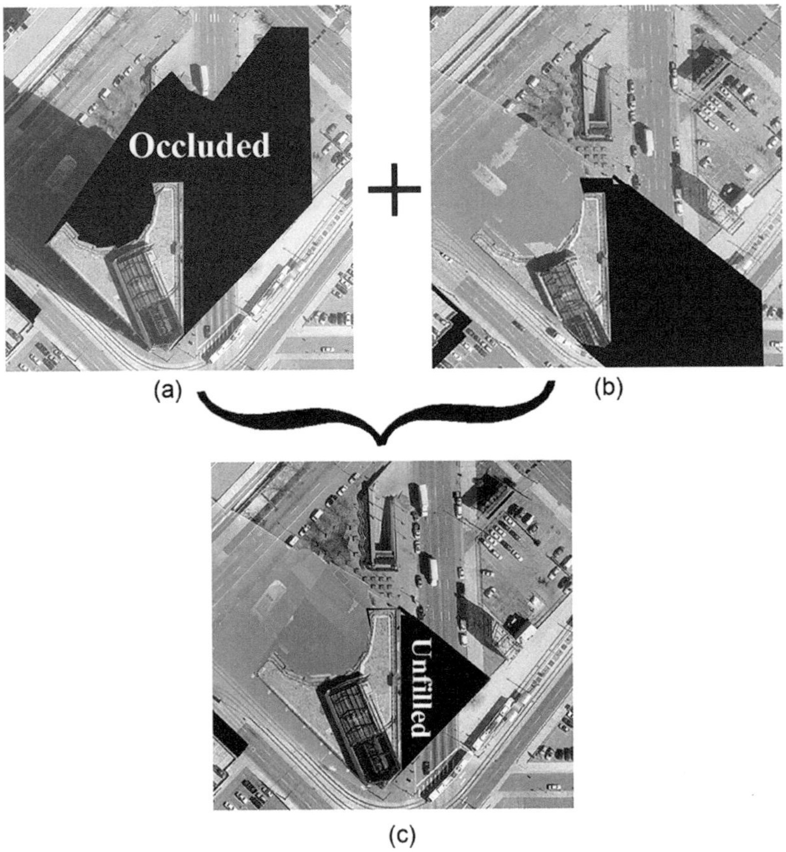

FIGURE 7.6 Building occlusion in a master image (a) and slave image (b), resulting in incomplete refilling in the resulting image (c).

and manually identified the occluded areas in the DV1119 orthoimage (master) and refilled the occluded areas from the DV1120 orthoimage (slave). As observed in Figure 7.6, it is not possible to find sufficient slave orthoimages to completely refill the occluded areas in the master orthoimage. This is especially true for those areas containing tall buildings located in the margin of images, where the occlusions are large due to the inherent property of perspective projection. The cause for incomplete refilling of the occluded areas is due to the low endlap of two adjacent images. Traditionally, a photogrammetric flying mission requires an approximately 65% endlap along the strip and 35% sidelap across the strip. This specification has had challenges in urban photogrammetry, where the tall buildings are major objects. Therefore how we design flight mission, such as endlap, sidelap, flying height, choose a camera with different focal length and IFOV, and the direction of flight need to be investigated.

7.2.2.6 Different Geometric Accuracies of Two Adjacent Orthoimages

Occlusion refilling requires that the rectified master and all slave orthoimages have consistent geometric accuracy. We checked the geometric accuracy of the orthoimages DV1119 and DV1120, both of which were generated using the same method (differential rectification model), the same DSM, the same GCPs, and the same iterative number. It can be found that the building in the DV1120 orthoimage was incompletely and inaccurately orthorectified to its planimetric position since the wall information is still visible in Figure 7.7b (building leaning over), whereas the wall information in the DV 1119 orthoimage is missing, which means the building is completely orthorectified in its upright position (see Figure 7.7a). This phenomenon is probably caused by the incomplete correction of lens distortion because the magnitude of the lens distortion is a function of object locations in the image plane. The identical ground building is imaged in different positions in the adjacent images, thus having different magnitudes of lens distortions. The unique method to overcome this problem is to find the exact lens distortion model to correct the distortions.

It can be imagined that if we employ the different geometric accuracy of slave orthoimages to refill the occluded area of the master orthoimage, the final orthoimage contains the geometric errors along with the seamlines, which probably are not seamless. Figure 7.8 demonstrates the existence of this type of error. As measured in Figure 7.8, the same feature in the two orthoimages has an obvious shift of about 7–20 pixels.

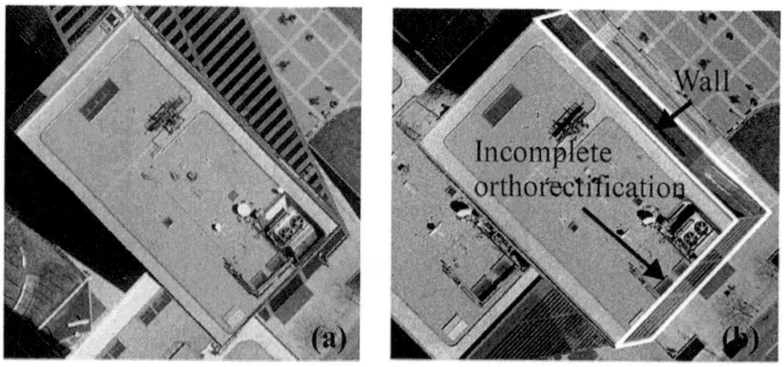

FIGURE 7.7 Comparison of orthorectification accuracy for the same building in two neighboring images using the same DSM, (a) DV1119 orthoimage, and (b) DV1120 orthoimage.

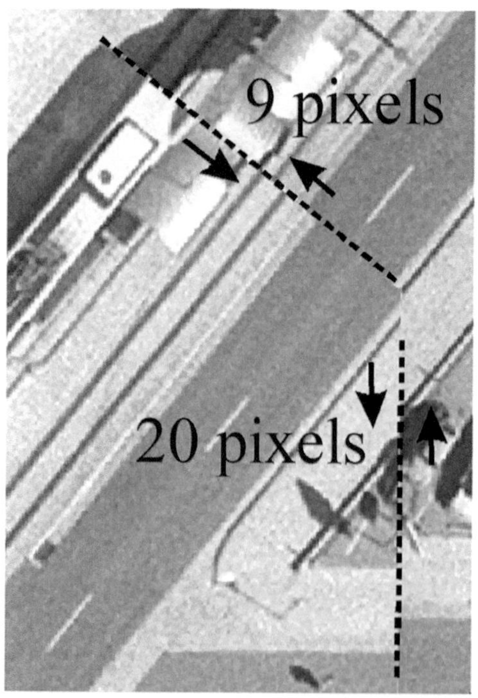

FIGURE 7.8 The shift of the same features in the neighbor images when mosaicked.

7.2.2.7 Radiometric Difference of Two Adjacent Orthoimages

The brightness difference between scenes is significant. This phenomenon in urban areas is more apparent due to occlusion and shadow. The generation of the true orthoimage in urban areas usually requires the occlusion refilling using the slave orthoimages, for which the radiometric difference in the different orthoimages should be kept to a minimum. Figure 7.9 illustrates an obvious radiometric contrast along the seamline, in which the gray difference between the master orthoimage and the slave orthoimage is large (Figure 7.9). An effective method to minimize this difference is to use those images acquired in the same time frame along an identical flight strip. Further, radiometric balancing is conducted before mosaicking (Zhou et al. 2002, 2005).

7.2.3 Problems Summary

The investigation for orthorectification of large-scale aerial images in urban areas using the traditional differential rectification model has discovered the following problems:

1. *The problems occurring in the single orthoimage*: The automatic identification and compensation of the building occlusion are a key step. Unsuccessful processing of this problem causes the phenomenon of "ghost images" in the resulting orthoimage. Another main problem is shadow detection and removal. Unsuccessful processing of this problem causes the low-quality orthoimage, i.e., visible seamlines, and obvious radiometric difference along seamlines. Therefore, identification and compensation of occlusion and shadow need to be conducted for urban orthoimage generation. Additionally, the inaccuracy of the DSM representation causes rough edges of buildings and inaccurate geometric positions of the edges of buildings. The lack of a detailed representation of small objects results in incomplete orthorectification for urban images. Therefore, an accurate model to represent urban buildings needs to be developed.
2. *The problems occurring between two orthoimages*: Difference in geometric and radiometric accuracy between two orthoimages can lead to the incomplete refilling of the occluded areas when mosaicked. To minimize the difference in geometric accuracy between two orthoimages, all distortions should be corrected, and to minimize the radiometric difference, the "best" patches in slave orthoimages and optimum seamlines should be chosen.

Overall, the existence of these problems indicates that the conventional orthorectification method is not able to orthorectify the objects into their correct and upright positions for a large-scale urban aerial image. Significant efforts are still needed to improve the methodology for the generation of true orthoimages for large-scale aerial images.

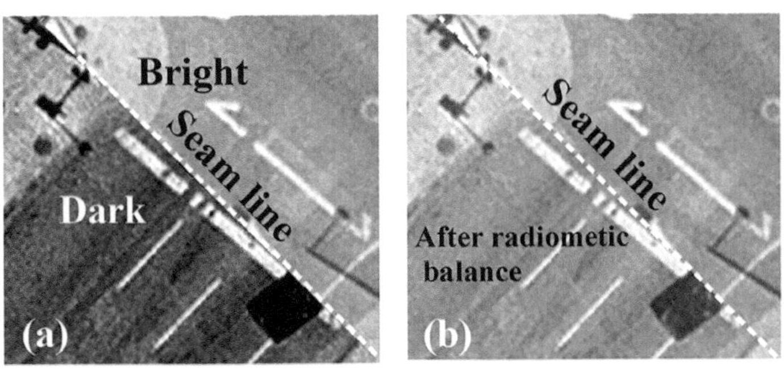

FIGURE 7.9 The radiometric contrast along the seam line between "master" and "slave" orthoimages: (a) before radiometric correction and (b) after radiometric correction.

7.3 PRINCIPLE OF TRUE ORTHOPHOTOMAP (TOM) GENERATION

The early studies for the generation of true orthoimages were conducted by a few investigators. Skarlatos (1999) and Joshua (2001) demonstrated that building occlusions do significantly influence not only image quality but also the accuracy of orthoimages. Amhar et al. (1998) and Schickler and Thorpe (1998) considered the hidden effects introduced by abrupt changes of surface height (e.g., buildings and bridges). Schickler and Thorpe (1998) and Mayr (2002) considered seamless mosaicking around fill-in areas in order to reduce gray-value discontinuities, but the radiometric aspect, or shadow enhancement to restore information within a building's shadow areas, has not been reported. Rau et al. (2002) treated enhancements of image radiometry, demonstrating a suitable enhancement technique to restore information within building shadow areas. Jauregui et al. (2002) presented a procedure for orthorectification of aerial photographs for producing and updating terrain surface maps. Vassilopoulou et al. (2002) used IKONOS images to generate orthoimages for volcanic hazard monitoring of Nisyros Island, Greece, and Siachalou (2004) used IKONOS images to generate the urban orthoimage. Cameron et al. (2000) analyzed orthorectified aerial photographs for measuring changes of native pinewood in Scotland, and Passini and Jacobsen (2004) analyzed the accuracy of orthoimages from very high-resolution imagery. Biason et al. (2004) further explored the automatic generation of true orthoimages.

7.3.1 Basic Steps of True Orthoimage Generation

To orthorectify a building to its correct, upright position, the building must be represented as part of the surface to be rectified. Although the DSM can represent the building surface (roof), the orthoimage generated using the DSM cannot achieve satisfactory accuracy, as described in Section 7.2.2.4 (see also Zhou et al. 2003). In addition, in order to compensate for the occluded areas, occlusion detection must be conducted through visible analysis. One such method is to trace the imaging rays from the top of the surface back to the projection center of the photograph. Only those objects whose rays do not intersect any other object before arriving at the projection center are not occluded, as a result, leading to a correctly rectified image pixel (Amhar et al. 1998). This method requires an exact digital building model (DBM), which describes the building structure, 3D coordinates, topological relationship, etc. On the other hand, the relief displacement caused by terrain should be corrected in a true orthoimage. For this reason, an exact representation of terrain should be given, i.e., digital terrain model (DTM), which does not contain buildings. Therefore, the basic steps of true orthoimage generation contain DTM-based orthoimage generation and DBM-based orthoimage generation as well as the emergence of the DTM- and the DBM-based orthoimages (see Figure 7.10).

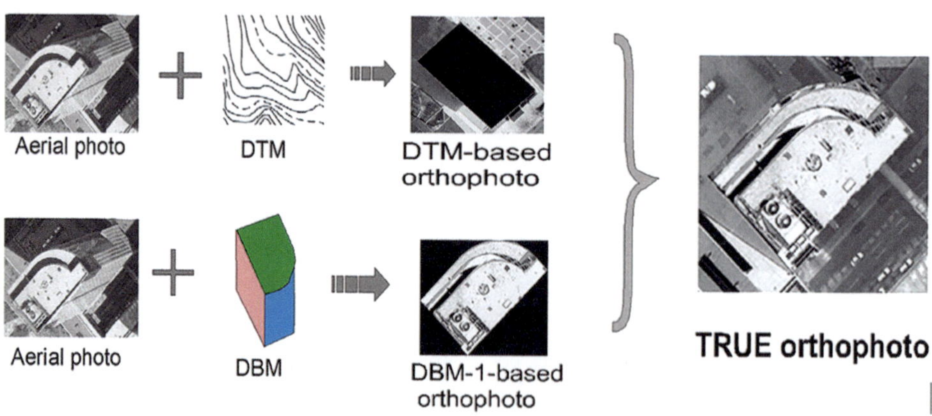

FIGURE 7.10 The procedure of true orthophoto generation.

7.3.2 DBM-Based Orthoimage Generation and Occlusion Detection

The process of DBM-based orthoimage generation only orthorectifies the displacement caused by buildings without considering the displacement caused by terrain. To this end, the occluded buildings need to be detected. At present, an effective and commonly used method for occlusion detection is to apply the DBM to calculate the distance between object surfaces to the projection center. This distance is called *Z distance*, and the method is called *Z-buffer* algorithm (Amhar et al. 1998). In this algorithm, a matrix is designed to store the Z-distances of each pixel, and another matrix is designed to record the location of an object point. Both of them are defined on an image plane. Based on the DBM, the Z-buffer is generated by projecting each DBM surface polygon onto the first image plane using collinearity equations. The projected polygon is rasterized and is filled with the *Z* distance and the polygon's identification (ID) code. The identification code is used to distinguish between walls and roofs. The filling processing via resampling the original image pixels will first check the ID codes and then rectify only those pixels that are from "roof," i.e., the wall will not be orthorectified. In this step, the displacements caused by visible buildings are orthorectified into their upright planimetric position.

Additionally, the occluded buildings cannot be orthorectified; consequently, holes are left in the orthoimage in this step. We can mark these occluded areas by simply back-projecting the individual DBM into the individual images and marking occluded areas as black, i.e., setting their brightness value to 0. This means that areas occluded by objects need to be filled with other orthoimages, called *slave orthoimage* (Rau et al. 2002; Amhar et al. 1998).

The above-generated orthoimage only containing buildings is thus called *DBM-based orthoimage* or *Index Map* (Rau et al. 2002).

7.3.3 DTM-Based Orthoimage Generation

DTM-based traditional orthoimage generation can be implemented by methods, called top-down or bottom-up processing (ray tracing) (Chen and Lee 1993a, 1993b). Ray tracing is an iterative method for tracing the intersection of a ray from the image plane to the terrain. For urban areas consisting of extremely tall buildings, this method has some inherent difficulties due to the occlusions of lower structures and objects by the nearby tall buildings. We recommend the bottom-up back projection method in this chapter. In this algorithm, once a DTM and the orientation parameters for an image are available, the collinearity equations are used to calculate the image pixel corresponding to a groundel, i.e., the gray values of the background area and visible roofs are calculated from the actual images. A detailed description can be found in Zhou et al. (2002). The basic procedures are as follows:

- For any point, $P\ (I, J)$ in the resulting image, (I, J) are its image coordinates in the image plane;
- Compute the planimetric coordinates of the point $P\left(X_S, Y_S\right)$ in the map-based coordinate system at the given scale;
- Interpolate the vertical coordinates, Z_S from the DTM using a bilinear interpolation algorithm;
- Compute the photo coordinate (x, y) and the image coordinate (i, j) of the point P in the original image by collinearity;
- Calculate the gray value g_{orig} by the nearest neighbor resampling algorithm;
- Assign the gray value g_{orig} as the brightness g_{result} of the resulting (rectified) image pixel.

The above procedure is then repeated for each pixel to be rectified.

7.3.4 Near True Orthoimage Generation

Based on the DBM-based orthoimage and the DTM-based orthoimage obtained above, the true orthoimage is capable of being created by merging the two types of orthoimages. Amhar et al. (1998) recommended a very simple operation, called *logical <OR> operation*, to carry out this merging operation when the background values are set at zero. We have tested this method and found that this logical <OR> operation is very effective.

The merging operation generates an orthoimage in which the buildings have been orthorectified to their correct positions. However, the resulting orthoimage still contains patches from those areas which were occluded by tall buildings. They need to be filled by using additional aerial images (sometimes called salve image). We call this intermediate orthoimage a *Near True Orthoimage (NTOI)*.

7.3.5 Occlusion Compensation

As mentioned above, the orthorectification process of an NDOI is capable of orthorectifying buildings into their proper locations, but also leaves blind spots. One of the most technically challenging steps is the mosaicking of conjugate areas of adjacent orthoimages to fill the occluded patches. For this purpose, a considerable overlap between aerial photographs is necessary.

1. *Optimization of seamlines*: The above occlusion identification can detect the boundary of the occluded areas. The boundary might fall on ground features not modeled in the DBM such as trees, cars, or shadows, which probably have different brightness values in the individual NTOI images (Schickler and Thorpe 1998) because the occlusion identification is based on the geometry of the imaging system, not on image radiometry. Therefore, the least square correlation algorithm is employed for the optimization of seamlines to obtain a seamless mosaic when compensating for the occluded area. The basic idea of this algorithm is to calculate the similarity of the image contents between the "master" and "slave" orthoimages. The pull range and search space are determined by the image texture and image quality. Based on this measure, the optimal seamline is along a path where the two neighbor images are mostly similar.
2. *Optimization of orthoimage patches*: The same ground object has different displacements in different images. Therefore, the qualities of orthorectification of the identical ground object in different NTOIs are different. An algorithm to automatically "pick up the best" orthoimage patch from the "slave" NTOIs for compensation of occluded areas has been developed. The basic idea of this algorithm is based on the fact that the relief displacement at image nadir is zero. Thus, this method first calculates the distances of each patch to its nadir and then chooses the patch with the shortest distance.

These two quality measures are combined for generating the so-called *weight image*, which is computed for each DOI. Based on these weight images, the best pixel will be picked up from the pixel with the highest weight.

7.3.6 Radiometric Balancing

Significant scene-to-scene radiometric variations are observed in tall building urban photos. These radiometric differences are primarily due to shadow and occlusion. To overcome this problem, radiometric balancing and blending operations are applied to individual scenes to prevent a patchy or quilted appearance in the final mosaic.

The radiometric balancing has been developed using the weights for blending individual scenes along the specified buffer zone. The weights are calculated by the cubic Hermite function (the detailed description can be found in Zhou and Jezek (2002)):

$$W = 1 - 3d^2 + 2d^3 \tag{7.1}$$

where W is the weighting function applied in the overlap area with values ranging from 0 to 1 and d is the distance of a pixel to the buffer line which is normalized from 0 to 1. The resulting brightness value (G) is then found from the equation:

$$G = W \cdot G_1 + (1 - W) \cdot G_2 \tag{7.2}$$

where G_1 and G_2 are the brightness of the neighboring images. In the buffer zone, large intensity values have lower weight, while small brightness values have high weight.

7.3.7 Optimization of Seamlines

The visibility analysis via DBM-based back projection also detects the boundary of the occluded areas in individual NDOI images. However, the boundary of the occluded areas is an initial seamline because the algorithm for occluded area identification is in effect based on the geometry of the imaging system and the ground objects (surfaces) and not on image radiometry. The initial seamlines might fall on ground features not modeled in the digital city model such as trees, cars, or shadows, and which can be different features in the individual DOIs. An optimal seaming process, to obtain the best seamlines from mosaicking two images, has been developed. The basic idea is to seek the similarity of the image contents from two different DOIs (Schickler et al. 1998). The criterion of similarity is the maximum correlation coefficient. Based on this measure, the optimal seamline is along a path where the two neighbor images are the most similar. The pull range and search area are determined upon the initial seamline selection.

7.4 THREE-DIMENSIONAL (3D) URBAN MODEL FOR ORTHORECTIFICATION

7.4.1 Data Structure

As described above, DBM-based orthoimage generation requires an effective representation of urban buildings. The generation of a 3D urban (building) model is a rather challenging task, from both a practical and a scientific point of view, because different applications (e.g., city planning, communication system design, control and decision-making, tourism, pollution distribution, military security operation, and visualization) require different data types and manipulation functions. Many models have been proposed, e.g., Breunig (1996), Graz (1999), Grün and Wang (1998), Zhou et al. (2000), and Zlatanoca (2000). It has been demonstrated that the maintenance and administration of this huge amount of urban data to be well organized need an all-powerful 3D urban model, which may be impossible to realize because a realistic presentation of the real world and the description of surface properties and material parameters are quite difficult (Grün and Wang 1998). For the purpose of true orthoimage generation, the data structure to be developed in this chapter requires:

- To generate the DBM-based high-quality orthoimage and
- To easily create, store, design, analyze, and query city objects for orthoimage-based urban applications.

The geospatial data to be used for true orthoimage generation includes (1) digital terrain data, (2) images, and (3) spatial objects, e.g., buildings. The vector Triangular Irregular Network data structure is capable of precisely representing the ground surface although the urban terrain is complex (Schickler and Thorpe 1998). The digital images including original images, generated orthoimages, and generated DTM or DBM images are usually represented by grid format with rows and

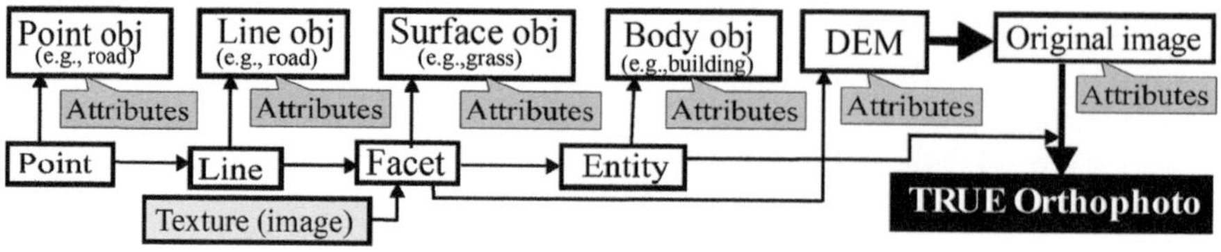

FIGURE 7.11 Data structure of the proposed 3D urban model for true orthophoto generation.

columns. These data structures are still adopted in this chapter. Spatial objects are abstractly understood as the following four types of geometric objects (Zhou et al. 2000, Grün and Wang 1998):

- *Point objects:* These are described by x and y planimetric coordinates, like benchmarks and wells.
- *Line objects:* These are one-dimensional objects with length as the only measurable one-dimensional spatial extension.
- *Surface objects:* These are two-dimensional or two-and-a-half-dimensional objects with area and perimeter as measurable spatial extensions.
- *Body objects:* These are three-dimensional objects with volume and surface area as measurable spatial extensions, which are bordered by facets.

For the above four types of objects, *Point* is the basic geometric element. For example, a *Point* can present a point object, or the starting or ending point of an *Edge*. An *Edge* is a line segment, which is an ordered connection between two points: beginning point and ending point. A *Facet* is completely described by the ordered edges that define the border of the facet. *Entity* is described by the ordered facets. Image data and thematic data can be attached to a facet. Each facet is related to an image patch through a corresponding link (see Figure 7.11). In the proposed data model, each object is identified by a defined attribute data, such as Type Identifier (TI), and thus four types of objects are referred to as PI (point identifier), LI (line identifier), FI (facet identifier), and EI (entity identifier), respectively. The other attribute data, e.g., thematic data and geometric data, can be attached to each type of object. The geometric data of 3D objects contains the information of position, shape, size, structure definition, and image index.

These defined object types are taken as the basic classes of primary geographic entities, from which other geographic entities are derived. For example, line and area types of objects frequently consist of several arcs, and each arc has two terminal points. Collectively, these four types of objects can represent most of the tangible natural and human phenomena that we encounter on a daily basis. This data model not only inherits primary operators but also defines its own special operators. Thus, a spatial object can be extracted into one of the object types according to its attributes (see Figure 7.11).

7.4.2 Topological Relationship

The topological relationships between geometrical elements are implicitly defined by the data structure. For example, a point object is presented by a distinct *Point* element. The line object is described by ordered *Edges*. Similarly, the body object is described by an *Entity* that is described by the ordered facets. Thus, the topological relationships between *Point* and *Edge*, *Edge* and *Facet*, and *Facet* and *Entity* are registered by the links between the geometrical elements (see Figure 7.12).

7.4.3 Implementation via Relational Database

The data structure described above is, in fact, a relational model, which can be implemented by relational database technology. In this relational model, each type of object is defined as a table.

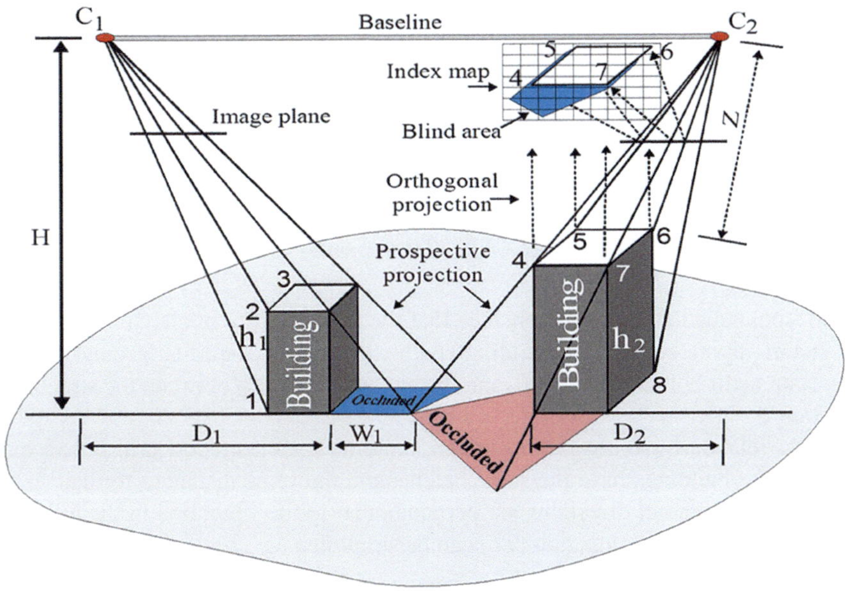

FIGURE 7.12 Visibility analysis for larger-scale true orthoimage generation.

For example, a table describing a point object includes three terms: Point Identification (PID), Attribute Linkage (AL), and *XYZ* coordinates. The PID is for the identification of a point object. The AL is a linkage to the attribute table describing point properties, such as point names and features. Different types of objects may have different attribute data. The *XYZ* coordinates are for geometric description of a point. Similarly, other types of objects, like line, can be referred. Based on the above relational structure, the query of a geometrical description for a type of object is easily realized. A detailed discussion of a relational database model of 3D city data for true orthoimage generation can be found in Zhou et al. (2003).

7.5 CITY STREET VISIBILITY ANALYSIS

Another important concern in urban orthoimage generation is that the high buildings occlude the streets. The traditional flight planning usually requires an approximate 65% endlap for the adjacent two photographs along the flight strips and 30% sidelap across the flight strips. It has been demonstrated that, from an orthoimage generation point of view, these early specifications for urban photogrammetric missions have encountered challenges due to the impacts of the occlusions, deep discontinuities, and shadows. To minimize their impacts on urban true orthophoto generation, we investigated and analyzed the relationship between occlusion and building height, street width, flying height, and AFOV (angular field of view) of the camera.

To ensure that the street is fully visible for occlusion compensation using the two contiguous images along a flight strip, $W_1 \sin \theta_1 + W_2 \sin \theta_2$ should be no more than the street width W (see Figure 7.12), i.e.,

$$W_1 \sin\theta_1 + W_2 \sin\theta_2 \le W \tag{7.3}$$

where W_1 and W_2 are the projected length for Building 1 and Building 2 on the ground and θ_1 and θ_2 are the angles between the street line and the ground radial of Building 1 and Building 2 from ground nadir points, respectively. Using the geometry depicted in Figure 7.12, it is not difficult to find that

$$W_1 = R_1 * h_1 / (H - h_1) \tag{7.4}$$

where R_1 is the distance between ground nadir, P_1, and Building 1; h_1 is the height of Building 1; and H is the flying height above the ground of the building. Similarly, for Building 2, we also have

$$W_2 = R_2 * h_2 \ /(H - h_2) \tag{7.5}$$

Rewriting equation (7.3) using equations (7.4) and (7.5), we obtain

$$R_1 * h_1 * \sin\theta_1 \ /(H - h_1) + R_2 * h_2 * \sin\theta_2 \ /(H - h_2) \leq W \tag{7.6}$$

As observed from equation (7.6), we conclude that the full visibility of city streets depends on the following factors: flying height (H), building heights (h_1 and h_2), building ground radial distance from the ground nadir points (R_1 and R_2), and the angles (θ_1 and θ_2) between the street line and the radial direction.

To model the relationship among flying height, building height, street width, and air base (P_1P_2), we assume that the buildings have the same heights and the same distances to their ground nadir points, and the ground radial directions are perpendicular to the street line in the following analyses. Under these assumptions, equation (7.6) can be simplified as

$$2 * R * h / (H - h) \leq W \tag{7.7}$$

Since $2*R = P_1P_2 - W$, equation (7.7) can be further rewritten as:

$$B * h / H \leq W \tag{7.8}$$

where B represents the length of the air base, P_1P_2. Equation (7.8) is the simplest relationship among flying height (H), building height (h), air base (B), and the street width (W).

With the model derived above, we simulated the three experiments to investigate the relationship among the four factors (flying height, air base, building height, and street width): (1) when the street width is assumed 20 m, the relationship between building height and flying height for different air bases (see Figure 7.13a); (2) when the street width is assumed 20 m, the relationship between the air base and flying height for different building heights (see Figure 7.13b); and (3) when the building height is assumed 30 m, the relationship between street width and flying height for different air bases (see Figure 7.13c). As observed from Figure 7.13, it is found that the required flying height increases dramatically as the building height or air base distance increases. For example, when the air base is 654 m long, the flying height of at least 3270 m must be achieved to ensure the complete visibility of both a street of 20 m wide and a building of 100 m high (Figure 7.13a). From equation (7.8) and Figure 7.13c, the required minimum flying height is inversely proportional to the street width, assuming that the air base and building height are predefined. Therefore, in practical urban photogrammetric flight planning, the minimum flying height should be determined by the width of the narrowest street to ensure full visibility of city streets. This condition will be able to meet the needs of 100% visibility of city streets for occlusion compensation in true orthoimage generation, in addition to the traditional photogrammetric requirements, such as stereo mapping, stereo compilation, and stereo visualization.

7.6 EXPERIMENTS AND ANALYSIS

7.6.1 The Flowchart and Basic Procedures

Figure 7.14 is a flowchart of the true orthoimage generation. The input data include aerial images from multiple strips with over 65% overlap, exterior orientation elements including three position

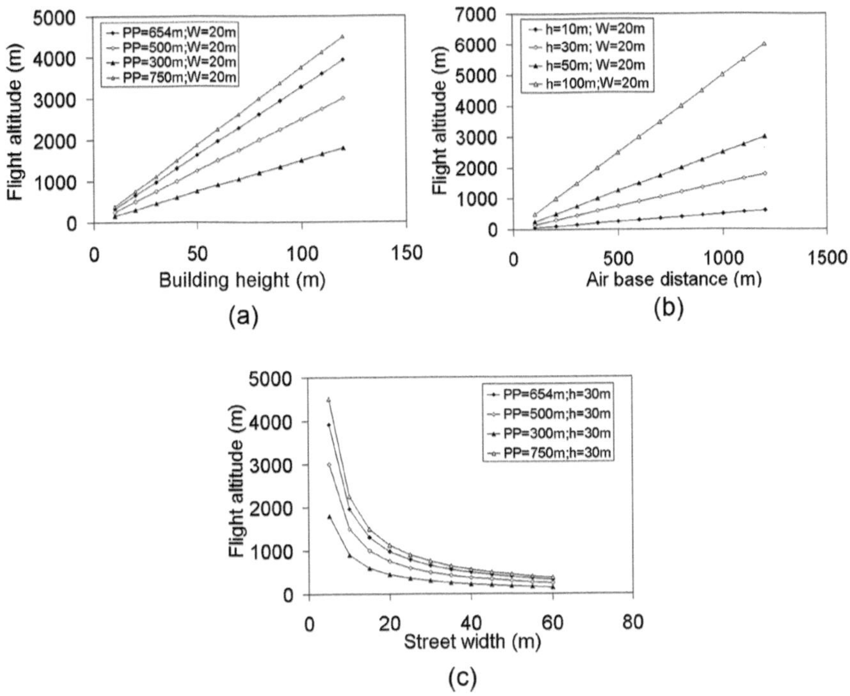

FIGURE 7.13 The relationships between flying height, (a) building height, (b) air base, and (c) street width. *PP* stands for air base, *W* stands for street width, and *h* stands for building height.

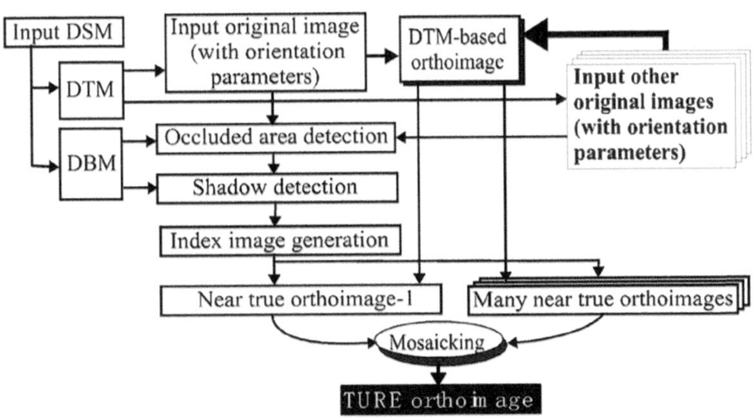

FIGURE 7.14 The procedure of true orthoimage generation.

elements and three attitude elements of each image, interior orientation elements including focal length, principal point coordinates, length of opposite fiducial marks, DBM, and DTM data.

The test field is selected in Downtown Denver, CO. Analytical Survey Inc. in Colorado Springs, CO provided us the digital surface data, five images from two strips, and corresponding parameters. The detailed descriptions for the dataset are given in Section 7.2.1. With the given DSM data, we used the ERDAS/Imagine software from Leica GeoSystem Inc. to manually extract each building and manage them using the developed data structure described in Section 7.4. Figure 7.15 is the visualization of the DBM model of Denver, CO. The DTM is generated by removing the digital buildings from the DSM.

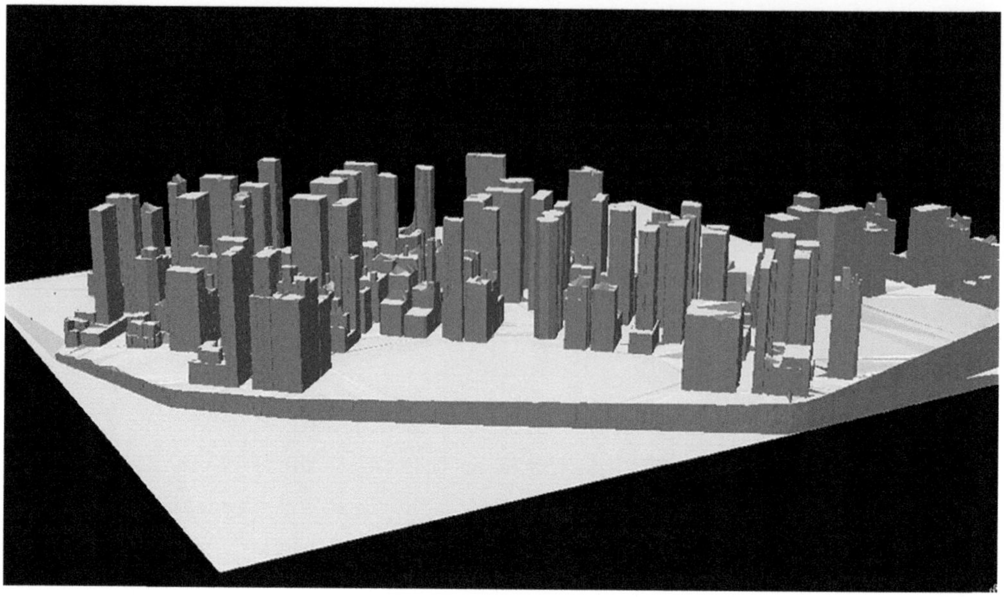

FIGURE 7.15 The city model of Denver created by our system.

The basic procedures for orthoimage generation include (Zhou et al. 2002)

1. *Determination of the output image size*: The size of output orthoimage is, in fact, a function of both the coverage of the original image and the size of DTM data. The intersection of two datasets will determine the size of the output orthoimage.
2. *Calculation of the positions of pixels in the output image and corresponding positions in the original image*: This process is calculated using the collinearity equation, in which the exposure station, the focal length, and the three exterior orientation angles of each image are known. If sufficiently well-distributed GCPs are provided, these parameters can be solved using collinearity.
3. *Brightness values assignment of the pixels of the output image via resampling the pixels in the original image*: Since each pixel in the output orthoimage rarely lies in an integrate pixel of the original image, gray resampling of the original pixels have to be conducted. The most commonly accepted methods are the nearest neighbor, bilinear interpolation, or cubic convolution methods. We employ the nearest neighbor method because this method consumes little computational time while not losing accuracy.

7.6.2 Shadow Detection and Restoration

7.6.2.1 Shadow Detection

Urban large-scale orthoimage generation requires detecting and removing shadow. Many efforts for detection and restoration of the shadowed areas from aerial images have been made in the past years, e.g., Salvador et al. (2001), Peng and Liu (2004), Liow and Pavlidis (1990), Noronha and Nevatia (2001), Stauder et al. (1999), Irvin et al. (1989), Jiang et al. (1992), Jaynes et al. (2004), and Tsai (2003). Most of these efforts did not consider the self-shadow and typically focused attention on the umbra, considering the penumbra as a particular case of umbra. Moreover, they only took advantage of the image gray information and 2D object geometric shapes when detecting the shadowed areas. This chapter presents an innovative method, which makes full use of DBM data for building shadow detection, but is also less interested in detecting self-shadow since the wall of buildings will be invisible in the true orthoimage. This proposed method is based on such a principle:

1. The length of shadow depends on both building height and the zenith of the light source (sun); the direction of shadow depends on the position of the sun relative to the buildings.
2. The zenith and the position of the light source relative to the buildings at the epoch of imaging remain constant. This means that the directions of shadow for all buildings within an aerial image are the same, and the length of shadow of each building can be measured using the building model if the zenith of the sun is known.

Based on the fact above, we first manually determine the relationship between the length and the direction of one building shadow and 3D building model, called *shadow relationship*. With the established shadow relationship, the other building shadows can be detected using the building model (DBM), which is stored in a DBM database.

Figure 7.16 illustrates the principle of DBM-based shadow detection. First, the direction (β) and the length (*L*) of the shadow are measured, and the relationship between the length of the shadow and the building height is established by

$$\tan(\alpha) = L / h \tag{7.9}$$

where *h* is the building height, and α is the zenith of the light source (sun). The solved zenith from equation (7.9) can be used to calculate the length of other buildings' shadow since the zenith is the same within an aerial image.

Further, the coordinates of the shadow boundary in a defined local coordinate system can be calculated by

$$x = L\ cos\ (\beta)$$

$$y = L\ sin\ (\beta) \tag{7.10}$$

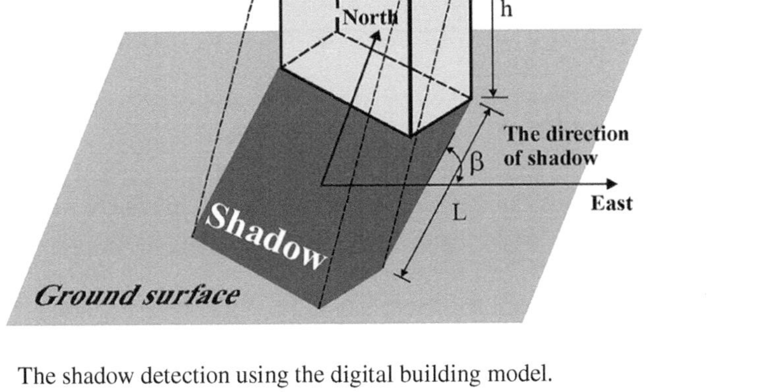

FIGURE 7.16 The shadow detection using the digital building model.

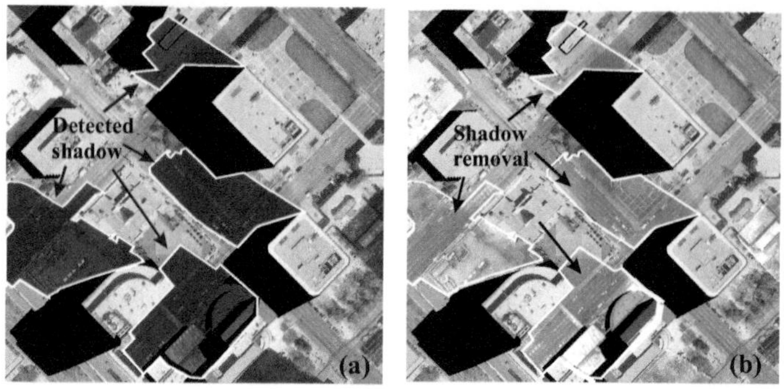

FIGURE 7.17 Comparison between (a) before shadow treatment and (b) after shadow treatment.

where x and y are coordinates of the shadow boundary and β is the direction of shadows in a given local coordinate system. The local coordinates of shadow boundary can be transformed to image coordinates; thus, the shadowed area can be detected.

The calculated zenith, α, and the measured direction, β, remain constant for all buildings within an aerial image. The DBM provides 3D geometric information of building, which will be used to calculate the length of the shadow and the boundary of the shadow. Ideally, this proposed method can achieve a satisfying result, but the real scene is more complex than imagined. We thus combine the proposed method and the region-growing method, which make full use of gray information, to detect the shadowed area. Figure 7.17 is a result of shadow detection from the proposed method.

7.6.2.2 Shadow Restoration

To restore the shadowed areas, Zhou et al. (2005) presented the use of a histogram adjustment algorithm for this purpose on the basis of changes of the mean and variance of the shadowed areas via matching the mean and variance of the image with no shadows. The mathematical model can be written as

$$Y = \sigma_l / \sigma_s * (X - \mu_s) + \mu_l \tag{7.11}$$

where Y is the output of the original pixel, X, in the shadowed areas after histogram adjustment; σ and μ are the standard deviation and mean; and the subscripts l and s stand for nonshadow and shadowed areas, respectively.

Figure 7.17 demonstrates the experimental results before and after the shadow treatment. As seen from Figure 7.17, the shadowed objects caused by tall buildings in Figure 7.17a have been substantially restored and the image quality of these shadowed areas is now comparable to that of the ones around them. The effectiveness of shadow removal can also be demonstrated by their histograms, as shown in Figure 7.18. It can be observed that the mean and dynamic ranges of the data in the shadowed areas become comparable to those in the nonshadowed areas after shadow treatment.

7.6.3 Occlusion Detection and Compensation

In a true orthoimage, all objects like buildings and bridges should be presented in their true upright planimetric positions, i.e., what one can see in a true orthoimage is only the roofs of objects; hence, the walls should be invisible and the occluded areas should be refilled by conjugate orthoimage patches if they are visible.

To effectively generate orthoimages, a visibility analysis applying the Z-buffer techniques has been presented in Section 7.3.2 (see also Amhar et al. 1998; Rau et al. 2002). The Z-buffer in our study consists of two fictitious planes (or matrixes): one in the image plane to store the distance from

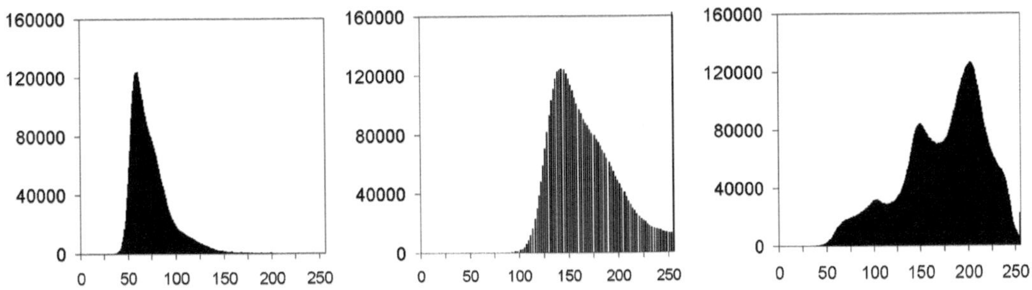

FIGURE 7.18 Comparison of histograms of the shadowed areas: (a) before adjustment, (b) after adjustment, and (c) no shadow in this area.

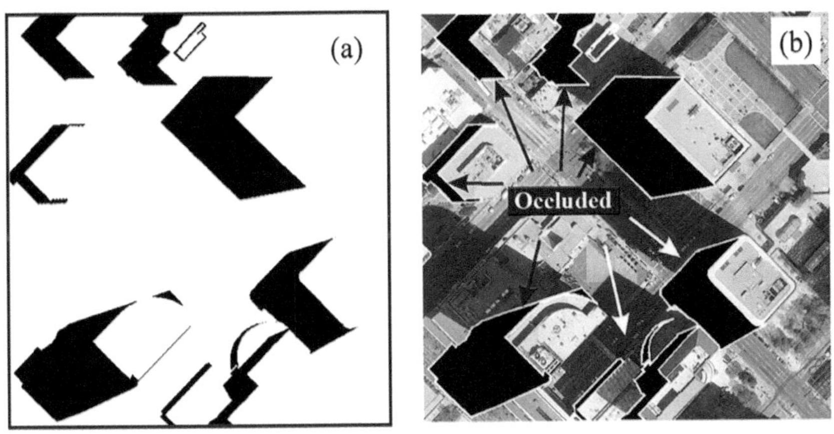

FIGURE 7.19 (a) Index map generated using the Z-buffer algorithm and (b) orthoimage with the occlusion identified using the index map.

the projection center to the surface point, whose corresponding position in the image plane is first calculated with the imaging system parameters (interior and exterior orientations) using the collinearity condition, and the other one in the ground surface plane to be used as a binary index map with the same resolution, dimension, and projection properties as DBM. When an area is occluded, it means that more than one surface points have the same coordinates in the image plane. Under these circumstances, we will select the shorter distance to store and mask the index map at the corresponding location where the distance to the projection center is longer. With this method, all the hidden areas can be identified in the index map (Figure 7.19a), and when the actual orthorectification begins, we only need to find the brightness values for the surface points that are not masked in the index map, all masked areas will therefore can be set as blank or other background values, indicating they are invisible or occluded in the resulting orthophoto (Figure 7.19b).

The occlusion compensation is implemented by refilling the occluded areas from neighbor slave orthoimages. This process is implemented by mosaicking, for which the automatic selection of seamlines from slave orthoimages has been described in Section 7.3.3. Based on this point, the occluded area must be visible in one, even several slave orthoimages in order to completely refill an occlusion. To reduce the radiometric discrepancy, it is better to use images acquired on the same flight mission due to their similarity in imaging conditions. Figure 7.19 shows that the master image was compensated by conjugate orthoimage patches.

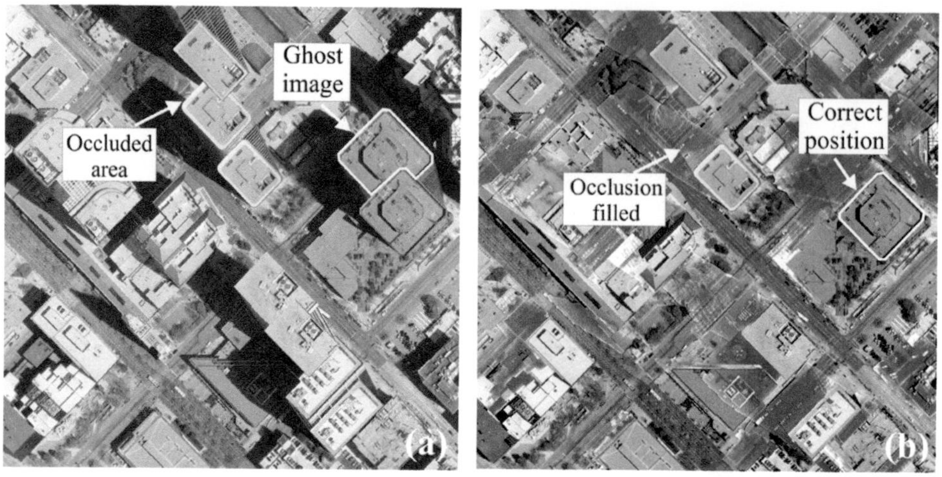

FIGURE 7.20 The comparison of orthorectified large-scale aerial images by (a) the traditional differential method and (b) our method.

7.6.4 Radiometric Balancing

Significant scene-to-scene radiometric variations are observed in urban aerial images. To overcome this problem, radiometric balancing and blending operations are applied to individual scenes to prevent a patchy or quilted appearance in the final mosaic. We have developed an Arc/Info-based tool in combination with a number of available GIS functions, including line and polygon buffering, distance calculation, and grid algebraic operations, to perform the radiometric balancing. In this tool, the weights for blending individual scenes along the specified buffer zone are calculated by the cubic Hermite function (the detailed description can be found in Zhou et al. (2002)).

7.6.5 True Orthoimage Generation

After the above procedures are implemented, a true orthoimage can be created. Figure 7.20b depicts the results of a true orthoimage that is generated using our method. For the comparison purpose, Figure 7.20a is an orthoimage generated by the conventional differential model. As seen in Figure 7.20a, the occlusions of buildings still exist in the resulting orthoimage as evidenced by the "doubling" of the building roofs. The building occlusions in Figure 7.20b have been completely removed and refilled. The buildings are correctly orthorectified in their true planimetric positions, and the shadows of buildings have been removed. The results in Figure 7.20b demonstrate the capability of the proposed methodology to generate true orthoimages from large-scale images in urban areas. Even though there remains room for improvement, such as to identify conjugate areas of multiple slave images, and to compensate for the spectral difference between refilled and surrounding areas.

7.7 CONCLUSIONS

A methodology for the generation of true orthoimages from large-scale aerial images in urban areas has been investigated in this chapter. The core technique of the methodology involves four key parts. First, a DBM and a DTM that accurately model the urban buildings and terrain must be obtained. Second, DTM-based and DBM-based orthoimages will be generated independently, and then the true orthoimage is generated by merging the DTM-based and the DBM-based orthoimages (i.e., orthorectification must be done by merging DTM-based orthorectification and DBM-based orthorectification). Third, shadowed areas and occluded areas must automatically be detected.

The shadowed areas should be restored and the lost information due to the occlusion should be compensated by mosaicking or re-filling the conjugate patches from adjacent slave orthoimages. Fourth, radiometric differences caused by scene-to-scene must be balanced and seamlines for mosaicking should be optimized for high-quality orthoimage products.

Various photogrammetric flight plannings based on test field data have been simulated for further understanding of the relationship between occlusion and flying height, building height, and street width. An accurate mathematical model, which can be used to plan the endlap and sidelap and flying height for various building heights, street widths, types of cameras, and other factors (e.g., street directions, building distribution pattern), has been developed. The experimental results are very helpful for assisting the urban photogrammetric flying planning mission.

REFERENCES

Amhar, F., Josef, J.; Ries, C., The generation of true orthophotos using a 3D building model in conjunction with a conventional DTM. *International Archives of Photogrammetry and Remote Sensing*, 32(4), 1998, pp. 16–22.

Biason, A.; Dequal, S.; Lingua, A., A new procedure for the automatic production of true orthophotos. *International Archives of the Photogrammetry, Remote Sensing and Spatial Information Sciences*, 2004, vol. 35, pp. 1682–1777. XXXV, 2004.

Breunig, M., *Integration of Spatial Information for Geo-information Systems*, Springer Corp.: Berlin, 1996.

Cameron, A. D.; Miller, D. R.; Ramsay, F.; Nikolaou, I.; Clarke, G. C., Temporal measurement of the loss of native pinewood in Scotland through the analysis of orthorectified aerial photographs. *Journal of Environmental Management*, 58, 2000, pp. 33–43.

Chen, L. C.; Lee, L. H., Rigorous generation of digital orthophotos from SPOT images. *Photogrammetric Engineering & Remote Sensing*, 59(3), 1993a, pp. 655–661.

Chen, L. C.; Rau, J. Y., A unified solution for digital terrain model and orthoimage generation from SPOT stereopairs. *IEEE Transactions on Geoscience and Remote Sensing*, 31(6), 1993b, pp. 1243–1252.

Graz, M. G., Managing large 3D urban databases. *Photogrammetric Week*, Stuttgart University, 1999, pp. 341–349.

Gruen, A.; Wang, X., CC-Modeler: A topology generator for 3-D city models. *ISPRS Journal of Photogrammetry & Remote Sensing*, 53, 1998, pp. 286–295.

Irvin, R. B.; McKeown Jr., D. M., Methods for exploiting the relationship between buildings and their shadows in aerial imagery. *IEEE Transactions on Systems, Man, and Cybernetics*, 19, 1989, pp. 1564–1575.

Jauregui, M., Vílchez, J.; Chacón, L., A procedure for map updating using digital mono-plotting. *Computers & Geosciences*, 28(4), 2002, pp. 513–523.

Jaynes, C., Webb, S.; Steele, R., Camera-based detection and removal of shadows from interactive multiprojector displays. *IEEE Transactions on Visualization and Computer Graphics*, 10(3), 2004, pp. 290–301.

Jiang, C.; Ward, M. O., Shadow identification. In: *Proceedings of IEEE International Conference on Computer Vision and Pattern Recognition*, Champaign, Illinois, USA, 1992, pp. 606–612.

Joshua, G., Evaluating the accuracy of digital orthophotos quadrangles (DOQ) in the context of parcel-based GIS. *Photogrammetric Engineering & Remote Sensing*, 67(2), 2001, pp. 199–205.

Liow, Y.-T., Pavlidis, T., Use of shadows for extracting buildings in aerial images. *Computer Vision, Graphics, and Image Processing*, 49(2), 1990, pp. 242–277.

Mayr, W., True orthoimages. *GIM International*, 37, 2002, pp. 37–39.

Noronha, S.; Nevatia, R., Detection and modeling of buildings from multiple aerial images. *IEEE Transactions on Pattern Analysis and Machine Intelligence*, 23(5), 2001, pp. 501–518.

Passini, R.; Jacobsen, K., Accuracy analysis of digital orthophotos from very high resolution imagery. *International Archives of the Photogrammetry, Remote Sensing and Spatial Information Sciences*, XXXV, 2004pp. 695–700..

Peng, J.; Liu, Y., The role of context and model in urban aerial image interpretation focusing on buildings. In: *IEEE International Conference on Networking, Sensing and Control*, Taipei, Taiwan, vol. 1, March 21–23, 2004, pp. 1–6.

Rau, J. Y.; Chen, N. Y.; Chen, L. C., True orthophoto generation of built-up areas using multi-view images. *Photogrammetric Engineering and Remote Sensing*, 68(6), 2002, pp. 581–588.

Salvador, E.; Cavallaro, A.; Ebrahimi, T., Shadow identification and classification using invariant color models. *In 2001 IEEE International Conference on Acoustics, Speech, and Signal Processing*, Salt Lake City, UT, USA. 2001, Vol. 3, pp. 1545–1548.

Schickler, W.; Thorpe, A., Operational procedure for automatic true orthophoto generation. *International Archives of Photogrammetry and Remote Sensing*, 32(4), 1998, pp. 527–532.

Siachalou, S., Urban orthoimage analysis generated from IKONOS data. *International Archives of the Photogrammetry, Remote Sensing and Spatial Information Sciences*, XXXV, 2004.

Skarlatos, D., Orthophotograph production in urban areas. *Photogrammetric Record*, 16(94), 1999, pp. 643–650.

Stauder, J.; Mech, R.; Ostermann, J., Detection of moving cast shadows for object segmentation. *IEEE Transactions on Multimedia*, 1(1), 1999, pp. 65–76.

Tsai, J. D., Automatic shadow detection and radiometric restoration on digital aerial images. In: *International Geoscience and Remote Sensing Symposium (IGARSS): Learning from Earth's Shapes and Colours*, Toulouse, France, vol. 2, July 21–25, 2003, pp. 732–733.

USGS, Digital orthophoto program.https://www.usgs.gov/publications/digital-orthophoto-quadrangles-0USGS. U.S. Department of the Interior, U.S. Geological Survey, 1998.

Vassilopoulou, S.; Hurni, L.; Dietrich, V.; Baltsavias, E.; Pateraki, M.; Lagios, E.; Parcharidis, I., Orthophoto generation using IKONOS imagery and high-resolution DEM: A case study on volcanic hazard monitoring of Nisyros Island (Greece). *ISPRS Journal of Photogrammetry and Remote Sensing*, 57(1–2), 2002, pp. 24–38.

Zhou, G.; Chen, W.; Kelmelis, J. A.; Zhang, D., A comprehensive study on urban true orthorectification. *IEEE Transactions on Geoscience and Remote Sensing*, 43, 2005, pp. 2138–2147. https://doi.org/10.1109/TGRS.2005.848417

Zhou, G.; Jezek, K.; Wright, W.; Rand, J.; Granger, J., Orthorectifying 1960's declassified intelligence satellite photography (DISP) of Greenland. *IEEE Geoscience and Remote Sensing*, 40(6), 2002, pp. 1247–1259.

Zhou, G.; Qin, Z.; Benjamin, S.; Schickler, W., Technical problems of deploying national urban large-scale true orthoimage generation. In: *The Second Digital Government Conference*, Boston, MA, May 18–21, 2003, pp. 383–387.

Zhou, G.; Xie, M.; Gong, J., Design and implementation of attribute database management system for GIS (GeoStar). *International Journal of Geographic Information Science*, 6(2), 2000, pp. 170–180.

Zlatanova, S., 3D GIS for urban development. ITC Dissertation Number 69. Doktor der technischen Wissenschaften an der Technischen Universitate Graz, 2000.

8 TOM Generation with Extremely High Buildings

8.1 INTRODUCTION

For urban areas, especially in very high building urban areas, the traditional method for digital orthophotomap is not capable of orthorectifying the displacements caused by high buildings. Therefore, the so-called true orthophotomap (TOM) generation has been investigated by researchers such as Amhar et al. (1998), Rau et al. (2002), Nielsen (2004), Biason et al. (2004), Mayr (2002), Palà and Arbiol (2002), Siachalou (2004), Habib et al. (2006), Zhou et al. (2005), Sheng et al. (2003), and Xie and Zhou (2006). Zhou et al. (2005) comprehensively discussed the TOM generation. Briefly, the TOM generation is composed of digital terrain model-based image orthorectification, DBM (digital building model)-based orthorectification, and their merging. Some associated work for TOM generation is the occlusion detection and refilling, shadow detection and compensation, and radiometric difference balance.

However, when we created a true orthoimage for those high buildings, such as over 150 m high, it can be found that some of the high buildings cannot be completely orthorectified into their upright and correct position, i.e., their walls can be viewed in the orthoimage (Figure 8.1). To further examine which type of errors causes this incomplete orthorectification, we orthorectified the same buildings in two different aerial images using the same DBM and method. It was found that the buildings are completely orthorectified in one image (Figure 8.1a), but not in the other image (Figure 8.1b). This fact has been discovered that the DBM is not a bane. In other words, the incomplete orthorectification is probably directly caused by the inexact exterior orientation parameters (EOPs) because the traditional orthorectification algorithm largely depends on both the accuracy and distribution of the ground control points (GCPs). Moreover, these GCPs are usually laid out on the ground. However, buildings occluding buildings and/or buildings occluding ground frequently happen in urban areas, resulting in that the good distribution of the GCPs is impractical. Thus, the orthorectified images must be incomplete.

For this reason, this chapter describes a method to increase the accuracy of orthorectifying high buildings using relative constraint conditions.

8.2 RELATIVE CONSTRAIN FOR ORTHORECTIFICATION

8.2.1 Traditional Orthorectification Model

The mathematical model of traditional orthorectification is called *photogrammetry differential model*, with which, for any given ground point such as G, it is expressed by

$$x_g - x_0 = -f\frac{a_1(X_G - X_S) + b_1(Y_G - Y_S) + c_1(Z_G - Z_S)}{a_3(X_G - X_S) + b_3(Y_G - Y_S) + c_3(Z_G - Z_S)} \tag{8.1a}$$

$$y_g - y_0 = -f\frac{a_2(X_G - X_S) + b_2(Y_G - Y_S) + c_2(Z_G - Z_S)}{a_3(X_G - X_S) + b_3(Y_G - Y_S) + c_3(Z_G - Z_S)} \tag{8.1b}$$

DOI: 10.1201/9781003632108-8

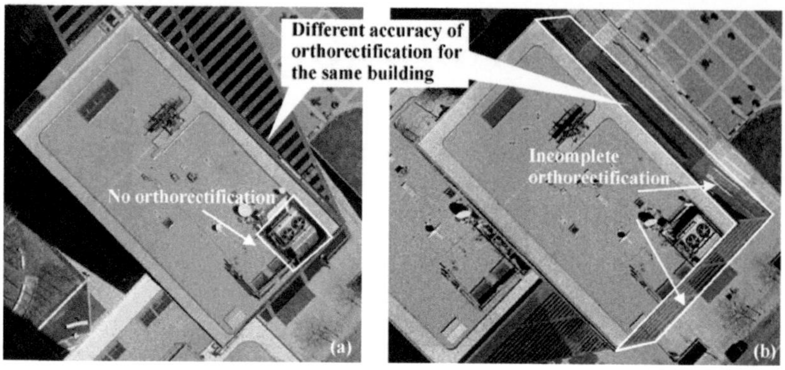

FIGURE 8.1 Incomplete orthorectification and different accuracies for the same building in the two aerial: (a) a building without orthorectification and (b) a building with incomplete orthorectification.

where f, x_0, y_0 are the camera's interior orientation parameters (IOPs); X_S, Y_S, Z_S are the position of exposure station; $a_1, \ldots, c_3$ are the elements of the rotation matrix, which is a function of three rotation angles ω, ϕ, κ; x_g, y_g are the image coordinates of the ground point G; X_G, Y_G, Z_G are the coordinates of the ground point G in the ground coordinate system.

Equation (8.1) can further be expressed by a linear equation after it is linearized using Taylor's series, i.e.,

$$V = AT - L \tag{8.2}$$

where $T = \begin{pmatrix} \phi & \omega & \kappa & X_S & Y_S & Z_S \end{pmatrix}^T$, $V = \begin{pmatrix} v_{gx1} & v_{gy1} & v_{gx2} & v_{gy2} \end{pmatrix}^T$, $L = \begin{pmatrix} l_{gx1} & l_{gy1} & l_{gx2} & l_{gy2} \end{pmatrix}^T$, and A is the coefficient (see Mikhail et al. 2001). With a number of GCPs, the nominal IOPs provided by the vendor can be calibrated, and the EOPs in equation (8.2) can be solved using least-squares estimation. Once the IOPs and EOPs were determined, any ground point can be orthorectified onto a predefined orthogonal plane. The detailed process of orthorectification can be referred to Zhou et al. (2002).

To increase the accuracy of the orthoimage, one uses to increase the number of GCPs and/or select well-distributed GCPs. This method has its limitations because the location and distribution of GCPs impact the accuracy of orthorectification, while most of the current GCPs are distributed on the ground, which means the top of high buildings has insufficient control information. Thus, this chapter plans to use relative constraint conditions, such as perpendicular and collinear conditions, to improve the accuracy. The mathematical models are below.

8.2.2 Perpendicular Control Condition

As shown in Figure 8.2, suppose that AB and BC are two edges of a flat house roof, and their corresponding edges in the image plane are ab and bc. Suppose that the coordinates of A, B, and C are (X_A, Y_A, Z_A), (X_B, Y_B, Z_B), and (X_C, Y_C, Z_C) in the object coordinate system. For a flat-roof cube house, the heights (i.e., Z coordinates) of A, B, and C are the same and the line segments AB and BC are perpendicular to each other.

If AB is not perpendicular to BC at an intersection angle θ, i.e., B deviates from its correct position at B', we draw a line from C to O, and make the line, CO is perpendicular to AB' (see Figure 8.2). l is the distance of B and O and can be expressed by

$$l = (X_C - X_B)\cos\theta + (Y_C - Y_B)\sin\theta = (X_C - X_B)\frac{(X_B - X_A)}{s_{AB}} + (Y_C - Y_B)\frac{(Y_B - Y_A)}{s_{AB}} \tag{8.3}$$

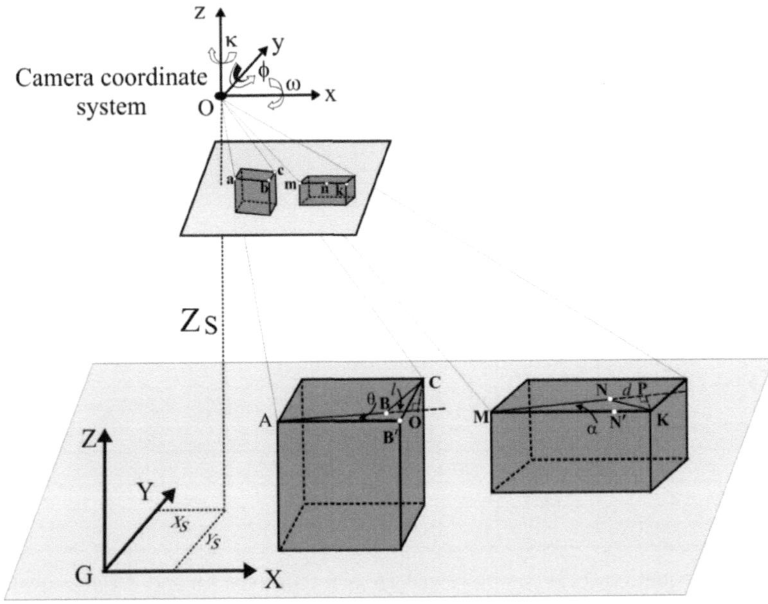

FIGURE 8.2 The geometry for the perpendicular constraint and collinear constraint condition.

where s_{AB} is the distance of the segment AB. Theoretically, the distance l should be zero. Thus, the differential form for equation (8.3) is

$$l_0 + \Delta l = \left[(X_C - X_B)\frac{(X_B - X_A)}{s_{AB}} + (Y_C - Y_B)\frac{(Y_B - Y_A)}{s_{AB}} \right] + \frac{1}{s_{AB}} \begin{bmatrix} (X_B - X_A)\Delta X_C + (Y_B - Y_A)\Delta Y_C + (X_B - X_C)\Delta X_A + (Y_B - Y_C)\Delta Y_A \\ +(X_C - 2X_B + X_A)\Delta X_B + (Y_C - 2Y_B + Y_A)\Delta Y_B \end{bmatrix} = 0 \tag{8.4}$$

Rewrite equation (8.4) in matrix form:

$$\begin{bmatrix} X_B - X_C \\ Y_B - Y_C \\ X_C - 2X_B + X_A \\ Y_C - 2Y_B + Y_A \\ X_B - X_A \\ Y_B - Y_A \end{bmatrix}^T \begin{bmatrix} \Delta X_A \\ \Delta Y_A \\ \Delta X_B \\ \Delta Y_B \\ \Delta X_C \\ \Delta Y_C \end{bmatrix} + \left[(X_C - X_B)(X_B - X_A) + (Y_C - Y_B)(Y_B - Y_A) \right] = 0 \tag{8.5}$$

Equation (8.5) is, in fact, a perpendicular constraint condition, which describes line AB as perpendicular to BC on the X–Y plane. Similarly, if the points B and C are on the plane X–Z, or Y–Z, the perpendicular relative constraint conditions can be expressed by, respectively,

$$\begin{bmatrix} X_B - X_C \\ Z_B - Z_C \\ X_C - 2X_B + X_A \\ Z_C - 2Z_B + Z_A \\ X_B - X_A \\ Z_B - Z_A \end{bmatrix}^T \begin{bmatrix} \Delta X_A \\ \Delta Z_A \\ \Delta X_B \\ \Delta Z_B \\ \Delta X_C \\ \Delta Z_C \end{bmatrix} + \left[(X_C - X_B)(X_B - X_A) + (Z_C - Z_B)(Z_B - Z_A)\right] = 0 \tag{8.6}$$

$$\begin{bmatrix} Y_B - Y_C \\ Z_B - Z_C \\ Y_C - 2Y_B + Y_A \\ Z_C - 2Z_B + Z_A \\ Y_B - Y_A \\ Z_B - Z_A \end{bmatrix}^T \begin{bmatrix} \Delta Y_A \\ \Delta Z_A \\ \Delta Y_B \\ \Delta Z_B \\ \Delta Y_C \\ \Delta Z_C \end{bmatrix} + \left[(Z_C - Z_B)(Z_B - Z_A) + (Y_C - Y_B)(Y_B - Y_A)\right] = 0 \tag{8.7}$$

The vector form for equation (8.5) can be rewritten as (equations 8.6 and 8.7 are similar)

$$\boldsymbol{C}_1 \hat{\boldsymbol{X}}_1 + \boldsymbol{W}_1 = 0 \tag{8.8}$$

where $\boldsymbol{C}_1 = \begin{bmatrix} X_B - X_C \\ Y_B - Y_C \\ X_C - 2X_B + X_A \\ Y_C - 2Y_B + Y_A \\ X_B - X_A \\ Y_B - Y_A \end{bmatrix}^T$, $\hat{\boldsymbol{X}}_1 = \begin{bmatrix} \Delta X_A \\ \Delta Y_A \\ \Delta X_B \\ \Delta Y_B \\ \Delta X_C \\ \Delta Y_C \end{bmatrix}$, $\boldsymbol{W}_1 = (X_C - X_B)(X_B - X_A) + (Y_C - Y_B)(Y_B - Y_A)$

8.2.3 Collinear Constrain Condition

Also, as depicted in Figure 8.2, suppose that the segments *MN* and *NK* are not collinear, at an intersection angle of α, we have

$$d = (Y_K - Y_N)\cos\alpha + (X_K - X_N)\sin\alpha = (Y_K - Y_N)\frac{(X_N - X_M)}{s_{MN}} + (X_K - X_N)\frac{(Y_N - Y_M)}{s_{MN}} \tag{8.9}$$

where d is the distance between K and P. Similarly, its linearized equation and matrix form are, respectively,

$$d_0 + \Delta d = \left[(Y_K - Y_N)\frac{(X_N - X_M)}{s_{MN}} + (X_K - X_N)\frac{(Y_N - Y_M)}{s_{MN}}\right] \tag{8.10}$$

$$+\frac{1}{s_{MN}}\begin{bmatrix} (Y_N - Y_M)\Delta X_K + (X_N - X_M)\Delta Y_K + (Y_N - Y_K)\Delta X_M + (X_N - X_K)\Delta Y_M \\ +(Y_K - 2Y_N + Y_M)\Delta X_N + (X_M - 2X_N + X_K)\Delta Y_N \end{bmatrix} = 0$$

$$\begin{bmatrix} Y_N - Y_K \\ X_K - X_N \\ Y_K - 2Y_N + Y_M \\ X_M - 2X_N + X_K \\ Y_M - Y_N \\ X_N - X_M \end{bmatrix}^T \begin{bmatrix} \Delta X_M \\ \Delta Y_M \\ \Delta X_N \\ \Delta Y_N \\ \Delta X_K \\ \Delta Y_K \end{bmatrix} + \left[(Y_K - Y_N)(X_N - X_M) + (X_K - X_N)(Y_N - Y_M)\right] = 0 \quad (8.11)$$

Equation (8.11) is, in fact, a collinear constraint condition, which describes line *MN* as collinear with *NK* on the *X–Y* plane. Similarly, if the points *M*, *N*, and *K* are on the plane *X–Z* or *Y–Z*, the corresponding collinear constraint conditions are, respectively,

$$\begin{bmatrix} Z_N - Z_K \\ X_K - X_N \\ Z_K - 2Z_N + Z_M \\ X_M - 2X_N + X_K \\ Z_M - Z_N \\ X_N - X_M \end{bmatrix}^T \begin{bmatrix} \Delta X_M \\ \Delta Z_M \\ \Delta X_N \\ \Delta Z_N \\ \Delta X_K \\ \Delta Z_K \end{bmatrix} + \left[(Z_K - Z_N)(X_N - X_M) + (X_K - X_N)(Z_N - Z_M)\right] = 0 \quad (8.12)$$

$$\begin{bmatrix} Z_N - Z_K \\ Y_K - Y_N \\ Z_K - 2Z_N + Z_M \\ Y_M - 2Y_N + Y_K \\ Z_M - Z_N \\ X_N - X_M \end{bmatrix}^T \begin{bmatrix} \Delta Y_M \\ \Delta Z_M \\ \Delta Y_N \\ \Delta Z_N \\ \Delta Y_K \\ \Delta Z_K \end{bmatrix} + \left[(Z_K - Z_N)(Y_N - Y_M) + (Y_K - Y_N)(Z_N - Z_M)\right] = 0 \quad (8.13)$$

The vector form for equation (8.11) can be written as (equations 8.12 and 8.13 are similar)

$$\boldsymbol{C}_2 \hat{\boldsymbol{X}}_2 + \boldsymbol{W}_2 = 0 \quad (8.14)$$

where $\boldsymbol{C}_2 = \begin{bmatrix} Z_N - Z_K \\ Y_K - Y_N \\ Z_K - 2Z_N + Z_M \\ Y_M - 2Y_N + Y_K \\ Z_M - Z_N \\ X_N - X_M \end{bmatrix}^T$, $\hat{\boldsymbol{X}}_2 = \begin{bmatrix} \Delta Y_M \\ \Delta Z_M \\ \Delta Y_N \\ \Delta Z_N \\ \Delta Y_K \\ \Delta Z_K \end{bmatrix}$, $\boldsymbol{W}_2 = (Z_K - Z_N)(Y_N - Y_M) + (Y_K - Y_N)(Z_N - Z_M)$

From Figure 8.2, if points *A*, *B*, and *C* and their corresponding imaged points *a*, *b*, and *c* are simultaneously GCPs, this means that one perpendicular or collinear constraint condition will add six additional unknown parameters (i.e., $\Delta X_A, \Delta Y_A, \Delta Z_A, \Delta X_B, \Delta Y_B, \Delta Z_B$ for perpendicular constrain). Thus, equation (8.2) can be extended into

$$\boldsymbol{V} = \boldsymbol{AT} + \boldsymbol{B}\hat{\boldsymbol{X}} - \boldsymbol{L} \quad (8.15)$$

where, $\boldsymbol{T} = \begin{pmatrix} \phi & \omega & \kappa & X_S & Y_S & Z_S \end{pmatrix}^T$, $\hat{\boldsymbol{X}} = \begin{pmatrix} \Delta X_A & \Delta Y_A & \Delta Z_A & \Delta X_B & \Delta Y_B & \Delta Z_B & \Delta X_M & \Delta Y_M & \Delta Z_M \\ \Delta X_N & \Delta Y_N & \Delta Z_N \end{pmatrix}^T$, $\boldsymbol{A}$ and $\boldsymbol{B}$ are coefficient matrix, $\boldsymbol{V}$ and L are similar to equation (8.2).

Combining equations (8.8), (8.14), and (8.15), we have

$$\begin{cases} \boldsymbol{V} = \boldsymbol{AT} + \boldsymbol{B}\hat{\boldsymbol{X}} - \boldsymbol{L} \\ \boldsymbol{C}_1\hat{\boldsymbol{X}}_1 + \boldsymbol{W}_1 = 0 \\ \boldsymbol{C}_2\hat{\boldsymbol{X}}_2 + \boldsymbol{W}_2 = 0 \end{cases} \tag{8.16}$$

Further, we rewrite equation (8.16) by

$$\begin{cases} \boldsymbol{V} = \boldsymbol{D}\delta\boldsymbol{X} - \boldsymbol{L} \\ \boldsymbol{C}_x\delta\boldsymbol{X} + \boldsymbol{W}_x = 0 \end{cases} \tag{8.17}$$

where $\boldsymbol{C}_x = (\boldsymbol{C}_1 \quad \boldsymbol{C}_2)^T$, and $\boldsymbol{W}_x = (\boldsymbol{W}_1 \quad \boldsymbol{W}_2)^T$, $\boldsymbol{D} = (\boldsymbol{A}|\boldsymbol{B})$

$$\delta\boldsymbol{X} = \begin{pmatrix} \Delta\phi & \Delta\omega & \Delta\kappa & \Delta X_S & \Delta Y_S & \Delta Z_S & \Delta X_A & \Delta Y_A & \Delta Z_A & \ldots & \Delta X_N & \Delta Y_N & \Delta Z_N \end{pmatrix}^T$$

With the least-squares establishment method (Lawson and Hanson 1995), equation (8.17) can be written as

$$\Phi = \boldsymbol{V}^T\boldsymbol{V} + 2\boldsymbol{K}_S^T(\boldsymbol{C}_x\delta\boldsymbol{X} + \boldsymbol{W}_x)$$

The corresponding normal equation is

$$\begin{cases} \boldsymbol{D}^T\boldsymbol{D}\delta\boldsymbol{X} + \boldsymbol{C}_x{}^T\boldsymbol{K}_S + \boldsymbol{D}^T\boldsymbol{L} = 0 \\ \boldsymbol{C}_x\delta\boldsymbol{X} + 0\boldsymbol{K}_S + \boldsymbol{W}_x = 0 \end{cases} \tag{8.18}$$

where $\boldsymbol{K}_S$ is an introduced unknown matrix. If the number of total observation equations is m, $\boldsymbol{K}_S$ is a $m \times 1$ matrix, i.e., $\boldsymbol{K}_S = (K_{S_1}, K_{S_2}, \ldots, K_{S_m})^T$.

Let $N_{dd} = D^T D$, equation (8.18) is rewritten as

$$\begin{pmatrix} \boldsymbol{N}_{dd} & \boldsymbol{C}_x{}^T \\ \boldsymbol{C}_x & \boldsymbol{0} \end{pmatrix} \begin{pmatrix} \delta\boldsymbol{X} \\ \boldsymbol{K}_S \end{pmatrix} + \begin{pmatrix} \boldsymbol{D}^T\boldsymbol{L} \\ \boldsymbol{W}_x \end{pmatrix} = 0 \tag{8.19}$$

The solution of unknown parameters would be

$$\begin{cases} \delta\boldsymbol{X} = -(\boldsymbol{Q}_{11}\boldsymbol{D}^T\boldsymbol{L} + \boldsymbol{Q}_{12}\boldsymbol{W}_x) \\ \boldsymbol{K}_S = -(\boldsymbol{Q}_{21}\boldsymbol{D}^T\boldsymbol{L} + \boldsymbol{Q}_{22}\boldsymbol{W}_x) \end{cases} \tag{8.20}$$

where

$$\begin{pmatrix} N_{dd} & C_x^T \\ C_x & 0 \end{pmatrix}^{-1} = \begin{pmatrix} Q_{11} & Q_{12} \\ Q_{21} & Q_{22} \end{pmatrix} \tag{8.21}$$

Equation (8.20) is a mathematical model to be used for orthorectification. As seen, this model combines the building relative controls and the traditional ground point control. Thus, higher accuracy of the orthoimage should be achieved.

In traditional photogrammetry, the number and distribution of GCPs significantly impact the accuracy of orthorectification. Similarly, the building size, for instance, the length and width of buildings, probably also impacts the accuracy of this proposed method. As seen in Figure 8.2, the longer the line *AB*, the smaller the error. On the other hand, the relief displacements caused by high buildings that occur along radial lines from the nadir point. Thus, the relief displacements are zero from imaged objects at the nadir point and increase with increased radial distances from the nadir.

8.3 VALIDATIONS AND ANALYSES

8.3.1 Experimental Data

A detailed description of the experimental data can be found in Zhou et al. (2005). Briefly, the experimental field is located in Downtown Denver, CO, where the highest building is 125 m and many others are around 100 m. Six original aerial images were acquired from two flight strips using an RC30 aerial camera at a focal length of 153.022 mm on April 1, 2000. The flying height was approximately 1650 m above the mean ground elevation of the imaged area. The aerial photos were originally recorded on film and later scanned into a digital format at a pixel resolution of 25 μm. Part of the scanned aerial images is shown in Figure 8.3a. Figure 8.3b is the 2D representation of the digital surface model (DSM).

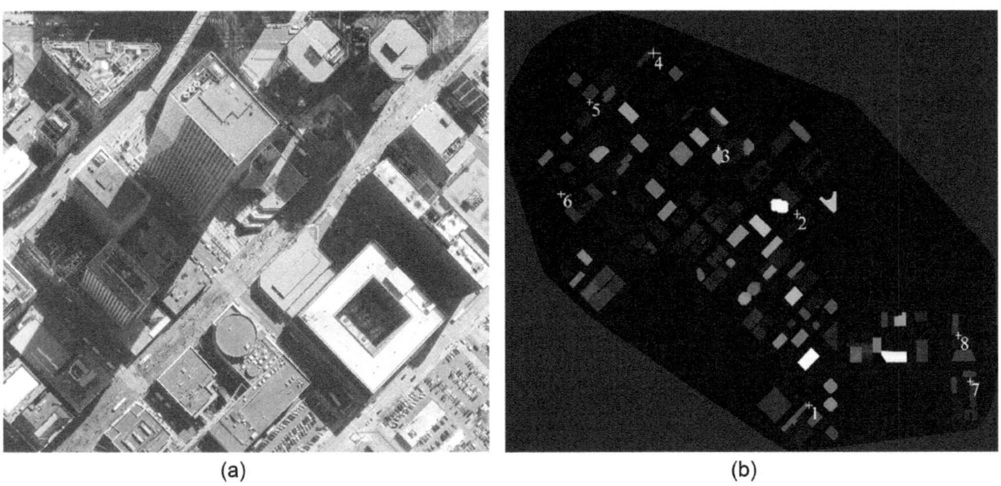

FIGURE 8.3 (a) Part of original aerial images and (b) digital surface model in Downtown Denver, CO, in which the brightness in the DSM represents the surface height.

8.3.2 Control Information

We manually measure 232 GCPs (i.e., 3D coordinates) from the DBM using Erdas/Imagine. The distribution of these GCPs covers the entire Downtown Denver (Figure 8.4). The measurement accuracy of the GCPs is approximately 1.48 pixels in the x direction and 1.27 pixels in the y direction averagely. The corresponding 2D image coordinates are measured from the original aerial images using the following method (Figure 8.5): We first selected a few (e.g., 8) GCPs including their 3D and 2D coordinates and then calculated the EOPs of the aerial image using space intersection. With the calculated EOPs, all other GCPs are back-projected onto the aerial image to obtain their approximate position in the aerial image plane. The precise locations (i.e., 2D image coordinates) are obtained by manually adjusting all these points in a magnified subwindow. The measurement accuracy of the 2D image coordinates is the subpixel level.

In addition to the 232 GCPs, we also measured 89 checkpoints from the DBM to evaluate the accuracy of the finally orthorectified image. Also, all the checkpoints are at the corners of the buildings. The measurement accuracy is consistent with the GCPs (Figure 8.4).

8.3.3 Constrain Line Extraction

With the measured GCPs, the corresponding lines to be taken as constraint controls are simultaneously extracted. In this experiment, the model of describing each building is realized using the parameterized Constructive Solid Geometry (CSG) method (Xie and Zhou 2006). In this model, each element of CSG primitive has been assigned its properties, such as wall, roof, and bottom, and each building model already contains its geometric topology, which describes the relationship of the building edges (Figure 8.6). Thus, the characteristics of topology can automatically be transferred into the relative controls. In other words, the relationship describing edges of each building, such as perpendicularity and collinearity, can be transformed into equation (8.8) and/or equation (8.14). For example, in Figure 8.7a and b, the faces F_a, F_b, and F_c, the edges L_1, L_2, and L_3, and their attributes are described in the CSG model. From the attributes of face data structure, L_1 and L_3 are automatically recognized as the edges of roofs and L_2 as the edge of wall. On the other hand,

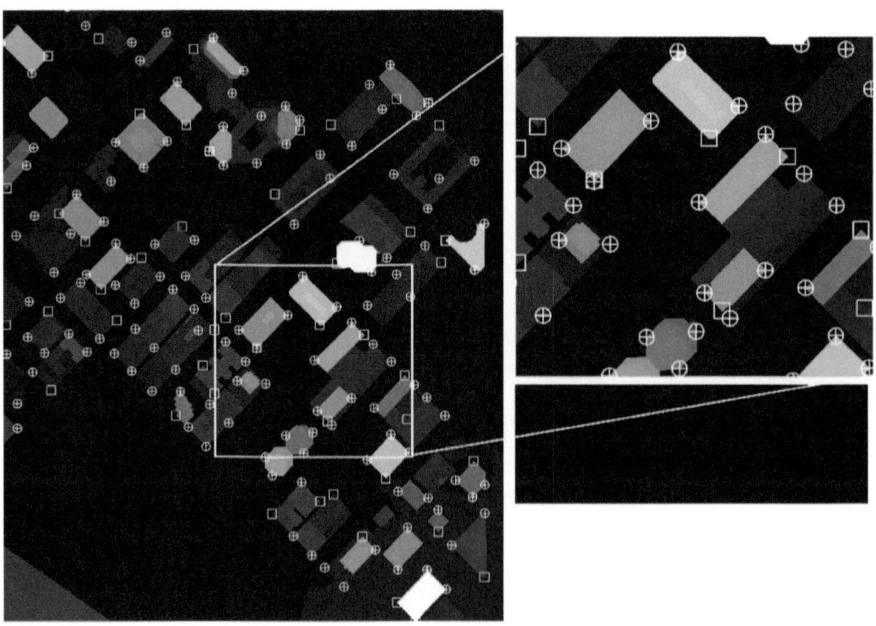

FIGURE 8.4 The measured ground control points and checkpoints as well as their image positions.

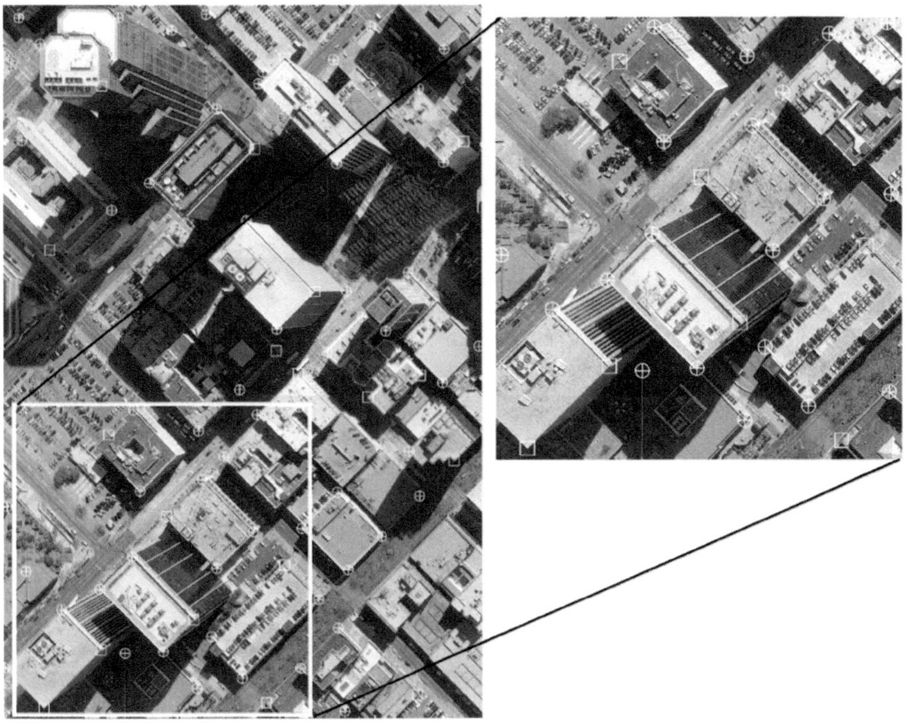

FIGURE 8.5 The 2D image coordinate measurement in the aerial image.

Single Building

CSG primitive ID	Operation	Attribute
CSG1	Union	A_1
CSG2	Intersection	A_2
...	...	...

CSG primitive

Component ID	Component Type
C_1	Quadrics
C_2	Polyhedron
...	...

Polyhedron

Polyhedron ID	Type	Face ID
PH1	Octahedron	$F_1, F_2, \ldots$
PH2	Hexahedron	$F_3, F_4, \ldots$
...	...	...

Quadrics

Quadrics ID	Type	Point ID	Radius	Height	Face ID
QD1	Cylinder	P_1	R_1	H_1	$F_5, F_6, \ldots$
QD2	Cone	P_2	R_2	H_2	$F_7, F_8, \ldots$
...	...	...	...	...	...

Face

Face ID	Type	Attribute	Edge ID
F_1	Triangle	Roof	$E_1, E_2, \ldots$
F_2	Rectangle	Wall	$E_3, E_4, \ldots$
F_3	Circle	Bottom	$E_5, E_6, \ldots$
...	...	...	...

Edge

Edge ID	Beginning Point	Ending Point
E_1	P_1	P_2
E_2	P_3	P_4
...	...	...

Point

Point ID	Space Coordinate		
	X	Y	Z
P_1	X_1	Y_1	Z_1
P_2	X_2	Y_2	Z_2
...	...	...	...

FIGURE 8.6 Data structure of the building model.

the control points P_1, P_3, and P_5 are extracted, and they can be automatically constructed lines l_1, l_2, and l_3. When L_1, L_2, and L_3 are back-projected onto the original aerial image, L_1, L_2, and L_3, whose topographic relationships have been described in the CSG model, will be matched with the lines l_1, l_2, and l_3, respectively (Figure 8.7c). Thus, the attributes (e.g., edges of roof and edges of wall) and topographic relationships (i.e., perpendicularity) of l_1, l_2, and l_3 can be inherited from L_1, L_2, and L_3. Thus, l_1 and l_3 are edges in the roof, and l_2 is a vertical line in the wall. Also, l_2 is perpendicular to l_1 and l_3 and l_1 is also perpendicular to l_3. Thus, when $l_1 \perp l_3$, equation (8.5) would be applied to construct a perpendicular relative control condition (equation) by replacing A, B, and C with P_5, P_3, and P_1 (see Figure 8.7).

Similarly, when point P_4 was measured on the line l_3, P_3, P_4, and P_5 should theoretically lie on a straight line of the building roof. Thus, a collinear constraint condition (equation 8.14) can be constructed by replacing M, N, and K with P_5, P_4, and P_3 (see Figure 8.7).

In this experiment, 106 buildings are back-projected onto the original aerial image. Seventy-six well-distributed lines are selected as relative controls. Figure 8.8 depicts these selected geometric control lines, in which the projected model wire is rendered with transparent mode, so that the occluded parts can be seen.

8.3.4 Accurate Comparison

Orthorectification was conducted through three methods. Method 1 only employs 8 GCPs; Method 2 uses 232 GCPs; and Method 3 (our method) employs 76 relative control lines and 56 GCPs. The IOPs were precisely calibrated and provided by the data vendor (Table 8.1). The EOPs are determined by space intersection, in which the three types of control data are employed. The results and accuracy are listed in Table 8.2. As seen from Table 8.2, our method has the highest accuracy at a standard deviation of 0.12 pixel.

After the above three groups of EOPs are determined, the orthoimages were generated using the procedures described by Zhou et al. (2002). The following two methods are employed to compare the accuracy of the three orthorectified images (orthoimages).

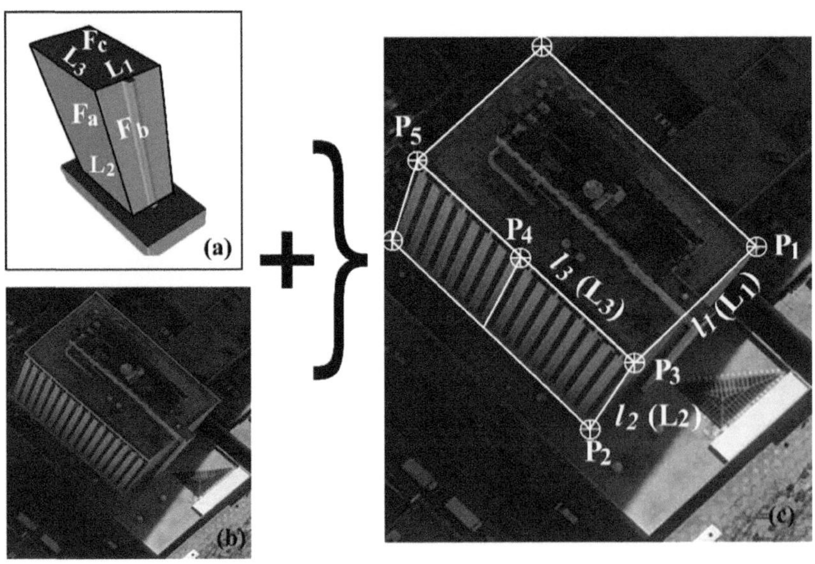

FIGURE 8.7 The extraction of building edges from the CSG model for relative controls: (a) the CSG model of buildings, (b) the building in the aerial image, and (c) the extracted building lines.

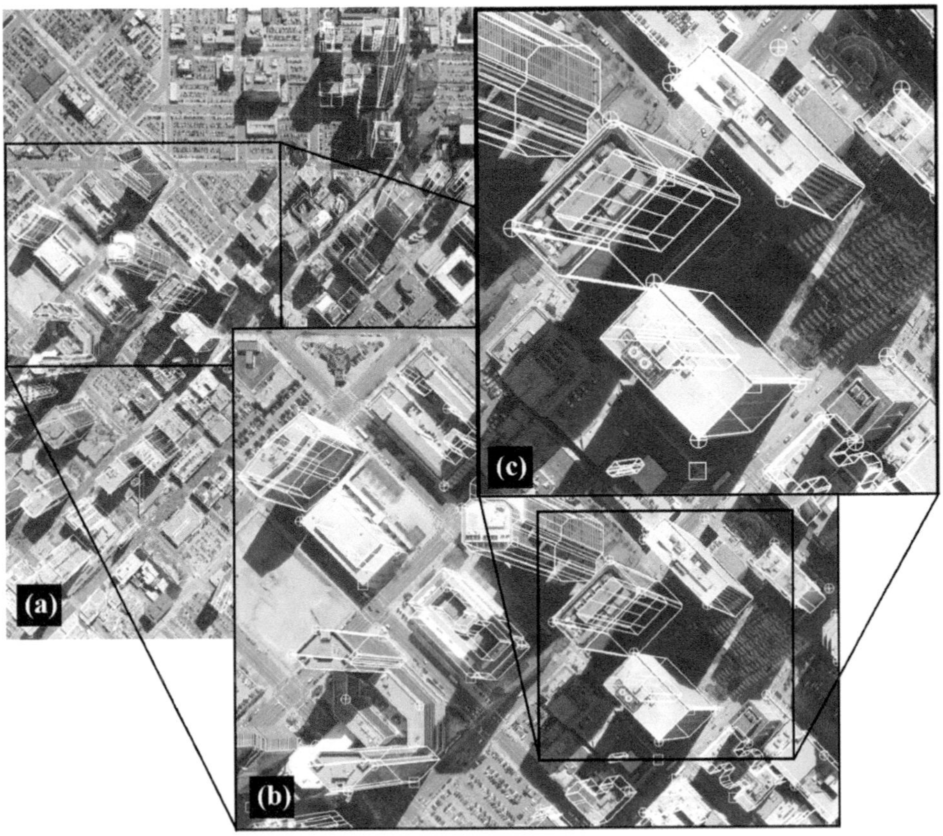

FIGURE 8.8 The selected building edges as constraint lines: (a) the original 3D model embedded in buildings, (b) the enlarged window from (a), and (c) a magnified window from (b).

TABLE 8.1
The Interior Orientation Parameter

Focal Length (mm)	Principal Point (mm)		Distortion Parameter		
	x_0	y_0	k_1	p_1	p_2
153.022	0.002	–0.004	0.2076×10^{-4}	-0.1142×10^{-6}	0.1982×10^{-6}

TABLE 8.2
Accuracy Comparison of the Three Methods for the Exterior Orientation Parameter Determination. Gray shade indicates accuracy index.

Methods		Exterior Orientation Parameter						
	GCPs	ϕ (arc)	ω (arc)	κ (arc)	X_S (ft)	Y_S (ft)	Z_S (ft)	δ_o (pixel)
M 1	8 pts	0.0104	0.0158	–1.5503	3,143,007.3	1,696,340.4	9032.0	0.65
M 2	232 pts	–0.0025	–0.0405	–1.5546	3,143,041.4	1,696,562.6	9070.7	0.49
M 3	56 pts + 76 lines	–0.0016	–0.0298	–1.5538	3,143,040.6	1,696,520.9	9072.3	0.12

1. *Visual check*: The wire lines derived from the DBM, which is taken as the "true" value, are superimposed onto the three orthorectified building roofs, respectively. Four orthorectified buildings with different heights and at different locations are selected for visually checking their accuracy (Figure 8.9). As seen, there are significant offsets between the building wire lines and the building edges in the orthoimage generated by Method 1. The average deviations for four buildings are approximately 15 and 10 pixels along the x and y directions. However, the average offsets from our method are only 1.0–1.5 pixels in both x and y directions. The results demonstrated that the accuracy of the orthoimage generated by our method has been greatly increased.
2. *Checkpoint check:* 89 checkpoints are employed to evaluate the absolute accuracy of the orthoimages orthorectified by three methods. The checkpoints are located in the corners of buildings and are considered the "true" value. The coordinates corresponding to the checkpoints in the three orthoimages are measured. The average deviation in both x and y directions is listed in Table 8.3. As seen from Table 8.3, the offsets from Methods 1 and 2 are approximately 3–5 ft and 2–4 ft, respectively, while they are 0.5–1.0 ft from our method. The results again demonstrated that our method can significantly improve the accuracy of the generated orthoimage.

As mentioned earlier, the relief displacements caused by high buildings that occur along radial lines from the nadir point, which means that the relief displacements are zero for imaged objects at the nadir point and increase with increased radial distances from the nadir. To examine how much the proposed method could improve the accuracy for the marginal and central objects, we measured the errors for the margin and central orthoimage and found that the proposed method could

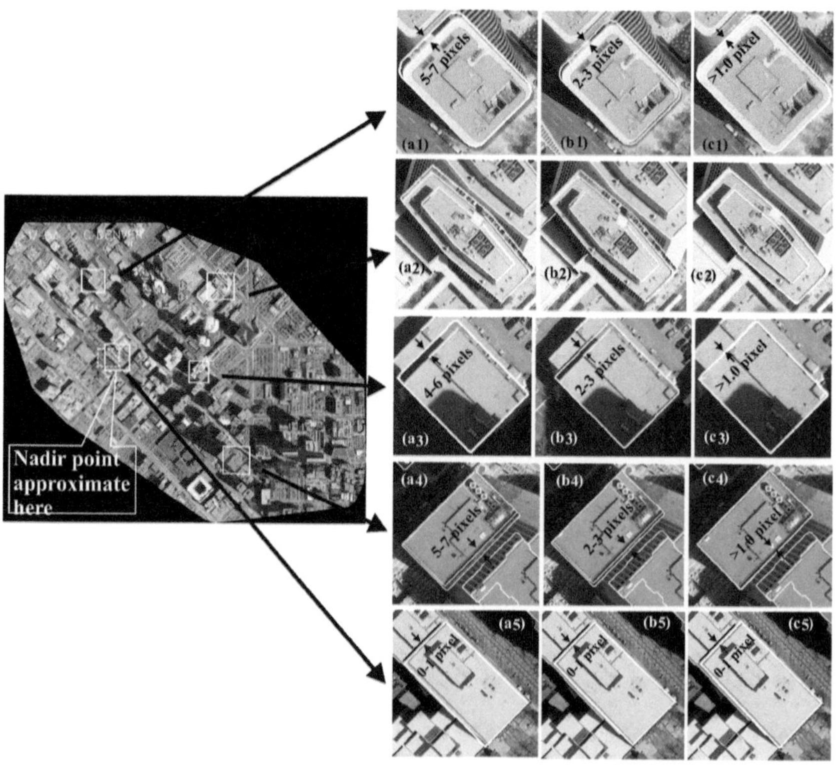

FIGURE 8.9 The comparison of building's orthorectification using three control conditions: (a) 8 control points, (b) 232 ground control points, and (c) 56 control points plus 76 control lines.

TABLE 8.3
Accuracy Comparison of Orthorectified Image

	Method 1 (ft)		Method 2 (ft)		Our method (ft)	
	ΔX	ΔY	ΔX	ΔY	ΔX	ΔY
Mean	3.621	4.874	2.918	3.977	0.78	1.125

improve the accuracy of the orthoimage approximately 5–7 ft for those objects in the image margin, but approximately 0–1 pixel for those high objects surrounding the nadir point (see Figure 8.9).

8.4 CONCLUSION

This chapter describes a method for increasing the accuracy of orthorectification for extremely high buildings. This proposed method considers the building geometric features, including the perpendicularity and collinearity of building edges, and then models these geometric features as constraint conditions. These modeled geometric features are merged into the orthorectification model. A test field located in Downtown Denver, CO, has been used to evaluate our methods. The experimental results of comparing the accuracy achieved by our method and other methods are presented. The experimental results demonstrated that the proposed method can improve the accuracy by 2–5 ft for those buildings of over 100 m high and 5–7 ft for those buildings with over 100 m high in the margin of imagery.

REFERENCES

Amhar, F.; Josef, J.; Ries, C., The generation of true orthophotos using a 3D building model in conjunction with a conventional DTM. *International Archives of the Photogrammetry, Remote Sensing*, 32(4), 1998, pp. 16–22.

Biason, A.; Dequal, S.; Lingua, A., A new procedure for the automatic production of true orthophotos. *International Archives of the Photogrammetry, Remote Sensing and Spatial Information Sciences*, 35, 2004, pp. 1682–1777.35,

Habib, A.; Kim, E.; Kim, C., New methodologies for true ortho-photo generation. *Photogrammetric Engineering & Remote Sensing*, 2006 (In press).

Lawson, C. L.; Hanson, R. J., *Solving Least Squares Problems*, New Ed ed.; Society for Industrial & Applied Math; Prentice Hall: Englewood Cliffs, NJ, 1995.

Mayr, W., 2002. Bemerkungen zum thema, true ortho-image. *Photogrammetrie – Fernerkundung – Geoinformation*, 4, 2002, pp. 237–244.

Mikhail, E. M., Bethel, J. S., and McGlone, J. C. (2001). Introduction to Modern Photogrammetry. Wiley, New York, USA.

Nielsen, M. Ø., *True orthophoto generation, informatics and mathematical modelling*. Thesis, Technical University of Denmark, Lundtofte, Denmark, 2004.

Palà, V.; Arbiol, R., True orthoimagery of urban areas. *GIM International*, 16, 2002, pp. 50–51.

Rau, J. Y.; Chen, N. Y.; Chen, L. C., True orthophoto generation of built-up areas using multi-view images. *Photogrammetry Engineering and Remote Sensing*, 68(6), 2002, pp. 581–588.

Sheng, Y. W.; Gong, P.; Biging, G. S., True orthoimage production for forested areas from large-scale aerial photographs. *Photogrammetry Engineering and Remote Sensing*, 69(3), 2003, pp. 259–266.

Siachalou, S., Urban orthoimage analysis generated from IKONOS data. *International Archives of the Photogrammetry, Remote Sensing and Spatial Information Sciences*, XXXV, 2004.

Xie, W.; Zhou, G., Urban 3D building model applied to true orthoimage generation. In: *First Workshop of the EARSeL Special Interest Group on Urban Remote Sensing*, Berlin, Germany, March 2–3, 2006.

Zhou, G.; Chen, W.; Kelmelis, J.; Zhang, D., A comprehensive study on urban true orthorectification. *IEEE Transactions on Geoscience and Remote Sensing*, 43(9), 2005, pp. 2138–2147.

Zhou, G.; Jezek, K.; Wright, W.; Rand, J.; Granger, J., Orthorectifying 1960's disclassified intelligence satellite photography (DISP) of Greenland. *IEEE Transactions on Geoscience and Remote Sensing*, 40(6), 2002, pp. 1247–1259.

9 On-Board Orthorectification

9.1 INTRODUCTION

The traditional orthorectification methods correct images and remove distortion pixel-by-pixel using a personal computer (PC)-based platform on the basis of a digital elevation model (DEM). It is difficult to achieve a demand for (near) real-time performance because the processing unit is a pixel and there is a great amount of image data. On the other hand, since the algorithm complexity of orthorectification is very high, serial instruction processing systems take much time to perform the orthorectification algorithms. Thus, how to improve the speed of the orthorectification process has become an urgent issue when applied in on-board processing of a spacecraft.

An effective solution for the (near) real-time processing of image orthorectification is to perform the orthorectification on hardware. In recent decades, the field programmable gate array (FPGA) has been widely used in the image processing (such as imaging compression (Escamilla-Hernández et al. 2008; Kate 2012), filtering (Pal et al. 2016; Wang et al. 2016; Zhang et al. 2016), edge detection (Ontiveros-Robles et al. 2016, 2017), real-time processing of video images (Li et al. 2015; Li et al. 2016), and motion estimation (González et al. 2012, 2013; Botella et al. 2010)) to make real-time processing come true. González et al. (2012, 2013) optimized matching-based motion estimation algorithms using an Altera custom instruction-based paradigm and a combination of synchronous dynamic random access memory and on-chip memory in Nios II processors and presented a low-cost system. Botella et al. (2010) proposed an architecture for a neuromorphic robust optical flow based on an FPGA, which was applied in a difficult environment. In addition to a very-high-speed integrated circuit hardware description language and Verilog HDL, OpenCL is usually used to design an FPGA (Waidyasooriya et al. 2016; Rodriguez-Donate et al. 2015). Waidyasooriya et al. (2016) proposed an FPGA architecture for three-dimensional (3D) finite difference time domain acceleration applying OpenCL, which solved the problem of designing time. Rodriguez-Donate et al. (2015) evaluated the use of a convolution operator in signal processing disciplines focused on FPGA evaluation under different optimizations with respect to thread and memory level exploitation, in which OpenCL was used. In the remotely sensed (RS) community, although there is also some research on applying FPGA to the real-time processing of RS images, it is still not enough to meet the requirement in practice. For example, Thomas et al. (2008) and Kalomiros and Lygouras (2008) proposed an image processing system by combining software and hardware that can improve the speed of image correction and mosaicking. Kuo and Gordon (2010) presented a processing method whereby the computing process is migrated to an FPGA, which aimed at solving the problem when the number of pixels in an image was huge and the transformation calculation of the floating-point matrix was complex. Halle et al. (2001) designed an on-board bispectral infrared detection system based on the neural network processor NI1000, a digital signal processor (DSP), and an FPGA. The system can perform on-board radiometric correction, geometric correction, and texture extraction. Malik et al. (2014) built a quick process hardware platform using an FPGA. The hardware platform could process 390 frames of 640 × 480 images per second. Tomasi et al. (2012) researched a stereo vision algorithm based on an FPGA to perform the correction of video graphics array images (57 fps). Greisen et al. (2011) applied the pipeline method of an FPGA to correct the color of stereo video images. Kumar and Sridharan (2007) realized the real-time correction of images using an FPGA under a dynamic environment.

This chapter describes a hardware platform based on an FPGA for RS image orthorectification. Through decomposing the orthorectification algorithm, several basic algorithms of image processing can be obtained, reducing the complexity of the algorithm and reaching the purpose of (near)

DOI: 10.1201/9781003632108-9

real-time orthorectification. The proposed FPGA-based orthorectification platform integrates three modules: a memory module, a coordinate transformation module (including the transformation from geodetic coordinates to photo coordinates and the transformation from photo coordinates to scanning coordinates), and an interpolation module (IM).

9.2 FPGA IMPLEMENTATION FOR THE ORTHORECTIFICATION ALGORITHM

9.2.1 A Brief Review of the Orthorectification Algorithm

Many orthorectification models have been proposed over the past few decades. According to the type of image, terrain of the covered area, and geomorphic features, an appropriate model can be chosen to orthorectify the RS images. Generally, orthorectification models contain a rigorous correction model based on the collinearity condition equation, a rational function model, or an affine-transformation-based correction model. In this chapter, the collinearity condition equation is used to implement RS image orthorectification on the proposed platform. The collinearity condition equation model is based on the corresponding relationship of space geometry between the point of image space and the point of object space. The collinearity condition equation is the model that is applied to simulate and solve the position and posture of a sensor at the time of imaging. The collinearity condition equation is very suitable for various resolutions of images and the situation where the parameters of the orbit are known or unknown. In this section, the collinearity condition equation model is briefly introduced.

The orthorectification method can be classified as the direct method and the indirect method. The direct method applies the image coordinates of the original image to compute the coordinates of the orthophoto, and the indirect method applies the image coordinates of the orthophoto to compute the image coordinates of the original image. In this chapter, the indirect method is used. The processes of the indirect method based on the collinearity condition equation include the following:

9.2.1.1 The Determination of the Geodetic Coordinates of the Pixels in the Orthophoto

Let (X_{g0}, Y_{g0}) be the geodetic coordinates of a marginal point on the left bottom of the orthophoto; (I, J) be the column and row coordinates of an arbitrary pixel in the orthophoto; Δx and Δy be the sample intervals of columns and rows, respectively; M be the scale denominator of the orthophoto; and (X_g, Y_g) be the geodetic coordinates of pixel G. The geodetic coordinates of pixel G can be obtained by

$$\begin{cases} X_g = X_{g0} + M \cdot (I + 0.5) \cdot \Delta x \\ Y_g = Y_{g0} + M \cdot (J + 0.5) \cdot \Delta y \end{cases} \tag{9.1}$$

After determining the geodetic coordinates of pixels in the orthophoto, the elevation (i.e., Z_g) of each pixel can be acquired through interpolating the digital surface model (DSM).

9.2.1.2 The Transformation from Geodetic Coordinates to Photo Coordinates

At the time of imaging, the ground point, the center of projection, and the photo point are on a line. According to the collinearity relationship among them, the photo coordinates of the ground point can be obtained by

$$\begin{cases} u = x_0 - f \dfrac{a_1(X_g - X_s) + b_1(Y_g - Y_s) + c_1(Z_g - Z_s)}{a_3(X_g - X_s) + b_3(Y_g - Y_s) + c_3(Z_g - Z_s)} \\ v = y_0 - f \dfrac{a_2(X_g - X_s) + b_2(Y_g - Y_s) + c_2(Z_g - Z_s)}{a_3(X_g - X_s) + b_3(Y_g - Y_s) + c_3(Z_g - Z_s)} \end{cases} \tag{9.2}$$

where (u, v) are the photo coordinates of the ground point; x_0, y_0, and f are the interior orientation elements; X_s, Y_s, and Z_s are the exterior orientation elements; and a_h, b_h, and c_h $(h = 1, 2, 3)$ are the elements of the rotation matrix R that can be obtained by equation (9.3).

$$R = \begin{pmatrix} a_1 & b_1 & c_1 \\ a_2 & b_2 & c_2 \\ a_3 & b_3 & c_3 \end{pmatrix} \tag{9.3}$$

$$= \begin{pmatrix} \cos\omega\cos\kappa - \sin\phi\sin\omega\sin\kappa & \cos\omega\sin\kappa & \sin\phi\cos\kappa + \cos\phi\sin\omega\sin\kappa \\ -\cos\phi\sin\kappa - \sin\omega\sin\phi\cos\kappa & \cos\omega\cos\kappa & -\sin\phi\sin\kappa + \cos\phi\sin\omega\cos\kappa \\ -\sin\phi\cos\omega & -\sin\omega & \cos\phi\cos\omega \end{pmatrix}$$

where ϕ, ω, and κ are three rotational angles along the x-, y-, and z-axes in coordinate transformation, respectively.

9.2.1.3 The Transformation from Photo Coordinates to Scanning Coordinates

Because there is an affine deformation between the photo coordinate system and the scanning coordinate system, the following affine transformation is used to get the scanning coordinates for the pixels:

$$\begin{bmatrix} i' \\ j' \end{bmatrix} = \begin{bmatrix} m_1 & n_1 \\ m_2 & n_2 \end{bmatrix} \begin{bmatrix} u \\ v \end{bmatrix} + \begin{bmatrix} k_1' \\ k_2' \end{bmatrix} + \begin{bmatrix} i_0' \\ j_0' \end{bmatrix}$$
$$= \begin{bmatrix} m_1 & n_1 \\ m_2 & n_2 \end{bmatrix} \begin{bmatrix} u \\ v \end{bmatrix} + \begin{bmatrix} k_1 \\ k_2 \end{bmatrix} \tag{9.4}$$

where m_t, n_t, and k_t' $(t = 1, 2)$ are the coefficients of affine transformation; (i_0', j_0') are the scanning coordinates of the principal point; especially, $k_1 = k_1' + i_0'$ and $k_2 = k_2' + j_0'$; and (i', j') are the scanning coordinates of an arbitrary pixel.

9.2.1.4 Gray-Scale Bilinear Interpolation

The gray-scale of pixels in the orthophoto can be determined according to the acquired scanning coordinates. Because the obtained scanning coordinates may not only be in the center of a pixel, a gray-scale interpolation process is required. In this chapter, the bilinear interpolation method is applied.

$$\begin{aligned} f(i+p, j+q) &= (1-p)(1-q)f(i,j) + (1-p)qf(i,j+1) \\ &\quad + p(1-q)f(i+1,j) + pqf(i+1,j+1) \end{aligned} \tag{9.5}$$

where i and j are nonnegative integers; p and q are in the range of (0, 1); and $f(i, j)$ are gray values.

9.2.2 FPGA-Based Implementation for Orthorectification Algorithms

The parallel processing of an FPGA is a main research direction in the high-performance rapid calculation community. The calculation speed of an FPGA is affected by multiple factors (such as the amount of logical resources in the chosen FPGA and the optimal design of algorithms). Through analyzing the structure of an orthorectification algorithm and optimizing it, an FPGA-based hardware

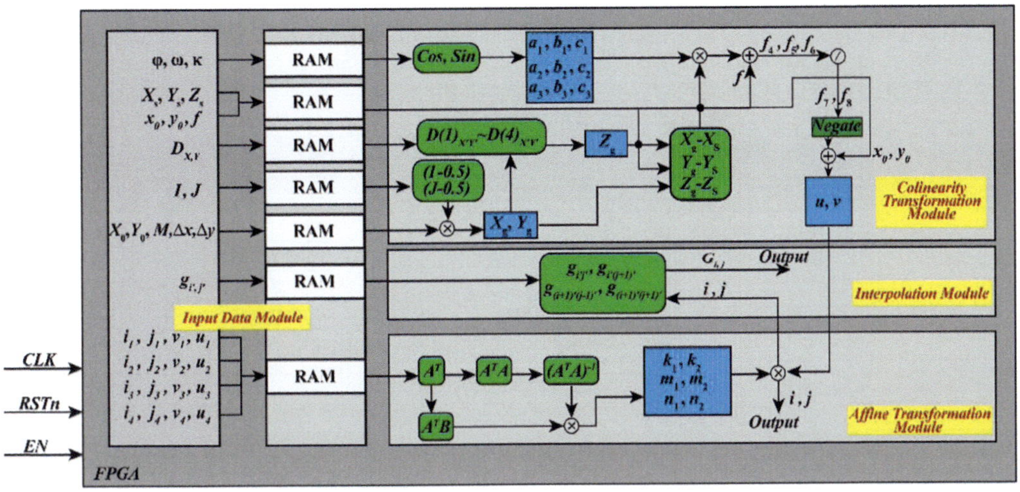

FIGURE 9.1 The proposed field programmable gate array (FPGA)-based architecture for the orthorectification of remotely sensed (RS) images. RAM, random access memory. (Courtesy of Zhou et al. 2017.)

architecture for orthorectification was designed (as shown in Figure 9.1). The FPGA-based hardware architecture contains three modules: (1) an input data module; (2) a coordinate transformation module that includes a collinearity equation transformation module (CETM) and an affine transformation module (ATM); and (3) an IM. The details of these modules are described as follows.

1. The original data and parameters are stored in the RAM of the input data module. These original data and parameters are sent to the CETM, ATM, and IM in the same clock cycle, when the enable signal is being received;
2. The coefficients of the collinearity conditional equation, geodetic coordinates of the orthophoto, and photo coordinates are calculated by the CETM. In the same clock cycle, the acquired photo coordinates are sent to the ATM;
3. The coefficients of the affine transformation are calculated in the ATM and then the coefficients and photo coordinates are combined to calculate the scanning coordinates, which are sent to the IM and output in the same clock cycle;
4. The gray-scale of the orthophoto is obtained by scanning the coordinates and cached gray-scale of the original image in the IM. In the same clock cycle, the obtained gray-scale of the orthophoto is output to the external memory.

9.2.2.1 FPGA-Based Implementation for a Two-Row Buffer

As is well known, to perform bilinear interpolation, several neighborhoods of a pixel are needed. However, an FPGA is not like software for storing the whole image in the internal storage and reading the value of the image pixel according to the index. Thus, according to the size of the neighborhood, several rows of image data should be cached in advance. Moreover, the cache should be built to store the required image data. Many research studies have made efforts to design the structure of a buffer, such as (Hsiao et al. 2010; Kazmi et al. 2017; Cao et al. 2012; Hu and Zhu 2010). Especially, the structure of a state machine (Hu and Zhu 2010) is a useful method for solving the issue of buffers.

According to Hu and Zhu (2010), a two-row buffer based on an FPGA is proposed. As shown in Figure 9.2, four states, i.e., IDEL, the beginning of all, the end of all, and the beginning of the line, are used for initialization and cache cleaning. In the "transing" state, according to the control signal and the address of the reading and writing, the corresponding operations are chosen, including storing the image data in the dual RAM from the external storage, waiting for image data to be taken

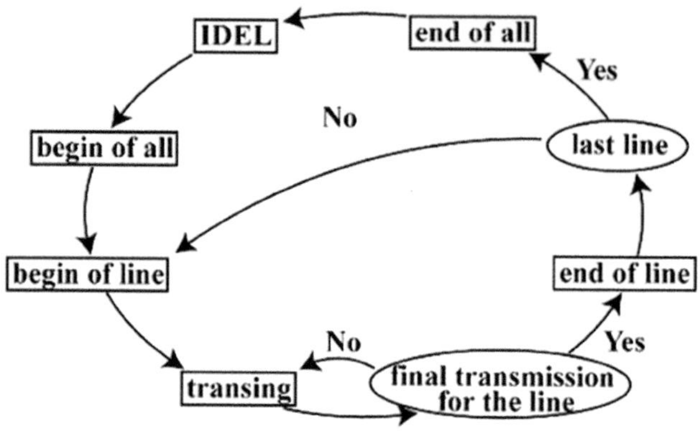

FIGURE 9.2 State diagram of cache for image input. (Hu and Zhu 2010.)

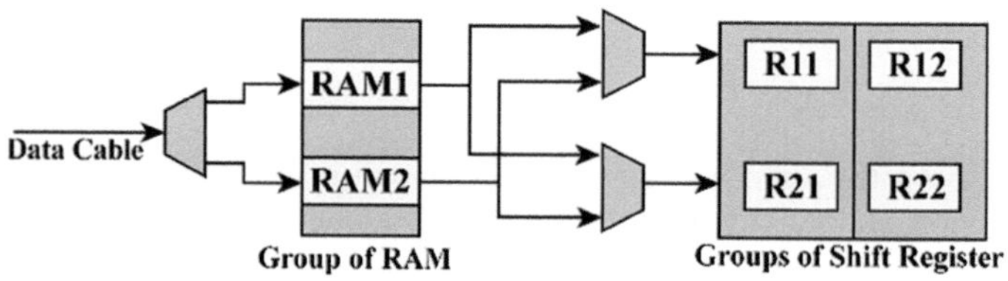

FIGURE 9.3 The model of the two-row buffer.

from the dual RAM, and turning to the "end of line" state. The number of dual RAM modules, i.e., n_{RAM}, can be determined by the number of rows of neighbors, i.e., n_{nb}, where n_{RAM} is equal to n_{nb}. To implement bilinear interpolation, 2×2 neighbors are needed. Thus, as shown in Figure 9.3, two dual RAMs are used as the cache media for the first-level cache of image input. When two dual RAMs are used to buffer data, a multiplexer is exploited to determine which dual RAM should store the data. For 2×2 neighbors, two groups of cycle shift registers are used to store the data from neighboring pixels of the input image in the second-level cache. For each group consisting of end-to-end registers, two registers, noted as R_{mn} (m, $n = 1, 2$), are contained. The data sent from the dual RAMs first are input selectors that ensure the row order is the same as the row order of the input image, and then the output of the selectors is sent to the groups of the cycle shift register. When storing the data from the dual RAMs into R_{m1} each time, the data stored in R_{mn} ($m = 1, 2$, $n = 1$) is assigned to $R_{m(n+1)}$ ($m = 1, 2$, $n = 1$), while the data stored in R_{m4} ($m = 1, 2$) is thrown out.

9.2.2.2 FPGA-Based Implementation for Coordinate Transformation

According to Section 9.2.1, the orthorectification method based on the collinearity condition equation needs two coordinate transformations. As shown from equations (9.1) to (9.4), these equations involve compound operations. However, the hardware implementation for an algorithm is based on the most basic logic operations. Thus, the orthorectification method based on the collinearity equation must be divided into several simple addition and multiplication operations, corresponding to the hardware components of the adder and multiplier, respectively. The details of the FPGA-based implementation for coordinate transformation modules are described as follows.

9.2.2.2.1 FPGA-Based Implementation for Calculating Geodetic Coordinates X_g, Y_g, and Z_g

To obtain the geodetic coordinates X_g, Y_g, and Z_g at part (1) of Section 9.2.1 using an FPGA chip, a parallel computation architecture is presented in Figure 9.4. As shown in Figure 9.4, four adders and four multipliers are used to compute X_g and Y_g in parallel by I, X_0, Δx, M, J, Δy, and Y_0, respectively.

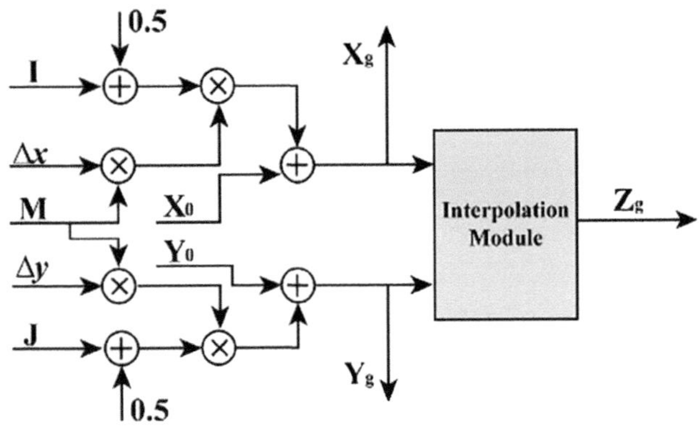

FIGURE 9.4 FPGA-based parallel computation architecture for calculating geodetic coordinates X_g, Y_g, and Z_g.

After obtaining X_g and Y_g, they are sent to the IM for calculating Z_g. To ensure that X_g, Y_g, and Z_g are output in the same clock cycle, the delay units are utilized for computing process. The details of the IM are presented in Section 9.2.2.3, because its computing process is the same as gray interpolation.

9.2.2.2.2 FPGA-Based Implementation for the Transformation from Geodetic Coordinates to Photo Coordinates

For an FPGA-based implementation in parallel computing for the transformation from geodetic coordinates to photo coordinates, a modification of equation (9.2) is needed. The intermediate variables produced in the computing process can be divided into three levels. The goal of the first level is to implement the margin calculation of the geodetic coordinates between any point and the principal point of the photograph. The first level contains

$$f_1 = X_g - X_s; \quad f_2 = Y_g - Y_s; \quad f_3 = Z_g - Z_s \tag{9.6}$$

The second level (see equation (9.7)) is to calculate the products between the transformation coefficients and intermediate variables f_1, f_2, and f_3.

$$f_4 = a_1 f_1 + b_1 f_2 + c_1 f_3; \quad f_5 = a_2 f_1 + b_2 f_2 + c_2 f_3; \quad f_6 = a_3 f_1 + b_3 f_2 + c_3 f_3 \tag{9.7}$$

The computation of the third level (see equation 9.8) is based on the first and the second levels to obtain the photo coordinates,

$$f_7 = f_4 / f_6; \quad f_8 = f_5 / f_6 \tag{9.8}$$

To compute f_4, f_5, and f_6 in equation (9.7), the elements of the rotation matrix R (i.e., a_i, b_i, and c_i, $i = 1, 2, 3$) should first be calculated based on equation (9.3). To compute the rotation matrix R using an FPGA chip, a parallel computation module is presented in Figure 9.5a, in which a_i, b_i, and c_i ($i = 1, 2, 3$) are calculated by the sin and cos functions of the three rotational angles ϕ, ω, and κ. Through using the CORDIC IP core, the sin and cos functions can be implemented by an FPGA. To ensure that a_i, b_i, and c_i ($i = 1, 2, 3$) are obtained in the same clock cycle, the delay units should be exploited in the computing process, where there are 12 multipliers and four adders.

To implement the transformation from geodetic to photo coordinates, i.e., equations (9.6)–(9.8) using an FPGA chip, a parallel computation module is presented in Figure 9.5b. As shown in

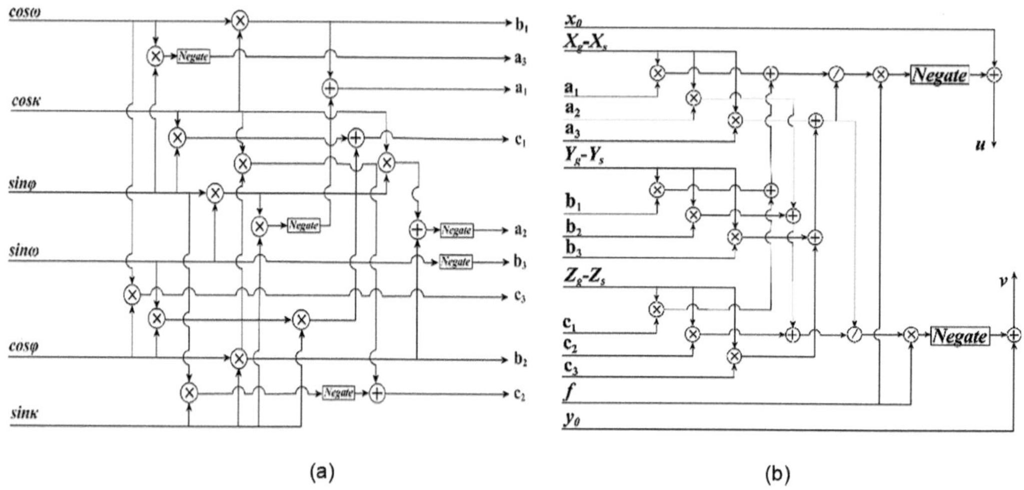

FIGURE 9.5 (a) FPGA-based computation for elements of rotation matrix R and (b) the FPGA implementation architecture for coordinate transformation from geodetic to photo coordinates. (Courtesy of Zhou et al. 2017.)

Figure 9.5b, the initial data (i.e., x_0, y_0, f, X_S, Y_S, and Z_S) is stored in the input data module, the rotation matrix is calculated by using the sin and cos functions and rotation angles, and the geodetic coordinates (X_g, Y_g, and Z_g) calculated by Figure 9.4 are used to compute the photo coordinates (u, v). In the computing process, eleven multipliers, eight adders, and two dividers are employed. Moreover, the delay units are also utilized to ensure that the outputs (u, v) are sent into the next module in the same clock cycle.

9.2.2.2.3 FPGA-Based Implementation for the Transformation from Photo to Scanning Coordinates

To realize the transformation from photo to scanning coordinates using an FPGA chip, equation (9.4) is divided into two levels. The first level is

$$g_1 = A_1B; \quad g_2 = A_2B \tag{9.9}$$

Additionally, the second level includes

$$g_3 = g_1 + k_1; \quad g_4 = g_2 + k_2 \tag{9.10}$$

$$i = g_3 + i_0'; \quad j = g_4 + j_0' \tag{9.11}$$

where $A_1 = (m_1, n_1)$ and $A_2 = (m_2, n_2)$ are the affine transformation coefficients and $B = (u, v)^T$ are the photo coordinates. The transformation coefficients, i.e., m_s, n_s, and k_s ($s = 1, 2$), should be calculated first based on the following least squares method.

To compute the coefficients of affine transformation, four control points are needed. Let H be the matrix of the control point scanning coordinates, Q be the matrix of the photo coordinates of the known fiducial points, and C be the matrix of the affine transformation coefficients,

$$H = \begin{bmatrix} i_1 & j_1 & i_2 & j_2 & i_3 & j_3 & i_4 & j_4 \end{bmatrix}^T \tag{9.12}$$

$$Q = \begin{bmatrix} 1 & 0 & u_1 & 0 & v_1 & 0 \\ 0 & 1 & 0 & u_1 & 0 & v_1 \\ 1 & 0 & u_2 & 0 & v_2 & 0 \\ 0 & 1 & 0 & u_2 & 0 & v_2 \\ 1 & 0 & u_3 & 0 & v_3 & 0 \\ 0 & 1 & 0 & u_3 & 0 & v_3 \\ 1 & 0 & u_4 & 0 & v_4 & 0 \\ 0 & 1 & 0 & u_4 & 0 & v_4 \end{bmatrix} \tag{9.13}$$

$$C = \begin{bmatrix} k_1 & k_2 & m_1 & m_2 & n_1 & n_2 \end{bmatrix}^T \tag{9.14}$$

$$H = QC \tag{9.15}$$

According to the least squares method, equation (9.15) can be solved by

$$C = (Q^T Q)^{-1} \left(Q^T H\right) \tag{9.16}$$

9.2.2.2.3.1 FPGA-Based Implementation for Q^TQ and Q^TH To implement the matrix multiplication Q^TQ based on an FPGA chip, it can be rewritten using an optimum method. As shown in equation (9.17), the elements of the matrix contain three formats, i.e., "$a_1 + b_1 + c_1 + d_1$", "$a_1^2 + b_1^2 + c_1^2 + d_1^2$", and "$a_1a_2 + b_1\ b_2 + c_1c_2 + d_1d_2$". Moreover, the matrix of $G = Q^TQ$ is symmetric, i.e., the elements of the lower triangular matrix are the same as the elements of the upper triangular matrix. To save the resources of the FPGA chip, the upper triangular matrix is computed only in parallel based on the FPGA architecture of Figure 9.6. In the presented architecture, 12 multipliers and 15 adders are used in the parallel computing process. The delay units are applied to ensure that the results are output into the next module in the same clock cycle.

Because the format of matrix E's elements is similar to the format of matrix G (see equations 9.17 and 9.18), the FPGA-based architecture for the parallel computing process of Q^TH can be referenced from the architecture of b_{13} and b_{35} in Figure 9.6. The details of the parallel computation for Q^TH based on an FPGA are not repeated here.

$$G = Q^TQ = \begin{bmatrix} 1+1+1+1 & 0 & u_1+u_2+u_3+u_4 & 0 & v_1+v_2+v_3+v_4 & 0 \\ 0 & 1+1+1+1 & 0 & u_1+u_2+u_3+u_4 & 0 & v_1+v_2+v_3+v_4 \\ u_1+u_2+u_3+u_4 & 0 & u_1^2+u_2^2+u_3^2+u_4^2 & 0 & u_1v_1+u_2v_2+u_3v_3+u_4v_4 & 0 \\ 0 & u_1+u_2+u_3+u_4 & 0 & u_1^2+u_2^2+u_3^2+u_4^2 & 0 & u_1v_1+u_2v_2+u_3v_3+u_4v_4 \\ v_1+v_2+v_3+v_4 & 0 & u_1v_1+u_2v_2+u_3v_3+u_4v_4 & 0 & v_1^2+v_2^2+v_3^2+v_4^2 & 0 \\ 0 & v_1+v_2+v_3+v_4 & 0 & u_1v_1+u_2v_2+u_3v_3+u_4v_4 & 0 & v_1^2+v_2^2+v_3^2+v_4^2 \end{bmatrix} \tag{9.17}$$

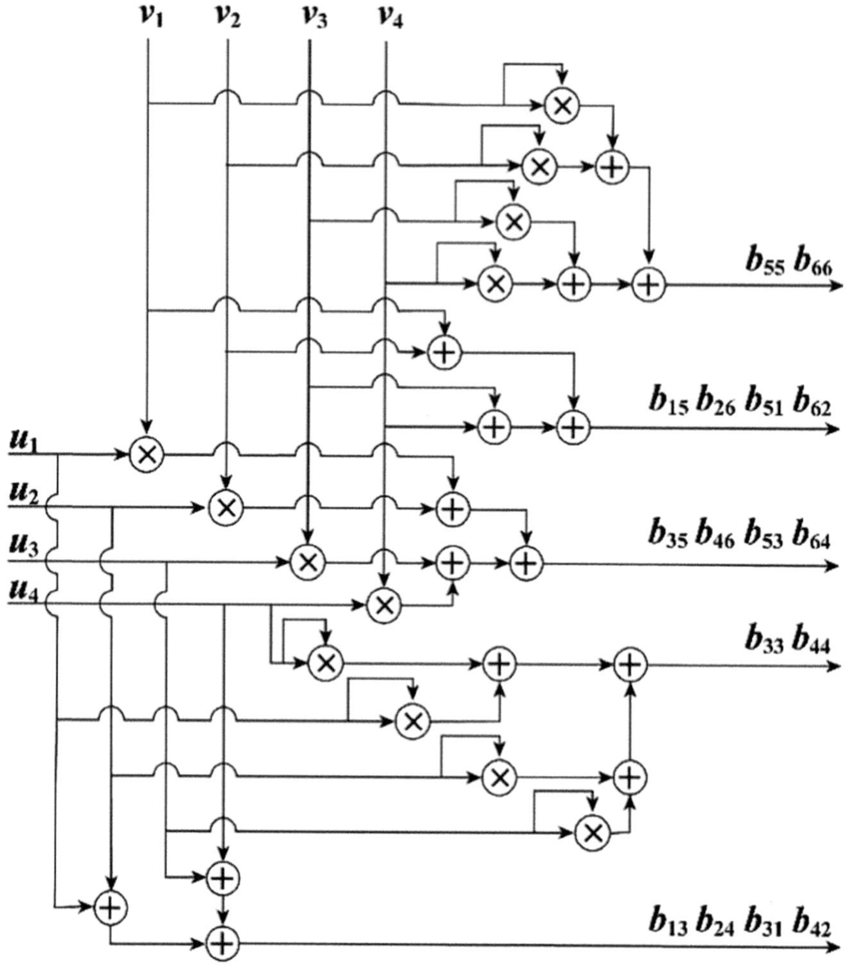

FIGURE 9.6 The FPGA implementation architecture of Q^TQ.

$$E = Q^T H = \begin{bmatrix} i_1 + i_2 + & j_1 + j_2 + & i_1u_1 + i_2u_2 + & j_1u_1 + j_2u_2 + & i_1v_1 + i_2v_2 + & j_1v_1 + j_2v_2 + \\ i_3 + i_4 & j_3 + j_4 & i_3u_3 + i_4u_4 & j_3u_3 + j_4u_4 & i_3v_3 + i_4v_4 & j_3v_3 + j_4v_4 \end{bmatrix}^T \tag{9.18}$$

9.2.2.2.3.2 FPGA-Based Implementation for G^{-1} To implement the inversion of matrix G (i.e., G^{-1}) based on the FPGA chip, it can be divided into two parts: (1) the implementation for decomposing matrix G using the LDL^T method and (2) the implementation of G^{-1}. The details of the implementation are described as follows.

To implement the decomposition of matrix G, the LDL^T method is used to modify the matrix G as follows (9.19):

$$G = \begin{bmatrix} g_{11} & g_{12} & \cdots & g_{1n} \\ g_{21} & g_{22} & \cdots & g_{2n} \\ \vdots & \cdots & \cdots & \vdots \\ g_{n1} & \cdots & \cdots & g_{nn} \end{bmatrix} = \begin{bmatrix} 1 & & & \\ l_{21} & 1 & & \\ \vdots & \cdots & \ddots & \\ l_{n1} & \cdots & l_{n(n-1)} & 1 \end{bmatrix} \begin{bmatrix} d_{11} & & & \\ & d_{22} & & \\ & & \ddots & \\ & & & d_{nn} \end{bmatrix} \begin{bmatrix} 1 & & & \\ l_{21} & 1 & & \\ \vdots & \cdots & \ddots & \\ l_{n1} & \cdots & l_{n(n-1)} & 1 \end{bmatrix}^T = LDL^T$$

(9.19)

where matrix L is a lower triangular matrix, matrix D is a diagonal matrix, and L^T is a transposed matrix of L. The elements of matrix L and matrix D, i.e., l_{ij} and d_{ii}, can be solved by equation (9.20):

$$\begin{cases} d_{ii} = g_{ii} - \sum_{k=1}^{i-1} l_{ik} d_{kk} l_{ik} \ (a) \\ l_{ij} = \dfrac{\left(g_{ij} - \sum_{k=1}^{j-1} l_{ik} d_{kk} l_{jk}\right)}{d_{jj}} \ (b) \end{cases}, \quad \begin{array}{c} j = 1, 2, \ldots, n \\ i = j+1, j+2, \ldots, n \end{array} \tag{9.20}$$

To reduce the number of multiplication, an intermediate variable $u_{ij} = l_{ij} d_{jj}$ is introduced. Therefore, equation (9.20) can be modified as equation (9.21):

$$\begin{cases} d_{ii} = g_{ii} - \sum_{k=1}^{i-1} u_{ik} l_{ik} \ (a) \\ u_{ij} = g_{ij} - \sum_{k=1}^{j-1} u_{ik} l_{jk} (b) \\ l_{ij} = \dfrac{u_{ij}}{d_{ii}} \quad (c) \end{cases}, \quad \begin{array}{c} j = 1, 2, \ldots, n \\ i = j+1, j+2, \ldots, n \end{array} \tag{9.21}$$

According to the characteristics of equation (9.21), the FPGA-based architecture for calculating l_{ij} and d_{ii} is shown in Figure 9.7. In the LDL^T method, the d_{ii} are first calculated based on equation (9.21a); subsequently, the elements of the same column of matrix L are calculated in parallel. Moreover, equation (9.21) shows that the elements of the later column of matrix L depend on the elements of the former column. As shown in Figure 9.7, five multipliers and twenty-five adders are used to calculate l_{ij} and d_{ii}. In the computing process, delay units are applied to ensure that the results are output into the next process in the same clock cycle.

After completing the computation of l_{ij} and d_{ii}, the inversion of matrix G can be calculated on the basis of l_{ij} and d_{ii}. According to equation (9.19), the inversion of matrix G can be rewritten as equation (9.22):

$$G^{-1} = (LDL^{T^{-1-1T^{-1-1}}} \tag{9.22}$$

To implement equation (9.22) based on the FPGA chip, an FPGA-based architecture is presented in Figure 9.8. In the presented architecture, five parts are contained. The details of the presented architecture are described as follows.

In the (i) part, MUX is applied to construct the column elements of matrix G.

In the (ii) part, the LDL^T method is used to calculate the elements of matrix L and matrix D, i.e., l_{ij} and d_{ii}.

In the (iii) part, the inversions of matrix D and matrix L are calculated in parallel. Moreover, in this part, the first MUX is used to construct the vector of L, and L^{-1} is calculated using a systolic array architecture that is applied for the fast inversion of dense matrices (El-Amawy 1989). The second MUX of this part is utilized to construct the vector of $(L^{-1})^T$. Through calculating the reciprocal of the elements of the diagonal matrix D, the D^{-1} can be obtained and output. For D^{-1}, a row vector is constructed consisting of the elements of matrix D.

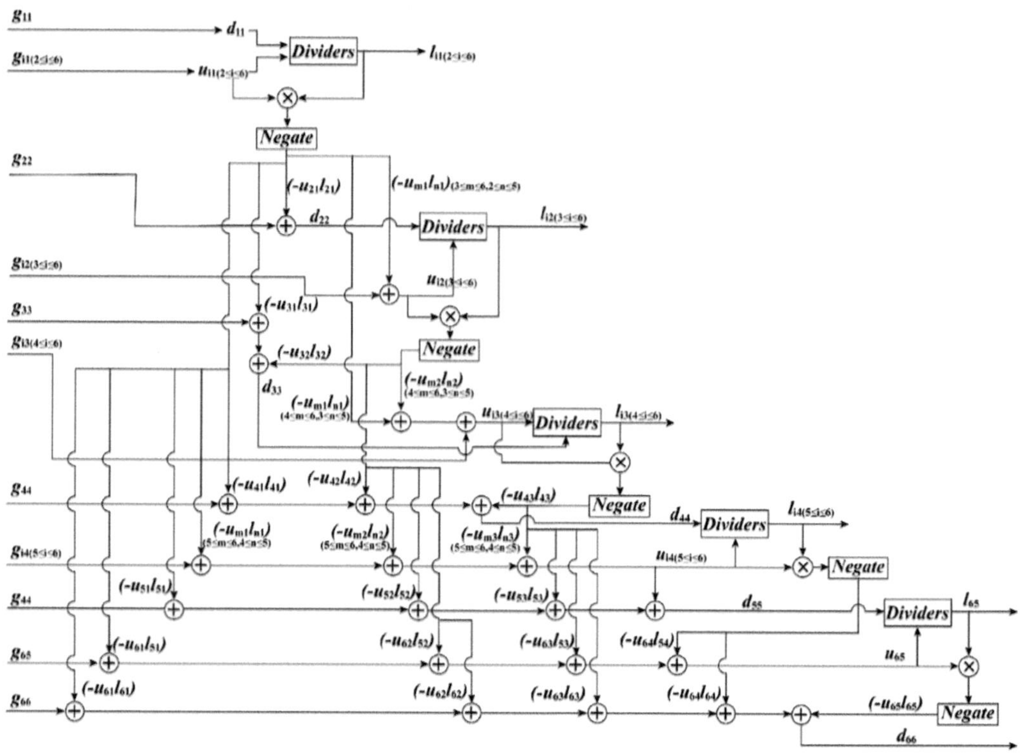

FIGURE 9.7 The FPGA-based architecture for calculating l_{ij} and d_{ii}. (Courtesy of Zhou et al. 2017.)

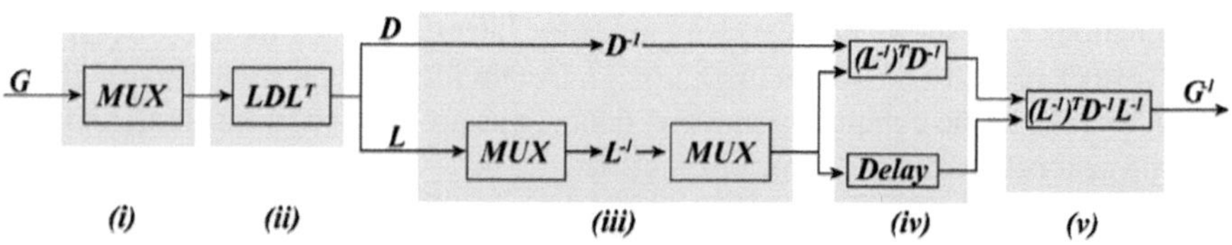

FIGURE 9.8 The FPGA-based architecture for the inversion of G.

In the (iv) part, the transposed matrix of L^{-1} is multiplied by D^{-1}, denoted as $P = (L^{-1})^T D^{-1}$, and delay units are used for delaying the output of L^{-1}.

In the (v) part, through multiplying P and L^{-1}, the inversion of matrix G, i.e., G^{-1}, is obtained and output.

9.2.2.2.3.3 FPGA-Based Implementation for the Transformation from Photo to Scanning Coordinates After obtaining the coefficients of the affine transformation based on the FPGA chip, the scanning coordinates can be calculated with these coefficients, and the photo coordinates based on the FPGA-based architecture are presented in Figure 9.9. As shown in Figure 9.9, four multipliers and four adders are used.

9.2.2.3 FPGA-Based Implementation for Bilinear Interpolation

In the whole process of orthorectification, the interpolation process is needed in two stages, i.e., the interpolation for geodetic coordinate Z_g and gray-scale. In Section 9.2.2.2.1, after obtaining the geodetic coordinates X_g and Y_g, Z_g can be obtained using X_g and Y_g to interpolate the DSM. In a

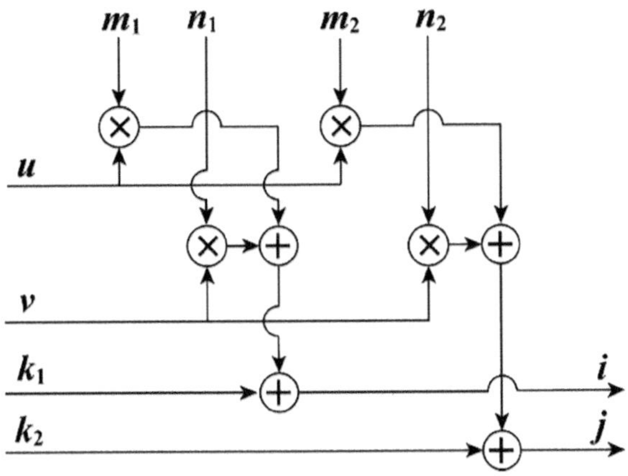

FIGURE 9.9 The FPGA-based architecture for the transformation from photo coordinates to scanning coordinates.

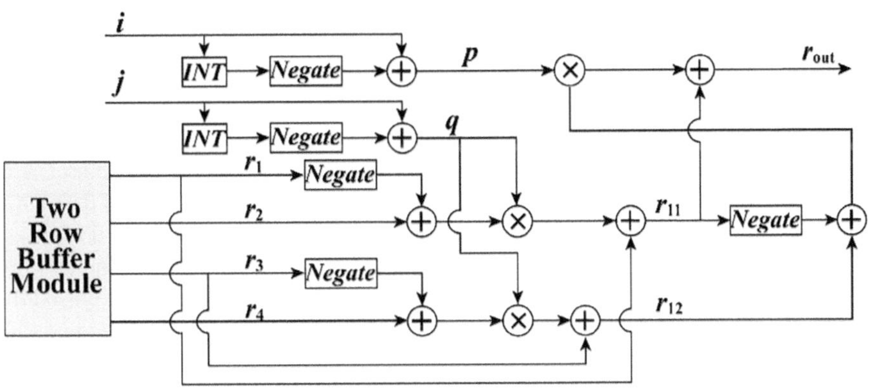

FIGURE 9.10 The FPGA implementation architecture for the bilinear interpolation algorithm.

similar way, after acquiring the scanning coordinates i and j, the gray-scale of the orthophoto can be acquired using i and j to interpolate the gray-scale of the original image. Because these two interpolation processes are similar, the FPGA-based architecture for interpolation is shared between them.

Considering the interpolation effect, the algorithm's complexity, and the resources of the FPGA, the bilinear interpolation method is used to implement the interpolation for Z_g and gray-scale. However, as shown in equation (9.5), the original bilinear interpolation method has eight times of multiplication, three times of addition, and two times of subtraction. Since the multiplication will take up many resources, equation (9.5) is rewritten as equation (9.23), which contains only three times of multiplication, three times of addition, and three times of subtraction.

$$\begin{cases} r_{11} = r_1 + q(r_2 - r_1) \\ r_{12} = r_3 + q(r_4 - r_3) \\ r_{out} = r_{11} + p(r_{12} - r_{11}) \end{cases} \tag{9.23}$$

where r represents the value of the DSM or gray-scale of the original image and $p = |i - INT(i)|$ and $q = |j - INT(j)|$ are intermediate variables. The FPGA implementation architecture of the bilinear interpolation algorithm is shown in Figure 9.10. The two-row buffer, in Figure 9.10, is an independent function module presented in Figure 9.3. The two-row buffer is packaged as an independent subsystem so that it can decrease the utilization of the buffer module. In this architecture for the bilinear interpolation algorithm, four multipliers and eight adders are utilized.

9.3 EXPERIMENT

9.3.1 The Software and Hardware Environment

The hardware platform used in this chapter is the AC701 Evaluation Kit of Artix-7 series produced by the Xilinx Company. The version of the FPGA is Xilinx Artix-7 XC7A200T FBG676ACX1349 D4658436A ZC. The design tool is ISE 4.7 and System Generator. The simulation tool is ModelSim SE10.1a. As shown in Figure 9.11, the FPGA Evaluation Kit uses the UART and JTAG ports to connect with the computer. The power port provides 250 V. The LCD panel and the screen show the results at the same time. To validate the proposed method, the orthorectification algorithm is also implemented using Matlab 2015a on PC with a Windows 7 (64 bit) operation system, which is equipped with an Intel(R) Core(TM) i7-4790 CPU @ 3.6 GHz and 8 GB RAM.

9.3.2 Data

To validate the proposed system based on an FPGA, two test datasets are used to perform the orthorectification. The first study area is located in central Denver, CO. The exploited aerial image (17,054 × 17,054), see Figure 9.12a, was collected on 17 April 2000 using an RC 30 aerial camera, which is the same as Zhou et al. (2005). The focal length is 153.022 mm, and the flying height is 1650 m above the mean ground elevation of the imaged area. The second dataset is acquired from an ERDAS IMAGINE example dataset, i.e., ps_napp.img (2294 × 2294) and ps_dem.img (see Figure 9.13). The known parameters, provided by the vendors, are listed in Table 9.1.

As described in Section 9.2.2.2.2, to obtain the scanning coordinates, the affine transformation coefficients must be solved. To this end, four fiducial points for each study area are used to acquire the affine transformation coefficients according to equation (9.16), and they are shown in Table 9.2.

After the above necessary parameters are acquired, they are taken as the constants and input to the proposed FPGA-based orthorectification system. The orthorectified results (orthophoto) using the proposed FPGA-based method are shown in Figures 9.14b and 9.15b. To validate the rectification's accuracy and speed, orthorectification for the same datasets was also implemented using the PC-based platform. The orthorectification results using the PC-based software are shown in Figures 9.14a and 9.15a.

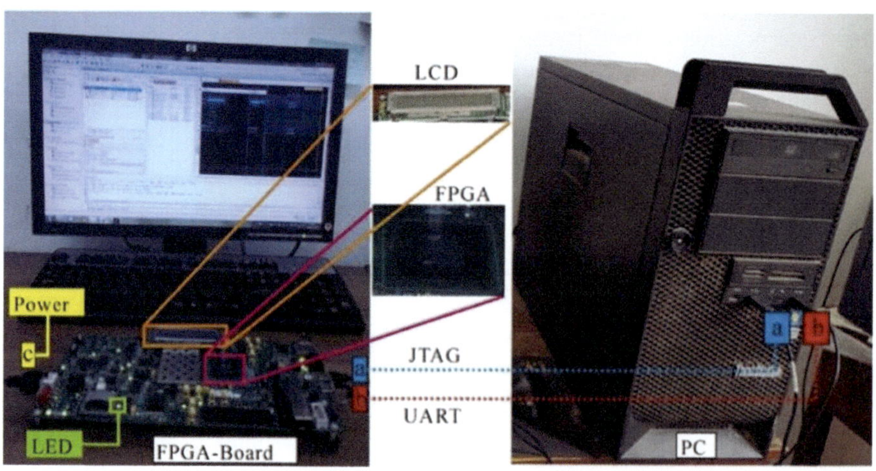

FIGURE 9.11 System diagram. (Courtesy of Zhou et al. 2017.)

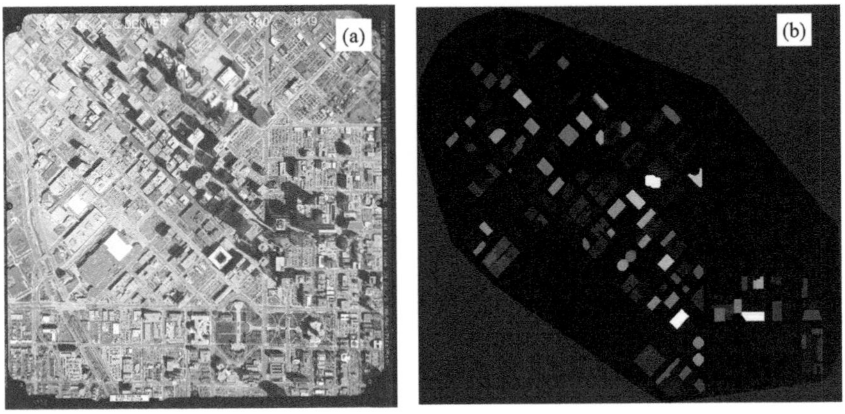

FIGURE 9.12 (a) The original aerial image covering the first study area and (b) digital surface model (DSM) covering the first study area. (Zhou et al. 2005.)

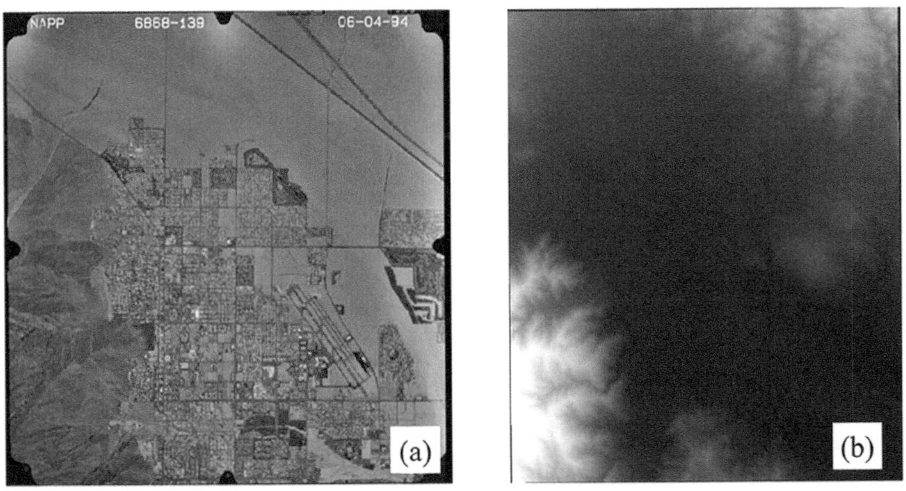

FIGURE 9.13 (a) The original aerial image covering the second study area and (b) digital elevation model (DEM) covering the second study area. (From ERDAS IMAGINE 9.2.)

TABLE 9.1
The Known Parameters for the Datasets of the Two Study Areas

Known Parameters	First Study Area	Second Study Area
x_0	0.002	–0.004
y_0	–0.004	0.000
f (mm)	153.022	152.8204
X_S (m)	3,143,040.5560	543,427.1886
Y_S (m)	1,696,520.9258	3,744,740.3247
Z_S (m)	9072.2729	6743.2730
ω (rad)	–0.02985539	0.63985182
ϕ (rad)	–0.00160606	–0.65999005
κ (rad)	–1.55385318	0.86709830

TABLE 9.2
Four Fiducial Points for Each Study Area

	The First Study Area				The Second Study Area			
#	*i*	*j*	*u*	*v*	*i*	*j*	*u*	*v*
FP_1	683.403	881.001	−196.100	191.150	87.500	88.501	−106.000	106.000
FP_2	15,820.521	835.103	182.325	192.300	2208.499	83.501	105.999	105.994
FP_3	15,868.602	15,970.971	183.525	−186.075	2213.503	2204.504	105.998	−105.999
FP_4	730.452	16,019.980	−194.925	−187.300	92.500	2209.501	−106.008	−105.998

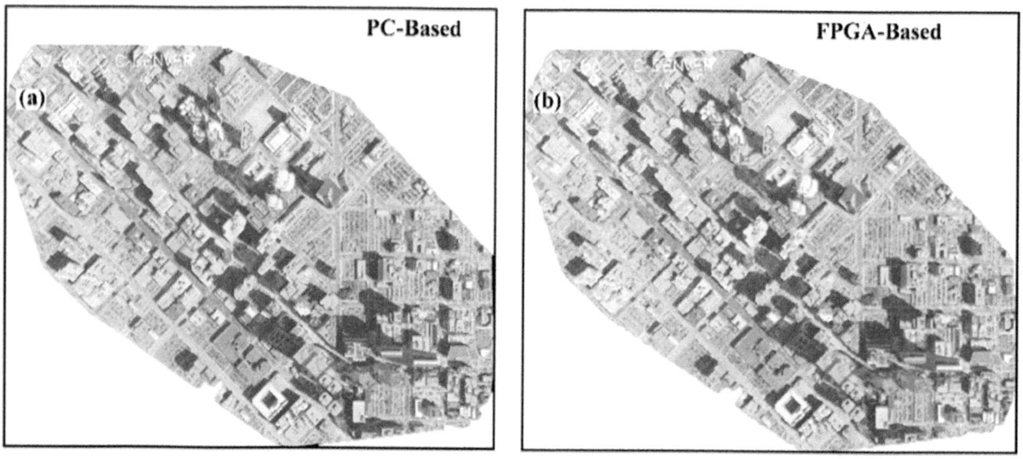

FIGURE 9.14 The orthophoto orthorectified by (a) the personal computer (PC)-based platform and (b) and the proposed method in the first study area.

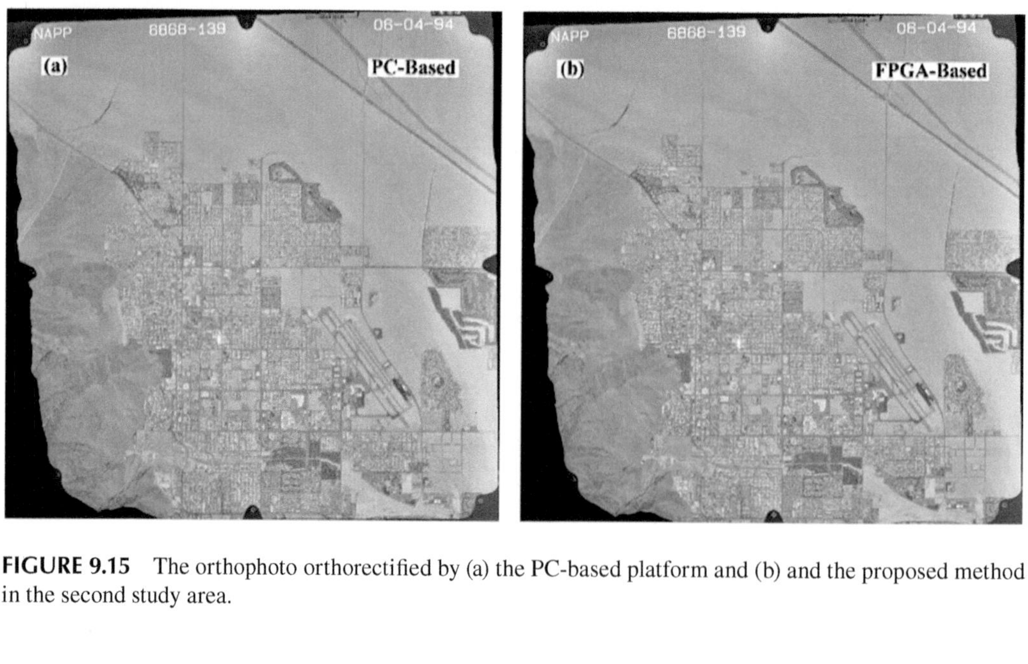

FIGURE 9.15 The orthophoto orthorectified by (a) the PC-based platform and (b) and the proposed method in the second study area.

9.4 DISCUSSION

9.4.1 Visual Check

To validate the rectified accuracy, the orthophoto results orthorectified by a PC-based platform are taken as the references. In each of the study areas, three subareas are chosen and zoomed into to visually check the accuracy (see Figures 9.16 and 9.17). As observed from Figures 9.16 and 9.17, the orthophotos orthorectified by the proposed method expose a one-pixel difference when compared to the results from the PC-based platform. Through the visual check, it can be concluded that the proposed method can meet the demand of orthorectification in practice.

9.4.2 Error Analysis

To further quantitatively evaluate the orthophoto accuracy obtained by the proposed method, the root-mean-square error (RMSE) (Shi and Shaker 2003; Reinartz et al. 2006) is applied to quantitatively analyze the rectification error of the proposed method. The RMSEs of the planimetric coordinates along the x- and y-axes, and distance (φ_X, φ_Y, and φ_S), are computed by, respectively,

$$\varphi_X = \sqrt{\frac{\sum_{k=1}^{n} (X_k' - X_k)^2}{n-1}} \tag{9.24}$$

$$\varphi_Y = \sqrt{\frac{\sum_{k=1}^{n} (Y_k' - Y_k)^2}{n-1}} \tag{9.25}$$

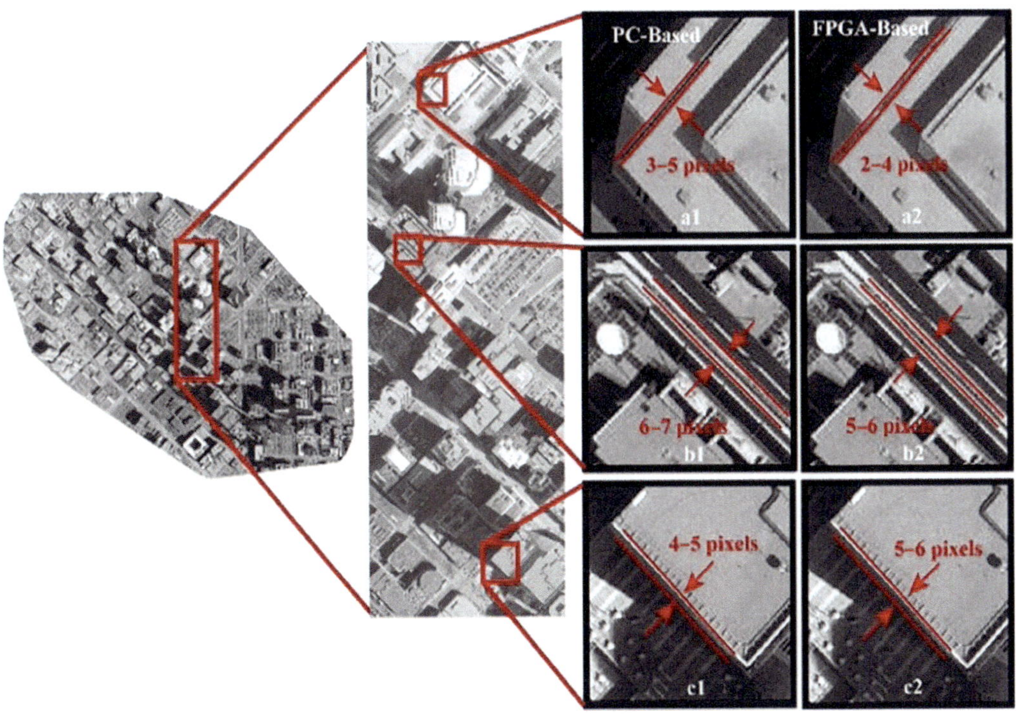

FIGURE 9.16 Visual check analysis for the orthorectified results in the three subareas of the first study area.

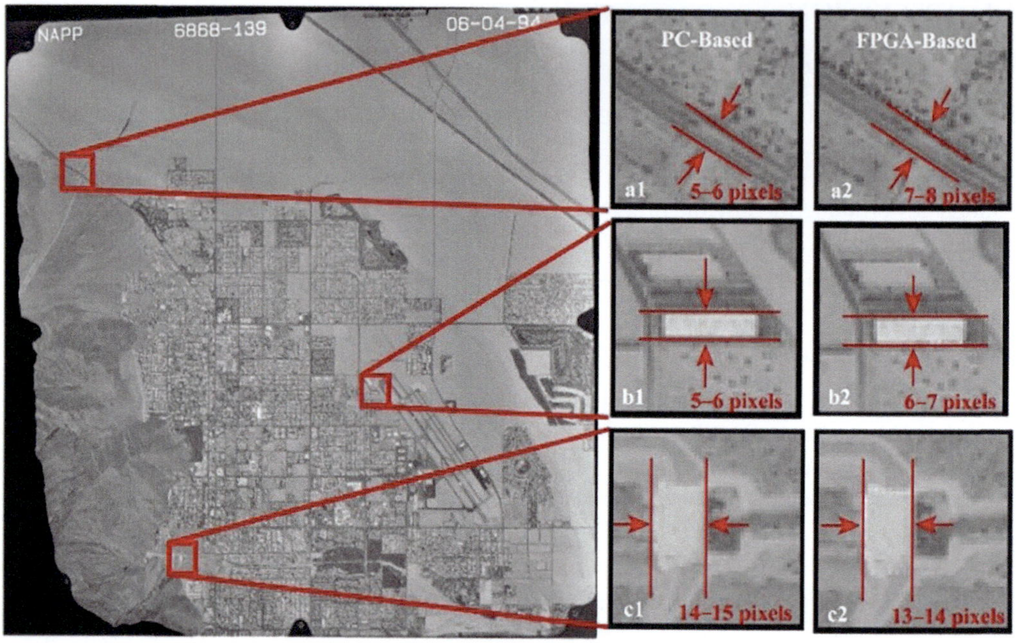

FIGURE 9.17 Visual check analysis for the orthorectified results in the three subareas of the second study area.

$$\varphi_s = \sqrt{\frac{\sum_{k=1}^{n} (X_k' - X_k)^2 + (Y_k' - Y_k)^2)}{n-1}} \tag{9.26}$$

where X_k' and Y_k' are the geodetic coordinates rectified by the proposed method; X_k and Y_k are the reference geodetic coordinates; and n is the number of checkpoints.

To this end, 90 checkpoints (see Figure 9.18) for the first study area were selected to validate the accuracy achieved by the proposed method. The differences in coordinates obtained between the proposed method and the PC-based platform are shown in Figure 9.19. According to equations (9.24)–(9.26) and Figure 9.19, the RMSEs of φ, φ_Y, and φ_S are 1.09, 1.61, and 1.93 m, respectively. In addition, other statistics, such as maximum value, minimum value, standard deviation, and mean of the difference value, are also computed and shown in Table 9.3. As shown in Table 9.3, the standard deviations of X and Y are very small. From Table 9.3, it can be found that the maximum errors of the X- and Y-coordinates are 1.16 and 1.89 m, respectively, and the standard deviation of the X- and Y-coordinates are 0.14 and 0.38 m, respectively. According to Zhang et al. (2006), the ultimate purpose of RS image rectification is to produce thematic maps from the rectified images. Whether the rectified images can satisfy the cartographic requirement of thematic maps depends on the scale of thematic maps. Because the minimum resolving distance on any map is only 0.1 mm, the tolerable errors on the ground distance vary with the scales of thematic maps. The tolerable error would be equivalent to 10 m on the ground if the scale of the thematic map is 1:100,000. Because the rectification error obtained by the proposed method ranges from ~1 to ~2 m, the correction accuracy level of the proposed FPGA-based platform is suitable for compiling 1:10,000–1:20,000 thematic maps.

However, it is also noted that differences between the proposed method and the PC-based platform in the X- and Y-coordinates still exist. The difference may be caused by the algorithms implemented through the FPGA hardware, such as fix-point computation, which propagate and accumulate. In addition, the proposed FPGA-based platform only applies two octaves, while the PC-based platform applies at least eight octaves.

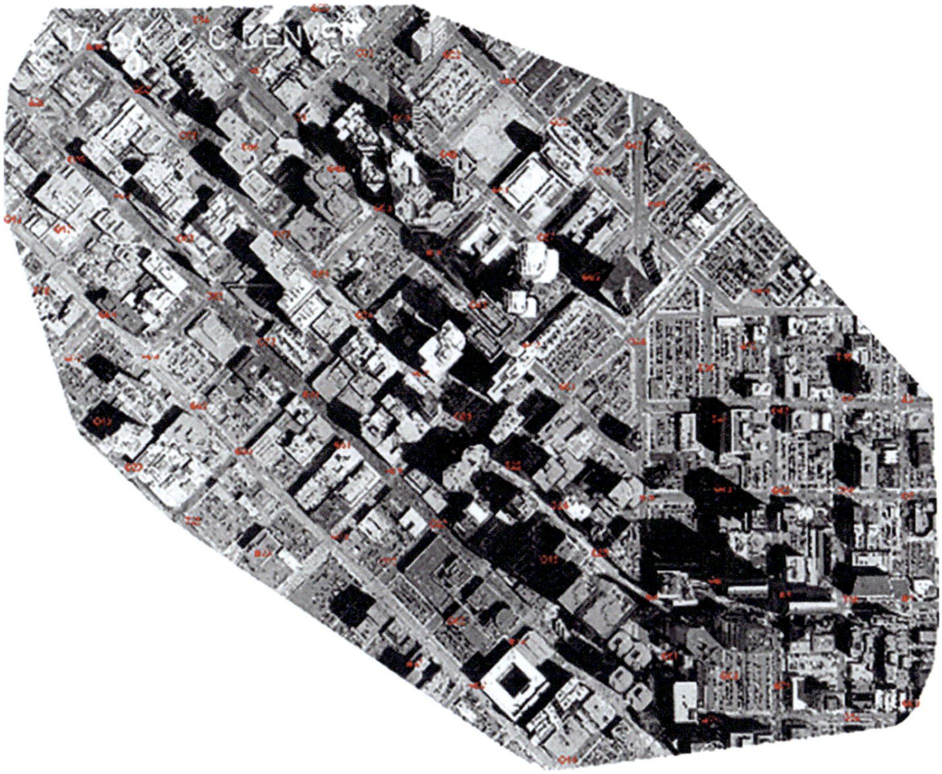

FIGURE 9.18 The distribution of the 90 checkpoints labeled as red in the first study area.

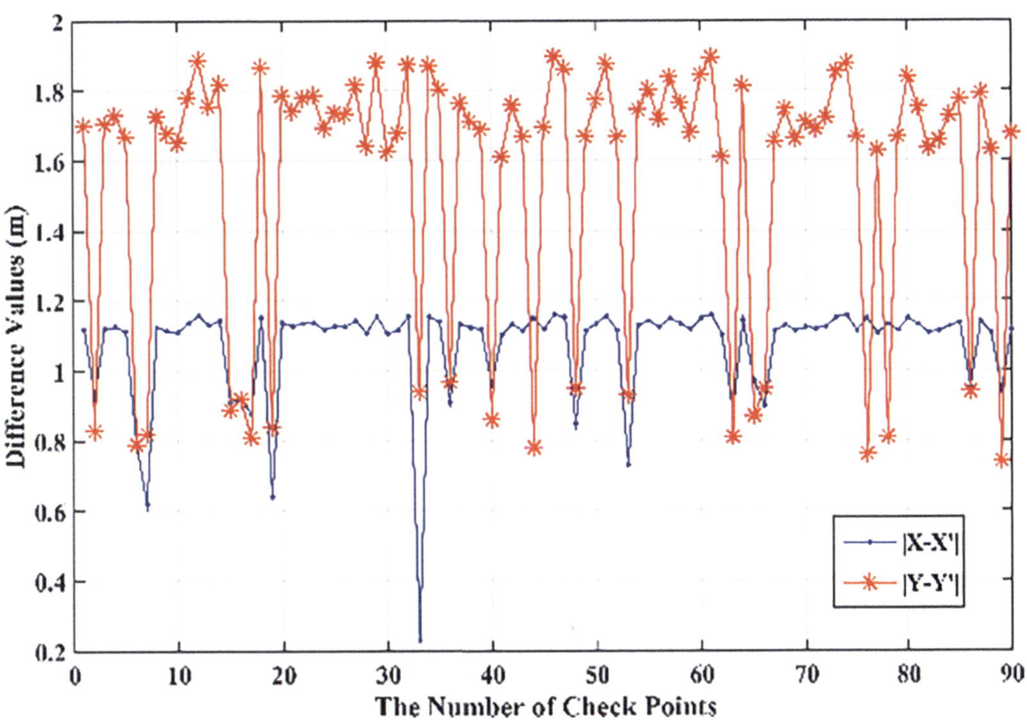

FIGURE 9.19 Different statistical analyses for orthophotos obtained by our method and the PC-based platform.

TABLE 9.3
Statistics of Difference Value of Geodetic Coordinate

#	Max	Min	Mean	Standard Deviation
X-coordinates	1.16 m	0.23 m	1.07 m	0.14 m
Y-coordinates	1.89 m	0.74 m	1.55 m	0.38 m

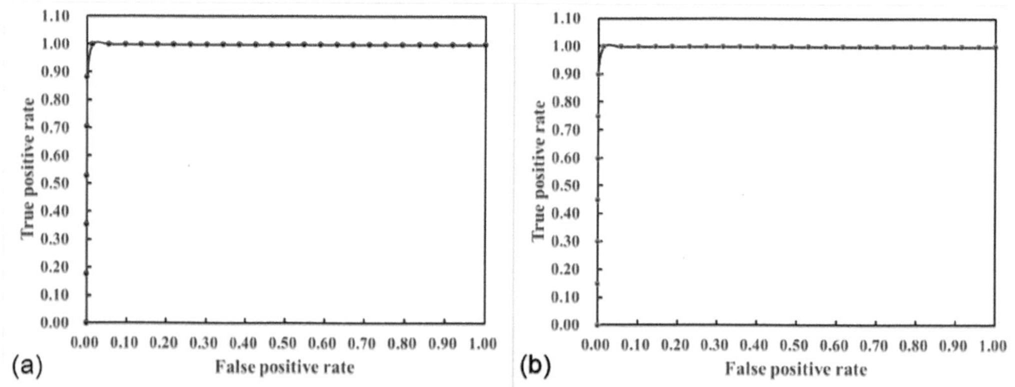

FIGURE 9.20 The receiver operating characteristics (ROC) curve analysis through the difference in the X-coordinates (a) and Y-coordinates (b). (Courtesy of Zhou et al. 2017.)

Moreover, another method (i.e., the receiver operating characteristics (ROC) curve) is used to evaluate the error of the proposed method. The ROC curve is useful for organizing classifiers and visualizing the performance of classifiers. The detailed information of the ROC curve can be found in Fawcett (2006). In an ROC graph, the vertical axis represents the true positive rate (TPR) acquired by equation (9.27) and the abscissa axis is the false positive rate (FPR) obtained by equation (9.28). Let the difference in X-coordinates and Y-coordinates, which are less than 1 m, be of a positive class, and the others be of a negative class. Then, the differences are sorted by descending order. Finally, three differences of X-coordinates (the same as Y-coordinates) are used as a group to calculate the TPR and FPR. The ROC curves of the X-coordinates and Y-coordinates are shown in Figure 9.20a and b, respectively. As shown in Figure 9.20, there are 17 X-coordinates and 20 Y-coordinates that are less than 1 m. The ROC curve of the X-coordinates point at (0.013, 1) produces the highest TPR. Additionally, the ROC curve of the Y-coordinates point at (0.014, 1) produces the highest TPR.

$$TPR = \frac{Positives\ correctly\ classified}{Total\ positives} \tag{9.27}$$

$$FPR = \frac{Negatives\ incorrectly\ classified}{Total\ Negatives} \tag{9.28}$$

9.4.3 Processing Speed Comparison

One of the most important factors in on-board orthorectification is the processing speed. To evaluate and compare the speed of the proposed method and the PC-based platform, a normalized metric, i.e., throughput representing the capacity of processing pixels per second, is used. For the proposed FPGA-based platform, the throughput is 11,182.3 kilopixels per second and the whole time of the orthorectification processing is 26.01 s for the first study area. However, for the PC platform, the throughput is 2582.9 kilopixels per second and the total time of the orthorectification processing is 112.6 seconds for the same image. That means the processing speed by the proposed FPGA-based platform is approximately 4.3 times faster than that by the PC-based platform.

9.4.4 Resource Consumption

In addition, this chapter takes the utilization ratio of each type of resource, such as input buffers (IBUF), output buffers (OBUF), and signal processors (DSP48E), to assess the proposed method.

First, the utilization ratios for the resources of the coordinate transformation module and the bilinear IM are analyzed independently. For implementing the coordinate transformation module of the proposed FPGA-based platform, the main hardware consists of 192 IBUF, 870 OBUF, and 78 DSP48E, as well as a few adding units (ADDSUB), multiplier units (MULT), and lookup tables (LUTs). The results of the utilization ratios of the logic unit for implementing the coordinate transformation module are shown in Table 9.4.

For the bilinear IM of the proposed FPGA-based platform, 129 slice resources, 291 IOs, and 75,779 LUTs are used. The utilization ratio of the IOs can reach 72% (291/400 ≈ 0.72). Additionally, the utilization ratio of LUT is 56% (75,779/134,600 ≈ 0.56) (see Table 9.5).

Generally, if the utilization ratio of a resource reaches 60%–80%, it shows that the selected device can meet the requirements of the design scheme. If the utilization ratio of a resource is too

TABLE 9.4
The Utilization Ratios of Logic Units for the Coordinate Transformation Module

#	Name of Logic Unit	Utilization Ratio (%)
Logic unit resource	Register	34
Distribution of logic unit	Flip flop	12
	LUT	27
	LUT-FF pairs	58
	Control sets	2
Input and output (IO)	IOs	78
	IOBs	54

LUT, lookup table; FF, flip flop.

TABLE 9.5
The Utilization Ratios of Logic Units for the Bilinear Interpolation Module

#	Name of Logic Unit	Utilization Ratio (%)
Use ratio of logic unit	Register	24
Distribution of logic unit	Flip flop	17
	LUT	56
	LUT-FF pairs	64
	Control sets	7
Input and output (IO)	IOs	72
	IOBs	65

low, it demonstrates that the selected device is wasted for implementing the design scheme. As shown in Tables 9.4 and 9.5, the utilization ratios of register used in the coordinate transformation module and the bilinear IM are only 34% and 24%, respectively. The utilization ratio of the register is relatively low in both models. The utilization ratios of LUT applied in these two models are 27% and 56%, respectively. The utilization ratio of LUT used in the bilinear IM is about twice higher than the utilization ratio of LUT applied in the coordinate transformation module. The reason is that the bilinear IM needs to store the gray values of neighbors. Additionally, the utilization ratios of the Flip Flop and control sets are low in both models. Moreover, the utilization ratios of LUT–FF pairs, IOs and IOBs, are relatively high in both models. In summary, according to the above comprehensive utilization ratios for resources, it can be demonstrated that the resources of the selected FPGA can meet the design requirement of the proposed FPGA-based orthorectification method.

9.5 CONCLUSIONS

In this chapter, an FPGA-based orthorectification method, which is intended to perform the orthorectification process on board spacecraft, is proposed for accelerating the speed of the orthorectification of RS images. The proposed FPGA-based orthorectification platform consists of a memory module, a coordinate transformation module (including the transformation from geodetic to photo coordinates and the transformation from photo to scanning coordinates), and a gray-scale IM based on a bilinear interpolation algorithm.

To validate the orthorectification's accuracy, an orthophoto orthorectified by a PC-based platform is taken as a reference. Two study areas, including three subareas, are chosen to validate the proposed method. The RMSE associated with maximum, minimum, standard deviation, and mean of the *X*- and *Y*-coordinates' differences is used. The experimental results demonstrated that the maximum errors of the *X*- and *Y*-coordinates are 1.16 and 1.89 m, respectively, and the standard deviations of the *X*- and *Y*-coordinates are 0.14 and 0.38 m, respectively. The RMSEs of the planimetric coordinates along the *X*- and *Y*-axes (φ_X, φ_Y) and the distance φ_S are 1.09, 1.61, and 1.93 m, respectively. Thereby, it can be concluded from these quantitative analyses that the proposed method can meet the demand of orthorectification in practice.

In addition, through analyzing the processing speed of orthorectification, it can be found that the processing speed by the proposed FPGA-based platform is approximately 4.3 times faster than that by the PC-based platform. In terms of resource consumptions, it can be found that the bilinear IM of the proposed method utilizes 129 slice resources and 291 IOs, whose utilization ratio of the IOs can reach 72% and the LUT achieves 56%.

REFERENCES

Botella, G.; Garcia, A.; Rodriguez-Alvarez, M.; Ros, E.; Meyer-Baese, U.; Molina, M. C., Robust bioinspired architecture for optical-flow computation. *IEEE Transactions on Very Large Scale Integration (VLSI) Systems*, 18, 2010, pp. 616–629.

Cao, T. P.; Elton, D.; Deng, G., Fast buffering for FPGA implementation of vision-based object recognition systems. *Journal of Real-Time Image* Processing, 7, 2012, pp. 173–183.

El-Amawy, A., A systolic architecture for fast dense matrix inversion. *IEEE Transactions on Computers*, 38, 1989, pp. 449–455.

Escamilla-Hernández, E.; Kravchenko, V.; Ponomaryov, V.; Robles-Camarillo, D.; Ramos, L. E., Real time signal compression in radar using FPGA. *Científica*, 12, 2008, pp. 131–138.

Fawcett, T., An introduction to ROC analysis. *Pattern Recognition Letters*, 27, 2006, pp. 861–874.

González, D.; Botella, G.; García, C.; Prieto, M.; Tirado, F., Acceleration of block-matching algorithms using a custom instruction-based paradigm on a NIOS II microprocessor. *EURASIP Journal on Advances in Signal Processing*, 2013, 2013, p. 118.

González, D.; Botella, G.; Meyer-Baese, U.; García, C.; Sanz, C.; Prieto-Matías, M., A low cost matching motion estimation sensor based on the NIOS II microprocessor. *Sensors*, 12, 2012, p. 13126.

Greisen, P.; Heinzle, S.; Gross, M.; Burg, A. P., An FPGA-based processing pipeline for high-definition stereo video. *EURASIP Journal on Image and Video Processing*, 2011, 2011, pp. 1–13.

Halle, W; Venus, H.; Skrbek, W., Thematic data processing on board the satellite BIRD. In: *Proceedings of SPIE Volume* 4540, *Sensors, Systems, and Next-Generation Satellites V (2001); International Symposium on Remote Sensing*, Toulouse, France, 2001, https://doi.org/10.1117/12.450683

Hsiao, P. Y.; Lu, C. L.; Fu, L. C., Multilayered image processing for multiscale Harris corner detection in digital realization. *IEEE Trans. Ind. Electron.* 57, 2010, pp. 1799–1805.

Hu, X.; Zhu, Y., Research on FPGA based image input buffer design mechanism. *Microcomputer Information, 26*, 2010, pp. 5–6+25.

Kalomiros, J. A.; Lygouras, J., Design and evaluation of a hardware/software FPGA-based system for fast image processing. *Microprocessors and Microsystems*, 32, 2008, pp. 95–106.

Kate, D., Hardware implementation of the Huffman encoder for data compression using altera DE2 board. *International Journal of Engineering, Science and Advanced Technology*, 2, 2012, pp. 11–15.

Kazmi, M.; Aziz, A.; Akhtar, P., An efficient and compact row buffer architecture on FPGA for real-time neighbourhood image processing. *Journal of Real-Time Image Processing*, 2017, https://doi.org/10.1007/s11554-017-0690-7

Kumar, P.R.; Sridharan, K. VLSI-efficient scheme and FPGA realization for robotic mapping in a dynamic environment. *IEEE Transactions on Very Large Scale Integration (VLSI)* Systems, 15, 2007, pp. 118–123.

Kuo, D.; Gordon, D., Real-time orthorectification by FPGA-based hardware acceleration. In: *Proceedings of SPIE Volume* 7830, *Image and Signal Processing for Remote Sensing, SPIE Remote Sensing*, Toulouse, France, XVI; 78300Y, 22 October 2010, https://doi.org/10.1117/12.864816

Li, H. H.; Liu, S.; Piao, Y., Snow removal of video image based on FPGA. In: *Proceedings of the 5th International Conference on Electrical Engineering and Automatic Control*, Weihai, China, 16–18 October 2015; Huang, B., Yao, Y., Eds.; Springer: Berlin/Heidelberg, Germany, 2016, pp. 207–215.

Li, H.; Xiang, F.; Sun, L., Based on the FPGA video image enhancement system implementation. *DEStech Transactions on Computer Science and Engineering*, 2016. https://doi.org/10.12783/dtcse/iceiti2016/6169

Malik, A. W.; Thornberg, B.; Imran, M.; Lawal, N., Hardware architecture for real-time computation of image component feature descriptors on a FPGA. *International Journal of Distributed Sensor Networks*, 2014, 2014, 14.

Ontiveros-Robles, E.; Gonzalez-Vazquez, J. L.; Castro, J. R.; Castillo, O., A hardware architecture for real-time edge detection based on interval type-2 fuzzy logic. In: *Proceedings of the IEEE International Conference on Fuzzy Systems*, Vancouver, British Columbia, Canada, July, 2016, pp. 24–29.

Ontiveros-Robles, E.; Vázquez, J. G.; Castro, J. R.; Castillo, O., A FPGA-based hardware architecture approach for real-time fuzzy edge detection. *Nature-Inspired Design for Hybrid Intelligent Systems*, 667, 2017, pp. 519–540.

Pal, C.; Kotal, A.; Samanta, A.; Chakrabarti, A.; Ghosh, R., An efficient FPGA implementation of optimized anisotropic diffusion filtering of images. *International Journal of Reconfigurable Computing*, 2016, 2016, 1.

Reinartz, P.; Müller, R.; Lehner, M.; Schroeder, M., Accuracy analysis for DSM and orthoimages derived from SPOT HRS stereo data using direct georeferencing. *ISPRS Journal of Photogrammetry and Remote Sensing*, 60, 2006, pp. 160–169.

Rodriguez-Donate, C.; Botella, G.; Garcia, C.; Cabal-Yepez, E.; Prieto-Matias, M., Early experiences with OpenCL on FPGAs: Convolution case study. In: *Proceedings of the 2015 IEEE 23rd Annual International Symposium on Field-Programmable Custom Computing Machines*, Vancouver, British Columbia, Canada, 2–6 May, 2015, pp. 235–235.

Shi, W.; Shaker, A., Analysis of terrain elevation effects on IKONOS imagery rectification accuracy by using non-rigorous models. *Photogrammetric Engineering & Remote Sensing*, 69, 2003, pp. 1359–1366.

Thomas, U.; Rosenbaum, D.; Kurz, F.; Suri, S.; Reinartz, P., A new software/hardware architecture for real time image processing of wide area airborne camera images. *Journal of Real-Time Image Processing*, 4, 2008, pp. 229–244.

Tomasi, M.; Vanegas, M.; Barranco, F.; Diaz, J.; Ros, E., Real-time architecture for a robust multi-scale stereo engine on FPGA. *IEEE Transactions on Very Large Scale Integration (VLSI) Systems*, 20, 2012, pp. 2208–2219.

Waidyasooriya, H.; Hariyama, M.; Ohtera, Y., FPGA architecture for 3-D FDTD acceleration using open CL. In: *Proceedings of the 2016 Progress in Electromagnetic Research Symposium (PIERS)*, Shanghai, China, 8–11 August, 2016, pp. 4719–4719.

Wang, E.; Yang, F.; Tong, G.; Qu, P.; Pang, T., Particle filtering approach for GNSS receiver autonomous integrity monitoring and FPGA implementation. *Telecommunication Computing Electronics and Control*, 14, 2016, https://doi.org/10.12928/telkomnika.v14i4.4196

Zhang, C.; Liang, T.; Mok, P. K. T.; Yu, W., FPGA implementation of the coupled filtering method. In: *Proceedings of the 2016 IEEE International Conference on Bioinformatics and Biomedicine (BIBM)*, Shenzhen, China, 15–18 December 2016, pp. 435–442.

Zhang, Y.; Zhang, D.; Gu, Y.; Tao, F., Impact of GCP distribution on the rectification accuracy of Landsat TM imagery in a coastal zone. *Acta Oceanologica Sinica (English Edition)*, 25, 2006, p. 14.

Zhou, G.; Chen, W.; Kelmelis, J.A.; Zhang, D., A comprehensive study on urban true orthorectification. *IEEE Transactions on Geoscience and Remote Sensing*, 43, 2005, pp. 2138–2147.

Zhou, G.; Zhang, R.; Liu, N., On-board orthorectification for images based on an FPGA. *Remote Sensing*, 9, 2017, 9. https://doi.org/10.3390/rs9090874

10 RPC-Based Orthorectification

10.1 INTRODUCTION

Chapter 9 described field-programmable gate array (FPGA)-based onboard orthorectification on the basis of traditional collinearity, for which the rigorous sensor physical models (interior orientation elements) must be known before orthorectifying an image. Alternatively, these elements are considered unknown parameters and solved using a number of high-quality ground control points (GCPs) with good distribution. However, not all of the sensor models are known, especially for those sensors used for reconnaissance. Thus, a "generic" or "rational function" sensor model that acts as an alternative to rigorous physical camera models is named "Rational Function Model (RFM)". Since an RFM is usually expressed by a ratio of polynomials—known as rational polynomial coefficients (RPCs), it acts as a universal replacement for complex, sensor-specific rigorous models to represent the relationship between satellite sensors and the ground.

Since the RFM is taken as the most popular generic model, independent of sensors themselves, and provides a mathematically simple and straightforward approach to describe the geometric mapping from the three-dimensional world (latitude, longitude, and height) to the corresponding two-dimensional image coordinate (pixel rows and columns). Moreover, the RPC model allows for rapid, accurate image orthorectification, georeferencing, and 3D reconstruction without needing the specific, often proprietary, parameters of the sensors (Yan et al. 2024); thus, this chapter describes an RPC-based onboard orthorectification algorithm and the FPGA-based implementation of the RPC algorithm.

10.2 RPC-BASED ORTHORECTIFICATION USING AN FPGA CHIP

High-resolution satellite sensors are different from conventional aerial frame perspective imaging and generally apply linear-array charge-coupled device (CCD) push-broom imaging technology. To deal with various types of images, many geometric processing models and algorithms are presented. One of the most widely used models is the RPC model, which is a general imaging model that is independent of the satellite sensor and platform. Many modern satellite images are always equipped with RPCs. Unlike rigorous physical models that are based on the collinear equation, which uses the ephemeris, attitude information, etc., to establish the acquisition geometry of the sensors, the RPC model does not require knowledge of the interior orientation elements and exterior orientation elements, which are sometimes not provided by vendors. The RPC model can produce uniform accuracy with a rigorous physical model and is a simple generalized model. The RPC model has been widely applied to orthorectify satellite images with the increasing utilization of high-resolution images, as in Fraser et al. (2002); Grodecki and Dial (2003); Wang and Ellis (2005); Hoja et al. (2008); Zhang et al. (2012); Yang and Zhu (2013); and Yang et al. (2017). The details of RPC orthorectification are given in Fraser et al. (2002) and Grodecki and Dial (2003).

In this chapter, the RPC algorithm is implemented using an FPGA. Usually, an FPGA chip can offer a highly flexible design, scalable circuits, and high efficiency in data processing for its pipeline structure and fine-grained parallelism. Moreover, an FPGA chip has advantages in size, weight, and power (SWaP) compared to GPU and CPU, which is helpful to integrate the FPGA into the on-board system.

However, the traditional RPC algorithm for orthorectification is computationally costly because of floating-point operations in the RPC algorithm. To implement on-board orthorectification using an FPGA chip in (near) real time, a fixed-point-based RPC (FP-RPC) algorithm is proposed, which can reduce the computation cost significantly. The details of the proposed method are given below.

DOI: 10.1201/9781003632108-10

10.2.1 Proposed RPC Algorithm

FP processing is a method that accelerates the calculation (Ferrer et al. 2005; Balster et al. 2011). To make the transformation between a fixed-point variable and a floating-point variable, multiplication by a constant is necessary to maintain the precision. When the constant is set to a power of 2, the multiplication can be seen as a single bit shift, i.e.,

$$F = 2^{\tau} F' \tag{10.1}$$

where F' is a floating-point variable, F is a fixed-point variable, and τ is a scale factor, which affects the binary accuracy of the resulting integer representation. A larger scale factor will produce a higher degree of binary accuracy.

In the proposed FP-RPC algorithm, all of the variables and constants are transformed to integers using equation (10.1). Table 10.1 gives the integer variables and their scale factors. In Table 10.1, a_i', b_i', c_i', and d' (i = 1–20) are multinomial coefficients. Generally, the values of b_1' and d_1' are 1. *Lon'*, *Lat'*, and *Hei'* are geodetic coordinates, which represent the longitude, latitude, and height, respectively. Lat'_{off}, Lat'_{scale}, Lon'_{off}, Lon'_{scale}, H'_{off}, H'_{scale}, $Line'_{off}$, $Line'_{scale}$, $Samp'_{off}$, and $Samp'_{scale}$ are the parameters for normalization. *Samp'* and *Line'* represent the image coordinates, sample, and line.

According to Fraser et al. (2002), Grodecki and Dial (2003), and equation (10.1), the normalized coordinates are converted into integers by

$$L = 2^{\tau_2} \frac{2^{-\tau_2} Lon - 2^{-\tau_3} Lon_{off}}{2^{-\tau_3} Lon_{scale}} \quad P = 2^{\tau_2} \frac{2^{-\tau_2} Lat - 2^{-\tau_3} Lat_{off}}{2^{-\tau_3} Lat_{scale}} \quad H = 2^{\tau_2} \frac{2^{-\tau_2} Hei - 2^{-\tau_3} H_{off}}{2^{-\tau_3} H_{scale}} \tag{10.2}$$

$$X = 2^{\tau_2} \frac{2^{-\tau_2} Samp - 2^{-\tau_3} Samp_{off}}{2^{-\tau_3} Samp_{scale}} \quad Y = 2^{\tau_2} \frac{2^{-\tau_2} Line - 2^{-\tau_3} Line_{off}}{2^{-\tau_3} Line_{scale}} \tag{10.3}$$

Moreover, the polynomials are converted into integers by

$$2^{-\tau_2} ND_{LS} = \left(2^{-\tau_1} C\right)\left(2^{-\tau_2} N^T\right) \tag{10.4}$$

$$ND_{LS} = 2^{-\tau_1} CN^T \tag{10.5}$$

where $ND_{LS} = \begin{bmatrix} Num_L & Den_L & Num_S & Den_S \end{bmatrix}^T$, $C = \begin{bmatrix} a_1 & a_2 & \dots & a_{19} & a_{20} \\ b_1 & b_2 & \dots & b_{19} & b_{20} \\ c_1 & c_2 & \dots & c_{19} & c_{20} \\ d_1 & d_2 & \dots & d_{19} & d_{20} \end{bmatrix}$, $N = \begin{bmatrix} 1 & L & P & H & LP & LH & PH & LL & PP & HH & PLH & LLL & LPP & LHH & LLP & PPP & PHH & LLH & PPH & HHH \end{bmatrix}$.

TABLE 10.1
Scale Factors and Integer Variables

Variable Name	Scale Factor	Integer Variable Name
a_i', b_i', c_i', and d_i' ($i = 1-20$)	τ_1	a_i, b_i, c_i, and d_i (i = 1–20)
Lon', *Lat'*, *Hei'*	τ_2	*Lon*, *Lat*, *Hei*
Lat'_{off}, Lat'_{scale}, Lon'_{off}, Lon'_{scale}, H'_{off}, H'_{scale}	τ_3	Lat_{off}, Lat_{scale}, Lon_{off}, Lon_{scale}, H_{off}, H_{scale}
$Line'_{off}$, $Line'_{scale}$, $Samp'_{off}$, $Samp'_{scale}$	τ_3	$Line_{off}$, $Line_{scale}$, $Samp_{off}$, $Samp_{scale}$

In addition, the normalized image coordinates (X', Y') are converted into integers by

$$2^{-\tau_2} Y = \frac{2^{-\tau_2} Num_L}{2^{-\tau_2} Den_L}, \quad 2^{-\tau_2} X = \frac{2^{-\tau_2} Num_S}{2^{-\tau_2} Den_S'} \tag{10.6}$$

$$Y = 2^{\tau_2} \frac{Num_L}{Den_L}, \quad X = 2^{\tau_2} \frac{Num_S}{Den_S} \tag{10.7}$$

Finally, the image coordinates ($Samp'$, $Line'$) are converted into integers by

$$\begin{cases} 2^{-\tau_3} Samp = \left(2^{-\tau_2} X\right)\left(2^{-\tau_3} Samp_{scale}\right) + 2^{-\tau_3} Samp_{off} \\ 2^{-\tau_3} Line = \left(2^{-\tau_2} Y\right)\left(2^{-\tau_3} Line_{scale}\right) + 2^{-\tau_3} Line_{off} \end{cases} \tag{10.8}$$

$$\begin{cases} Samp = \left(2^{-\tau_2} X\right) Samp_{scale} + Samp_{off} \\ Line = \left(2^{-\tau_2} Y\right) Line_{scale} + Line_{off} \end{cases} \tag{10.9}$$

10.2.2 Parallel Computation of Orthorectification Using an FPGA

Many factors affect the computation speed when an FPGA is adopted, such as the optimal design of algorithms and the logical resources of the utilized FPGA. By analyzing the structure of the FP-RPC algorithm and optimizing it, an FPGA-based hardware architecture for FP-RPC-based orthorectification is designed, as shown in Figure 10.1. As described in equations (10.2) to (10.9), their structures are similar. It is convenient for FPGAs to be implemented in parallel. As shown in Figure 10.1, the FPGA-based FP-RPC module can be divided into three submodules, that is, Read_parameter_mod (RPM), which is used to send parameters to other modules; Coordinate_Transform_mod (CTM), which is applied to transform geodetic coordinates to image coordinates; and Interpolation_mod (IM), which is utilized to perform bilinear interpolation. The details of these modules are given as follows.

a. For RPM, the coefficients of RPC can be calculated by least-squares adjustment (Tao and Hu 2001). According to Tao and Hu (2001), the computing processes of the RPC coefficients are as follows. Equation (10.7) can be rewritten as

$$F_X = Num_S\left(P, L, H\right) - 2^{\tau_2} X Den_S\left(P, L, H\right) = 0 \tag{10.10a}$$

$$F_Y = Num_L\left(P, L, H\right) - 2^{\tau_2} Y Den_L\left(P, L, H\right) = 0 \tag{10.10b}$$

So, the matrix form of the error equation can be expressed as

$$V = MA - R \tag{10.11}$$

where

$$M = \begin{bmatrix} \dfrac{\partial F_X}{\partial a_i} & \dfrac{\partial F_X}{\partial b_j} & \dfrac{\partial F_X}{\partial c_i} & \dfrac{\partial F_X}{\partial d_j} \\ \dfrac{\partial F_Y}{\partial a_i} & \dfrac{\partial F_Y}{\partial b_j} & \dfrac{\partial F_Y}{\partial c_i} & \dfrac{\partial F_Y}{\partial d_j} \end{bmatrix} \quad (i = 1, 2, \ldots, 20;\ j = 2, \ldots, 20)$$

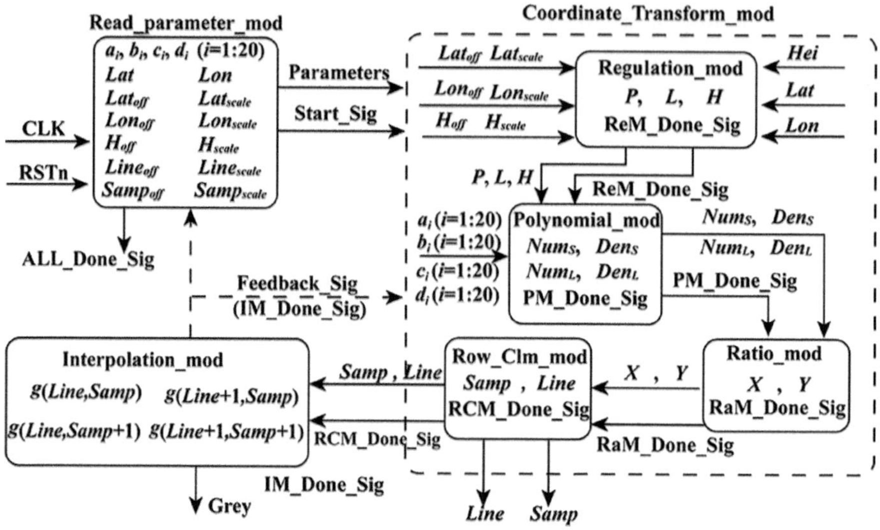

FIGURE 10.1 Flowchart showing implementation on an FPGA chip. (Courtesy of Zhang et al. 2018.)

$$R = \begin{bmatrix} -F_X^0 & -F_Y^{0T} \end{bmatrix} A = \begin{bmatrix} a_i & b_j & c_i & d_j \end{bmatrix}^T$$

Equation (10.11) is solved by the least-squares algorithm, and the solutions of RPC coefficients can be represented as

$$A = (M^T M)^{-1} M^T R \tag{10.12}$$

The solution for equation (10.12) is acquired by an iterative process. The entire algorithm of equation (10.12) has been implemented in our previous work (Zhou et al. 2018), in which the detailed implementation process can be found.

The normalization parameters can be calculated by the following equations:

$$Lat_{off} = \frac{\sum_{i=1}^{n} Lat_i}{n}, \quad Lon_{off} = \frac{\sum_{i=1}^{n} Lon_i}{n}, \quad H_{off} = \frac{\sum_{i=1}^{n} Hei_i}{n}, \tag{10.13a}$$

$$Samp_{off} = \frac{\sum_{i=1}^{n} Samp_i}{n}, \quad Line_{off} = \frac{\sum_{i=1}^{n} Line_i}{n}$$

$$Lat_{scale} = \max\left(\left|Lat_{\max} - Lat_{off}\right|, \left|Lat_{\min} - Lat_{off}\right|\right), \tag{10.13b}$$

$$Lon_{scale} = \max\left(\left|Lon_{\max} - Lon_{off}\right|, \left|Lon_{\min} - Lon_{off}\right|\right)$$

$$H_{scale} = \max\left(\left|Hei_{\max} - H_{off}\right|, \left|Hei_{\min} - H_{off}\right|\right), \tag{10.13c}$$

$$Line_{scale} = \max\left(\left|Line_{\max} - Line_{off}\right|, \left|Line_{\min} - Line_{off}\right|\right)$$

$$Samp_{scale} = \max\left(\left|Samp_{\max} - Samp_{off}\right|, \left|Samp_{\min} - Samp_{off}\right|\right) \tag{10.13d}$$

where n is the number of GCPs.

When the enable signal is being received, the geodetic coordinates (*Lon*, *Lat*, *Hei*) stored in the RAMs are read and sent to Coordinate_Transform_mod (CTM) with the attained parameters and the start signal (Start_Sig) in the same clock cycle.

b. When the Start_Sig, the constants, and the geodetic coordinates are being received in the CTM, the normalized coordinates (*P*, *L*, *H*) are first calculated in the regularization module (Regulation_mod, ReM); then, the normalized coordinates and the done signal of ReM (ReM_Done_Sig) are sent to the polynomial module (Polynomial_mod, PM) with a_i, b_i, c_i, and d_i (i = 1–20) in the same clock cycle to compute the numerators and denominators (Num_L, Num_S, Den_L, and Den_S) of equation (10.7); subsequently, when the done signal (PM_Done_Sig) of PM, Num_L, Num_S, Den_L, and Den_S are being received, the normalized coordinates (*X*, *Y*) of the image coordinates are calculated. Finally, when the normalized coordinates (*X*, *Y*) and the done signal (RaM_Done_Sig) of the ratio module (Ratio_mod, RaM) are being received, the image coordinates (*Samp*, *Line*) and the done signal (RCM_Done_Sig) of the image coordinate calculation module (Row_Clm_mod, RCM) are acquired and sent to the interpolation module (Interpolation_mod, IM) in the same clock cycle.
c. When the image coordinates (*Samp*, *Line*) and RaM_Done_Sig are being received in IM, the gray of pixel (*Samp*, *Line*) is obtained by interpolation, and the done signal (IM_Done_Sig) of IM is produced.
d. When the posedge clk of the signal, ALL_Done_Sig is being detected, the processing is finished.

10.2.2.1 Read Parameter Module

To ensure that the constants, geodetic coordinates, and the start signal (Start_Sig) are sent in the same clock cycle, a parallel module (i.e., the RPM) is designed (see Figure 10.2). In the RPM, the constants are assigned corresponding values, while the geodetic coordinates are stored in RAM. In this design, all values are expressed using a fixed point of 32 bits to ensure computational accuracy.

In the RPM, the geodetic coordinates are sent to the next module according to the order of the column. First, the address of RAM is initialized as 0. When the enable signal is detected, the first group of geodetic coordinates (Lat_0, Lon_0, Hei_0) is read from the RAM and sent to the next module with the constants and the Start_Sig in the same clock cycle. Starting from the second group of geodetic coordinates, the rules for reading and sending geodetic coordinates are changed. In other words, after the second group of geodetic coordinates (Lat_1, Lon_1, Hei_1), the geodetic coordinates will be read and sent unless the enable signal and the feedback signal (Feedback_Sig), which are sent by the interpolation module, are detected at the same time. After the final group of geodetic coordinates is read and sent, if the Feedback_Sig is received, the done signal (ALL_Done_Sig) of orthorectification is produced. When the ALL_Done_Sig is detected, the process of orthorectification is stopped.

10.2.2.2 Coordinate Transformation Module

As shown in Figure 10.1, for the CTM, the inputs contain the constants, the geodetic coordinates, and the Start_Sig; the outputs include image coordinates and the done signal of this module. The CTM can be divided into four submodules, namely ReM, PM, RaM, and RCM. Details regarding these four submodules are as follows.

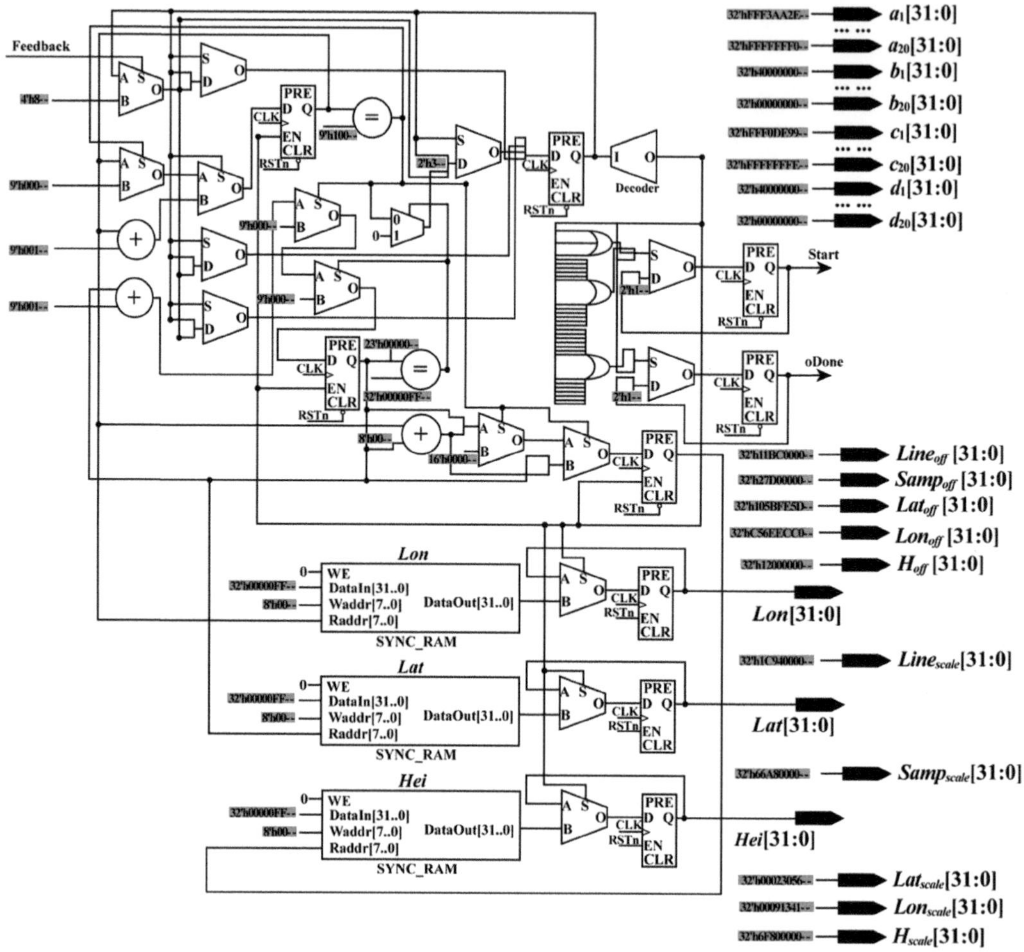

FIGURE 10.2 Schematic diagram of the read parameter module. (Courtesy of Zhang et al. 2018.)

10.2.2.2.1 Regulation Module

According to Section 10.2.1, the geodetic coordinates (*Lat*, *Lon*, *Hei*) should first be transformed into the normalized coordinates (*L*, *P*, *H*) based on equation (10.2) because this operation can minimize the introduction of errors during the computation of the numerical stability of equations. As shown in equation (10.2), the forms of these equations are uniform. In other words, they are suitable for implementation using an FPGA. To obtain the normalized coordinates (*L*, *P*, *H*) of the geodetic coordinates (*Lat*, *Lon*, *Hei*) using an FPGA chip, a parallel computation architecture is presented in Figure 10.3. In Figure 10.3, the structures of "Normalize *Lat*", "Normalize *Lon*", and "Normalize *Hei*" are similar. Thus, only the schematic diagram of "Normalize *Lat*" is presented (see Figure 10.4).

As shown in Figure 10.4, during the computation process, 1 divider, 10 adders, 10 flip flops, and 16 multiplexer units are mainly used to normalize the *Lat*. In this design, the relationship between "Normalize *Lat*", "Normalize *Lon*", and "Normalize *Hei*" is parallel. The normalized coordinates (*L*, *P*, *H*) are obtained in the same clock cycle as the done signal.

10.2.2.2.2 Polynomial Module

When the ReM_Done_Sig and (*L*, *P*, *H*) are being received by the PM module, the PM module starts to work. As shown in equations (10.4) and (10.5), these polynomials have a uniform form,

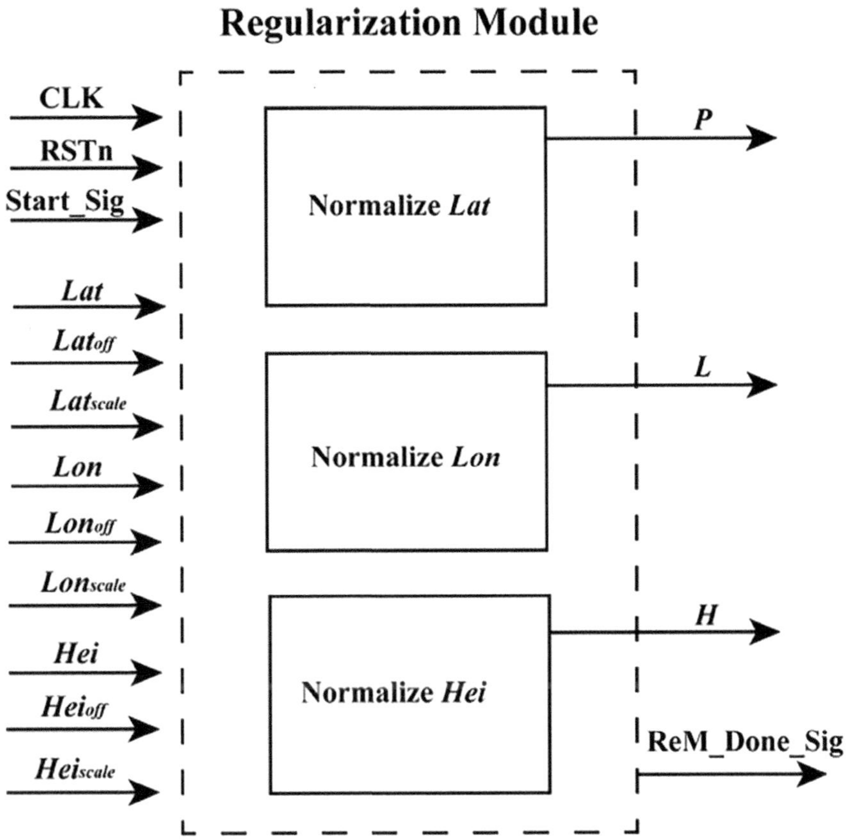

FIGURE 10.3 Schematic diagram of the ReM.

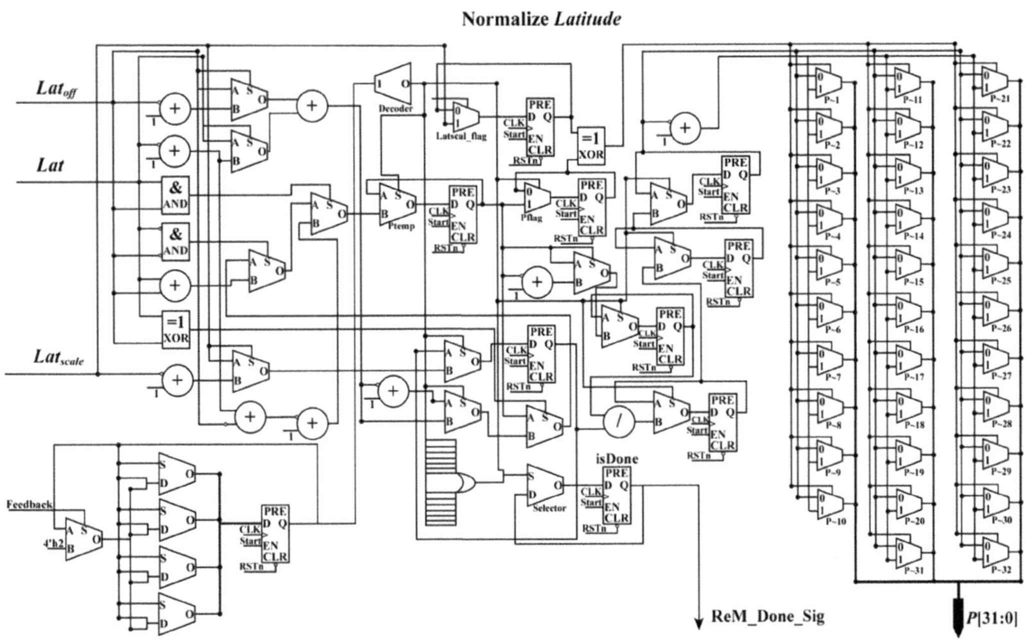

FIGURE 10.4 Schematic diagram of normalizing *Latitude*. (Courtesy of Zhang et al. 2018.)

which are suitable for the implementation of an FPGA chip in parallel. In these equations, variables such as *LH*, *LP*, and *PH* are shared. To implement these polynomials in parallel using an FPGA chip, a parallel computation architecture is proposed in Figures 10.5 and 10.6. As shown in those figures, the PM module is divided into two parts: one is used to perform multiplication and the other is applied to manipulate addition. When performing addition, some special operations about the

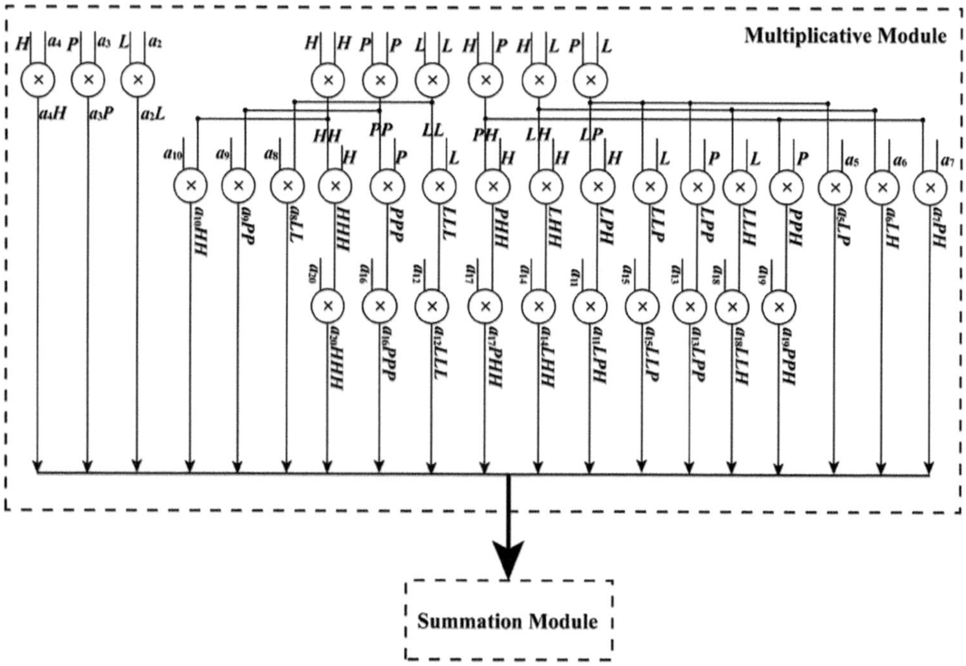

FIGURE 10.5 Schematic diagram of polynomial module.

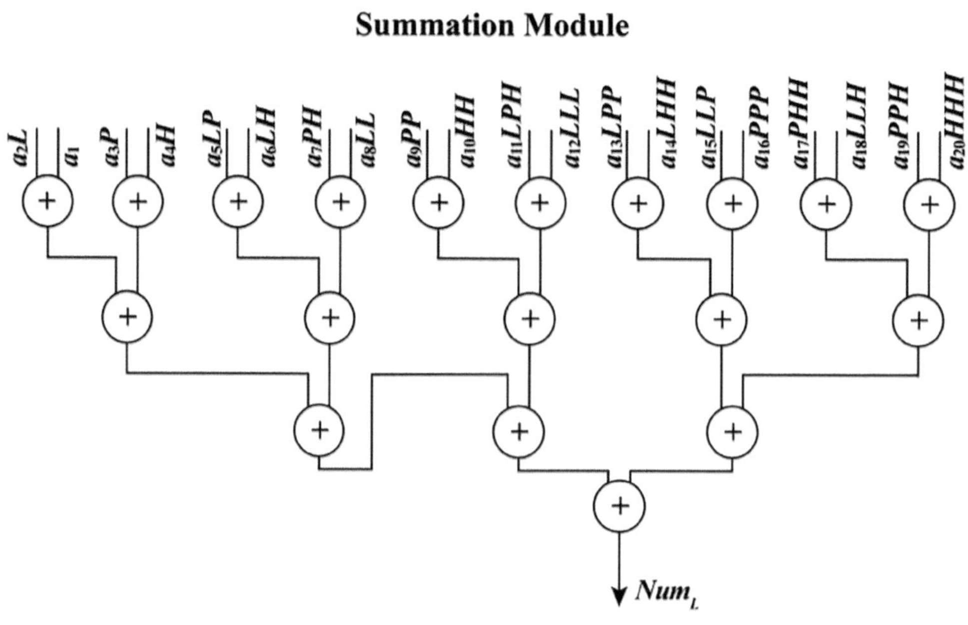

FIGURE 10.6 Schematic diagram of summation module.

positive and negative sets of data should be considered. Thus, for the additions in Figure 10.6, each of them is extended to a similar form, as shown in Figure 10.7, taking the addition between a_3P and a_4H as an example. In the example, three situations are considered: (1) a_3P and a_4H are both positive; (2) a_3P and a_4H are both negative; and (3) a_3P and a_4H have opposite signs. The details for an extended addition are shown in Figure 10.7.

To implement each polynomial, 35 multipliers are utilized in the multiplication and 19 extended additions are used. In each extended addition, 3 flip flops, 4 selectors, 7 adders, and 11 multiplexers are applied. After processing the PM module, four sums, i.e., Num_L, Num_S, Den_L, and Den_S are obtained with the done signal of the PM module, PM_Done_Sig, in the same clock cycle.

10.2.2.2.3 Ratio Module

When the PM_Done_Sig, Num_L, Num_S, Den_L, and Den_S are being received, the RaM module starts to calculate the normalized coordinates (X, Y) of the image coordinates. As shown in equation (10.6), the forms for the two equations are the same. It is convenient to calculate X and Y in parallel

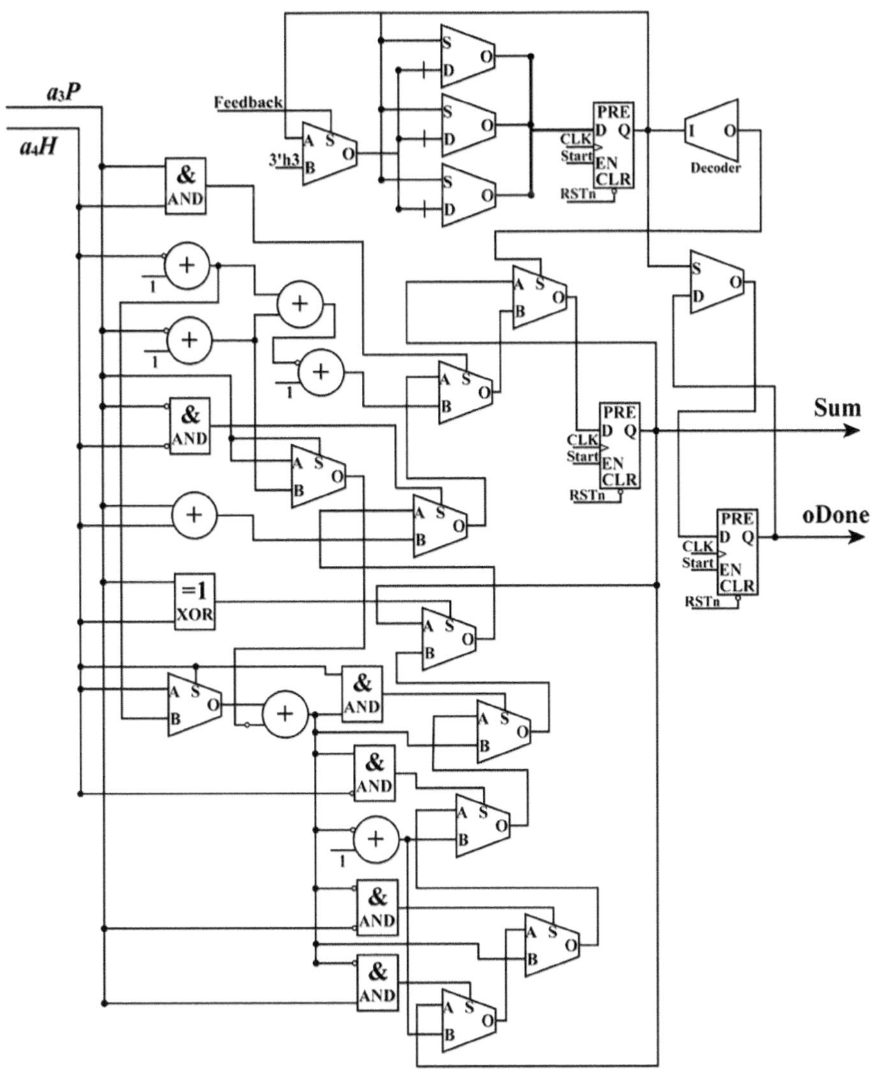

FIGURE 10.7 Details showing an example of extended addition. (Courtesy of Zhang et al. 2018.)

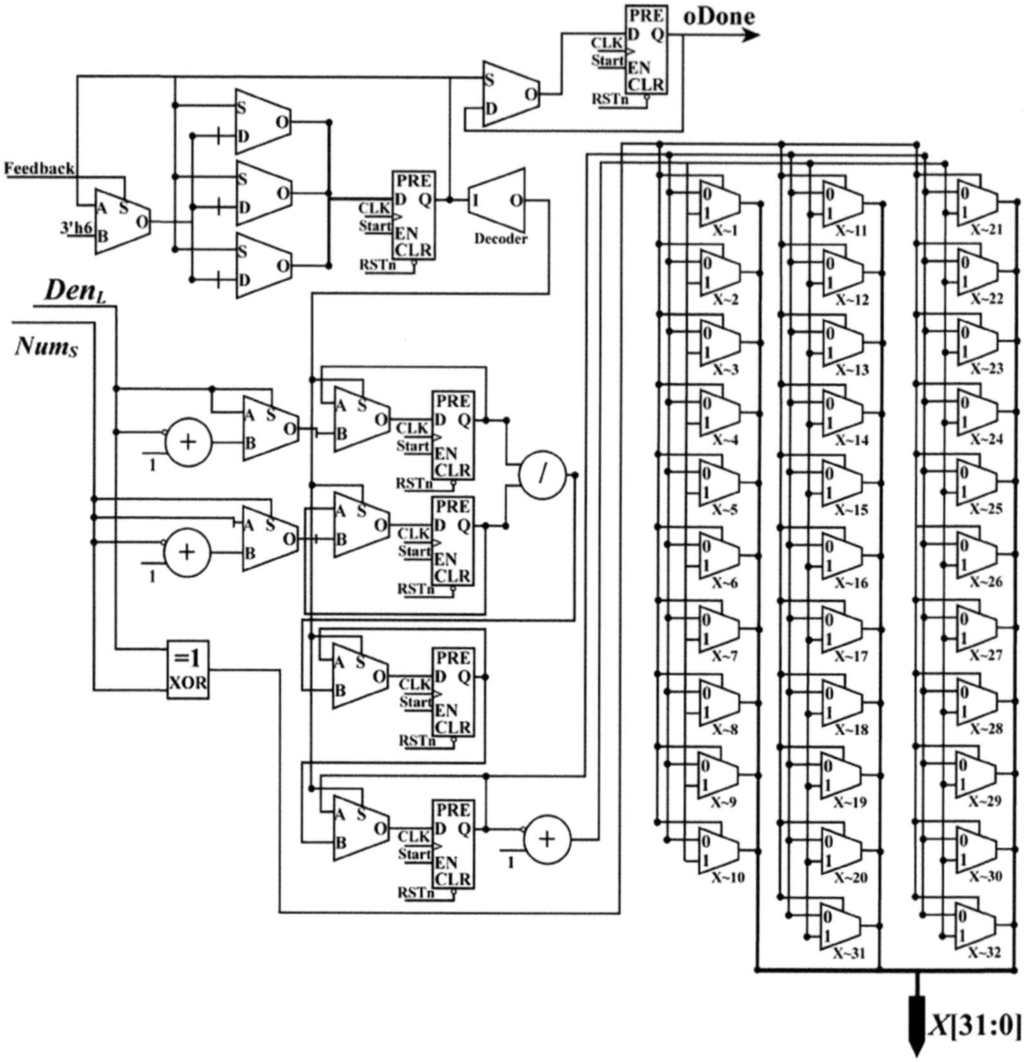

FIGURE 10.8 Schematic diagram of the ratio module. (Courtesy of Zhang et al. 2018.)

using an FPGA chip. As shown in Figure 10.8, a parallel-computing architecture that is used to calculate *X* is presented. In the same way, the *Y* coordinate can be obtained.

To obtain the *X* (or *Y*) coordinates, 1 divider, 3 adders, 6 multiplexers, 6 flip flops (2 flip flops are public), and 32 selectors are applied. After the processing of the RaM module, the *X* coordinate and *Y* coordinate are acquired with the done signal, RaM_Done_Sig, in the same clock cycle.

10.2.2.2.4 Image Coordinate Calculation Module

When the RaM_Done_Sig, *X*, and *Y* coordinates are being detected, the RCM module starts to calculate the image coordinates (*Samp*, *Line*), i.e., column and row indexes. As shown in equation (10.9), the equations give the relationship between the normalized coordinates (*X*, *Y*) and image coordinates (*Samp*, *Line*).

As shown in equation (10.9), the equations have a uniform form, which is helpful for implementation using an FPGA. To calculate the image coordinates (*Samp*, *Line*) in parallel, a parallel-computing hardware architecture is designed. Because the forms of the equations in equation (10.9)

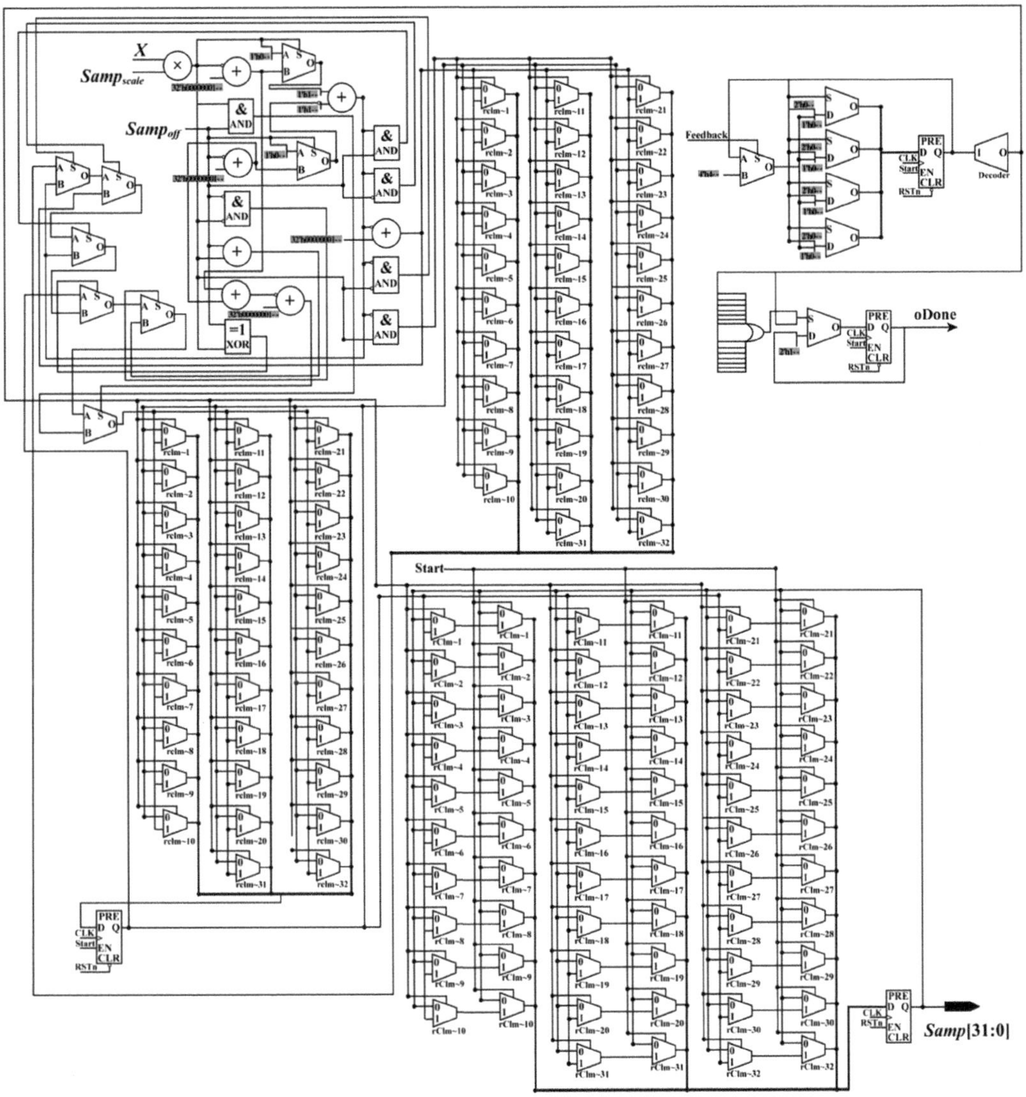

FIGURE 10.9 Schematic diagram used for calculating *Samp* coordinate. (Courtesy of Zhang et al. 2018.)

are similar, only the schematic diagram used for calculating the *Samp* coordinate is given. As shown in Figure 10.9, one multiplier, four flip flops (two of them are shared when calculating *Line* coordinate), five selectors (MUX) shared when calculating *Line* coordinate, seven adders, and 136 multiplexers (MUX21).

After the processing of the RCM module, the image coordinates, that is, the column and row indexes (*Samp*, *Line*), and the done signal (RCM_Done_Sig) are obtained in the same clock cycle. Up to this point, the whole processing of the coordinate transformation is done. The obtained image coordinates are sent to the interpolation module to interpolate the grayscale.

10.2.2.3 Interpolation Module

Because the obtained column and row indexes may not exist only at the center of a pixel, it is necessary to use the interpolation method to obtain the grayscale in the obtained column and row indexes. Considering the interpolation effect, the complexity of an interpolation algorithm, and the resources of an FPGA, the bilinear interpolation method is selected to implement the interpolation for grayscale. Mathematically, the bilinear interpolation algorithm can be expressed by the following equation:

$$g(i+p,j+q)=(1-p)(1-q)g(i,j)+(1-p)qg(i,j+1)$$
$$+p(1-q)g(i+1,j)+pqg(i+1,j+1) \quad (10.14)$$

where *i* and *j* are nonnegative integers; the intermediates $p = |i - int(i)|$ and $q = |j - int(j)|$ are within the range of (0, 1); and $g(i, j)$ represents the gray values.

To implement the bilinear interpolation algorithm in parallel using an FPGA chip, a parallel computation architecture was designed (see Figure 10.10). The designed hardware architecture contains four submodules/parts: (1) the subtract_mod, which is used to obtain the integer part (*iLine* and *iSamp*) and fractional part (*p* and *q*) of *Line* and *Samp* indexes, and to calculate the subtraction (1_*p* and 1_*q*) in equation (10.14); (2) the get_gray_addr_mod is applied to obtain the address of gray in RAM; (3) the multiplication part, which is utilized to calculate the multiplications in equation (10.14); (4) the calculate_sum_mod, which is used to compute the sum in equation (10.14). After the processing of the calculate_sum_mod, the results of interpolation in (*Samp*, *Line*) are obtained. The details of subtract_mod, get_gray_addr_mod, and calculate_sum_mod are described as follows.

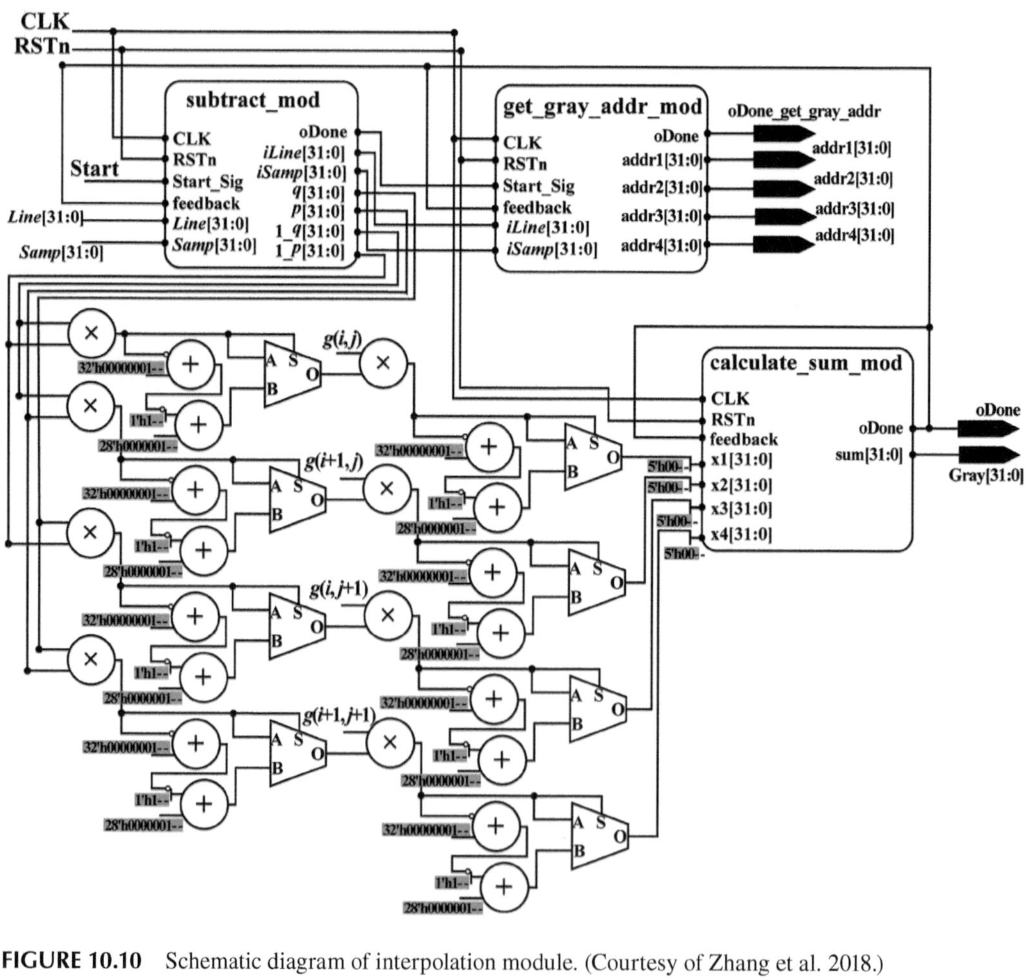

FIGURE 10.10 Schematic diagram of interpolation module. (Courtesy of Zhang et al. 2018.)

10.2.2.3.1 subtract_mod

As shown in equation (10.14), to perform the bilinear interpolation method, the gray values of four neighbors around the acquired column and row indexes are required. Thus, the acquired column and row indexes should be preprocessed to obtain the integer part and fractional part, which are used to calculate $(1 - q)$ and $(1 - p)$. To implement the function using an FPGA chip, a parallel-computing architecture is proposed, named subtract_mod. In subtract_mod, the methods used to acquire *iSamp*, q and 1_q are similar to those for obtaining *iLine*, p and 1_p, respectively. Thus, in this section, only the schematic diagram for obtaining *iSamp*, q and 1_q is given (see Figure 10.11).

As shown in Figure 10.11, to obtain *iSamp*, q and 1_q, three adders, seven multiplexers (MUX21), and nine flip flops, three of which are shared when *iLine*, p and 1_p are used. In addition, three MUXs are public. After the processing of the whole subtract_mod, *iSamp*, q, 1_q, *iLine*, p, and 1_p are acquired with the done signal in the same clock cycle. When obtaining these variables, *iSamp* and *iLine* are sent to the next submodule to retrieve the address of gray for four neighbors in RAM. Meanwhile, q, 1_q, *iLine*, p, and 1_p are sent to another part to perform multiplication.

10.2.2.3.2 get_gray_addr_mod

The grayscale of a pixel can be obtained according to the corresponding address. To obtain the gray values of four neighbors around the obtained column and row indexes in parallel, a parallel-computing hardware architecture is proposed (see Figure 10.12), called get_gray_addr_mod. In the get_gray_addr_mod, 3 LESS-THAN comparators, 9 adders, 10 MUX21, 12 flip flops, and 70 MUX are applied. After the processing of the get_gray_addr_mod, four addresses are obtained with the done signal in the same clock cycle. According to the obtained addresses, the gray values can be acquired from RAM. Then, they are sent to the multiplication part to perform the multiplication.

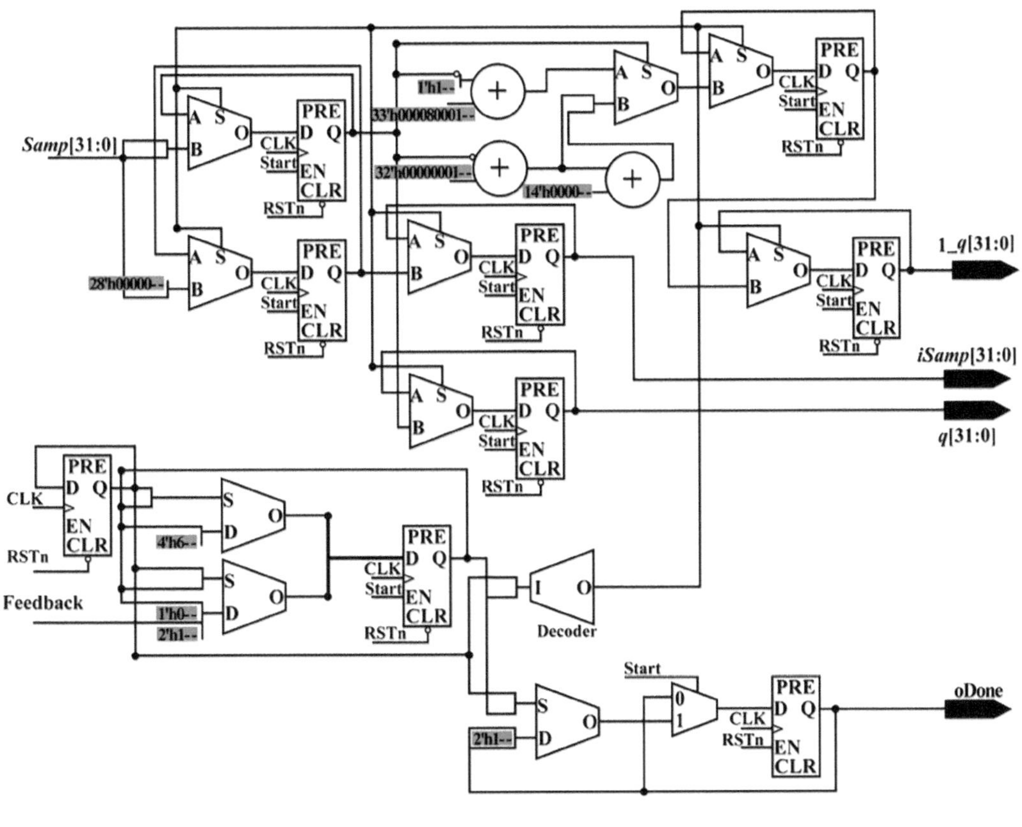

FIGURE 10.11 Schematic diagram of the subtract_mod module. (Courtesy of Zhang et al. 2018.)

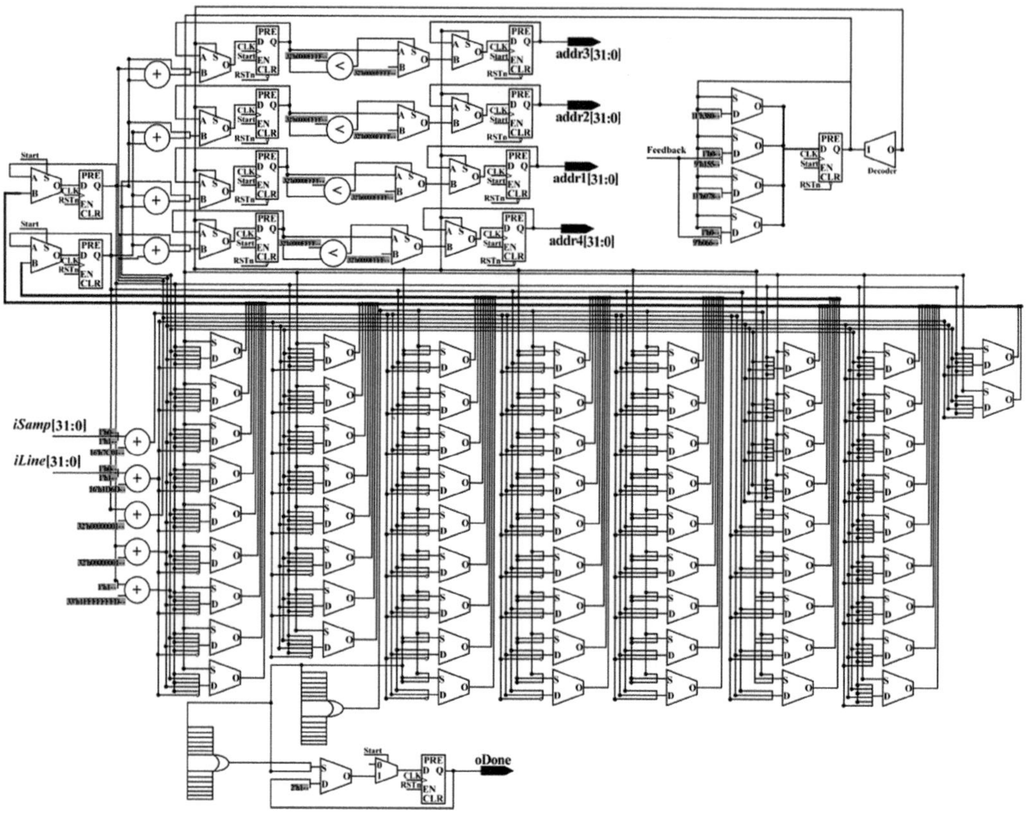

FIGURE 10.12 Schematic diagram of the get_gray_addr_mod module. (Courtesy of Zhang et al. 2018.)

10.2.2.3.3 calculate_sum_mod

As shown in Figure 10.10, after the multiplication process, four variables x_1, x_2, x_3, and x_4 are obtained in the same clock cycle. To implement the addition for four variables, two levels of additions are needed. Each addition corresponds to an extended addition that has an architecture that is similar to Figure 10.7. The details can be found in Section 10.2.2.2 and Figure 10.7. After the processing of the calculate_sum_mod, the result of interpolation in (*Samp*, *Line*) can be obtained.

10.2.3 Integration On-Board System

An FPGA device can be integrated into the on-board system as a part of the system, because an FPGA has advantages in size, weight, and power. After completing the proposed algorithm design using Verilog language, the designed algorithm can be programmed into the selected FPGA device.

10.3 EXPERIMENTS

10.3.1 Software and Hardware Environment

In this chapter, an Altera FPGA was used. The version of the FPGA is Kintex-7 XC7K325TFFG900-1 (see Figure 10.13), the design tool is Vivado 2016.4, and the simulation tool is ModelSim SE10.1d. The PC uses a Windows 7 (64 bit) operating system and has an Intel(R) Core(TM) i7-4790 CPU @ 3.6 GHz processor with 8 GB RAM. To validate the proposed method, the orthorectification algorithm is also implemented using MATLAB 2012a (MathWorks, 1 Apple Hill Drive, Natick, MA 01760-2098, USA).

FIGURE 10.13 Photograph of the FPGA platform. (Courtesy of Zhang et al. 2018.)

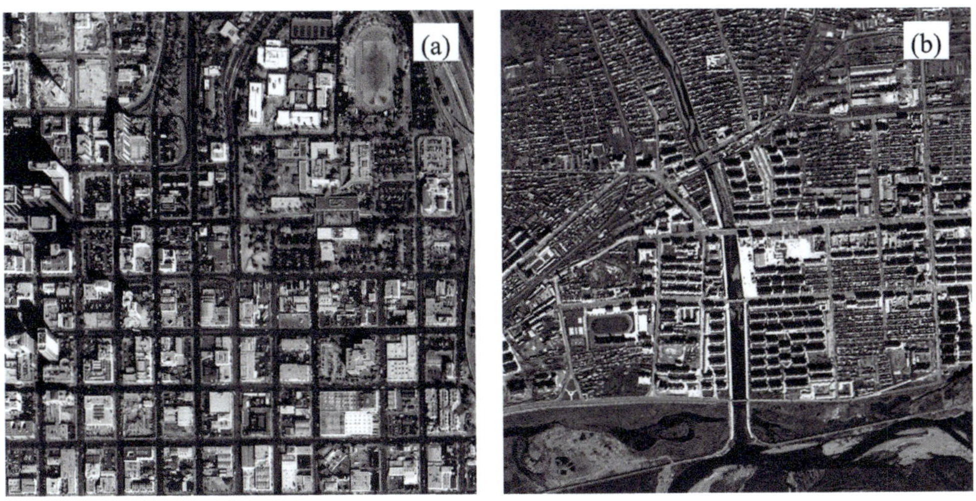

FIGURE 10.14 (a) Original IKONOS image and (b) original SPOT6 image.

10.3.2 Dataset

To validate the correction accuracy and processing speed of the proposed FPGA-based orthorectification method, two test datasets (as shown in Figure 10.14) are used in this chapter. The first study area is located in San Diego, California, US. The IKONOS-2 PAN image with the resolution of 1.0 m was collected on February 7, 2000. The wavelength range of IKONOS-2 PAN image is 450–900 nm. The second study area is located in Genhe, Inner Mongolia, China. The SPOT-6 PAN image with the resolution of 1.5 m was acquired on September 29, 2013. The wavelength range of SPOT-6 PAN image is 450–745 nm. The known parameters are listed in Tables 10.2–10.4.

According to the proposed method, input parameters should be transformed into fixed-point data. As shown in Tables 10.2–10.4, the values of the parameters of the two study areas are in different ranges. To ensure computation accuracy, all parameters are transformed into fixed-point of 32 bits using a different scale factor, τ. The details are given in Tables 10.5 and 10.6. In addition, the clock frequency is 100 MHz.

TABLE 10.2
Normalized Parameters

#	First Area	Second Area
$Line_{off}$ (pixels)	1135	24,874.5
$Samp_{off}$ (pixels)	2548	17,962.5
Lat_{off} (°)	32.718700	50.737358
Lon_{off} (°)	−117.13340	121.44648
H_{off} (m)	36.000	500
$Line_{scale}$ (pixels)	1829	24,874.5
$Samp_{scale}$ (pixels)	6570	17,962.5
Lat_{scale} (°)	0.01710000	0.41058849
Lon_{scale} (°)	0.07090000	0.45794952
H_{scale} (m)	223	500

TABLE 10.3
Rational Function Polynomial Coefficients of the First Study Area

#	Values	#	Values	#	Values	#	Values
a_1	−7.52883250E−4	b_1	1	c_1	−9.23491680E−4	d_1	1
a_2	4.60115225E−3	b_2	−1.68736561E−3	c_2	1.01134804	d_2	−1.68736561E−3
a_3	−1.03642070	b_3	1.88384395E−3	c_3	3.57115249E−4	d_3	1.88384395E−3
a_4	−3.93943040E−2	b_4	−6.55340329E−4	c_4	−1.17541741E−2	d_4	−6.55340329E−4
a_5	1.75874570E−3	b_5	2.11928788E−7	c_5	1.71162003E−3	d_5	2.11928788E−7
a_6	2.25762210E−4	b_6	−2.15886792E−7	c_6	−2.77384658E−4	d_6	−2.15886792E−7
a_7	6.47497342E−4	b_7	6.60194370E−8	c_7	−4.67286564E−5	d_7	6.60194370E−8
a_8	−1.22344418E−3	b_8	1.29969058E−6	c_8	−1.70712175E−3	d_8	1.29969058E−6
a_9	−1.95386510E−3	b_9	−6.96485750E−7	c_9	7.61699396E−7	d_9	−6.96485750E−7
a_{10}	2.70799645E−5	b_{10}	3.41030606E−7	c_{10}	2.98181047E−6	d_{10}	3.41030606E−7
a_{11}	5.10672113E−7	b_{11}	5.17265975E−10	c_{11}	8.68077245E−7	d_{11}	5.17265975E−10
a_{12}	2.05965613E−6	b_{12}	2.71171743E−10	c_{12}	1.42413537E−6	d_{12}	2.71171743E−10
a_{13}	−2.18726634E−7	b_{13}	−1.47633205E−10	c_{13}	−1.11312088E−6	d_{13}	−1.47633205E−10
a_{14}	5.40855097E−8	b_{14}	3.59570414E−10	c_{14}	2.35665930E−7	d_{14}	3.59570414E−10
a_{15}	−3.96996732E−6	b_{15}	2.28588675E−10	c_{15}	5.40107978E−7	d_{15}	2.28588675E−10
a_{16}	7.19308892E−7	b_{16}	−1.11864088E−10	c_{16}	−1.10872161E−10	d_{16}	−1.11864088E−10
a_{17}	−3.89372910E−7	b_{17}	−1.37823694E−10	c_{17}	−4.86264967E−10	d_{17}	−1.37823694E−10
a_{18}	−4.18443985E−6	b_{18}	−3.32951199E−9	c_{18}	−8.43531861E−7	d_{18}	−3.32951199E−9
a_{19}	4.50802285E−8	b_{19}	6.33689691E−10	c_{19}	−3.64346611E−8	d_{19}	6.33689691E−10
a_{20}	−1.57227534E−8	b_{20}	−5.49482473E−11	c_{20}	−2.38643375E−9	d_{20}	−5.49482473E−11

10.3.3 Results and Analysis

As shown in Figures 10.15a and 10.16a, after the processing of the proposed method, the orthorectified results (orthophoto) are obtained. To validate the accuracy and speed of the proposed rectification method, orthorectification for the same datasets was also implemented by applying the PC-based platform. On the PC-based platform, the proposed FP-RPC orthorectification is implemented using MATLAB codes.

TABLE 10.4
Rational Function Polynomial Coefficients of the Second Study Area

#	Values	#	Values	#	Values	#	Values
a_1	0.00207581	b_1	1	c_1	−0.01727220	d_1	1
a_2	0.05939323	b_2	5.06562471E−9	c_2	1.01955596	d_2	−2.43637767E−6
a_3	−1.06139835	b_3	−2.23329117E−9	c_3	0.00149223	d_3	1.76928700E−6
a_4	0.00300505	b_4	−2.21235982E−11	c_4	−0.00498582	d_4	−1.29701506E−7
a_5	9.99447200E−5	b_5	−3.85166222E−8	c_5	−0.01544990	d_5	−4.50015117E−5
a_6	4.48210655E−6	b_6	7.22875706E−11	c_6	0.00070933	d_6	1.03746933E−6
a_7	−0.00011601	b_7	1.96276656E−9	c_7	−0.00021711	d_7	−2.16896140E−6
a_8	−0.00285285	b_8	1.03082157E−8	c_8	0.01454522	d_8	2.27253718E−5
a_9	−0.00025849	b_9	4.58273834E−8	c_9	0.00146642	d_9	2.71848287E−5
a_{10}	8.66439267E−8	b_{10}	4.12786890E−12	c_{10}	−2.61417544E−6	d_{10}	−1.20465997E−7
a_{11}	1.94846265E−7	b_{11}	−8.77450090E−11	c_{11}	−2.38142758E−5	d_{11}	−8.45078827E−9
a_{12}	7.22766464E−7	b_{12}	−1.25925035E−9	c_{12}	4.37422992E−5	d_{12}	−2.25373082E−9
a_{13}	−2.45655192E−5	b_{13}	−1.41399102E−9	c_{13}	−0.00012373	d_{13}	1.66156835E−7
a_{14}	−2.15776820E−10	b_{14}	1.09626013E−12	c_{14}	3.1682470454E−7	d_{14}	−2.87815787E−10
a_{15}	3.53191253E−5	b_{15}	1.82192638E−9	c_{15}	0.00022374	d_{15}	−1.35493121E−7
a_{16}	3.58935305E−5	b_{16}	−8.74976255E−10	c_{16}	1.88815906E−5	d_{16}	−2.31068042E−8
a_{17}	−3.00220333E−9	b_{17}	2.05074275E−13	c_{17}	−1.37072926E−7	d_{17}	5.66675066E−10
a_{18}	−3.32028434E−7	b_{18}	8.02116204E−11	c_{18}	1.95362054E−5	d_{18}	−2.37161908E−9
a_{19}	−1.44250161E−7	b_{19}	−1.15845372E−10	c_{19}	2.68897003E−6	d_{19}	6.41890372E−9
a_{20}	1.88636696E−12	b_{20}	5.66352056E−16	c_{20}	−1.15446432E−9	d_{20}	−7.23372539E−12

TABLE 10.5
Scale Factors for Parameters of First Area

	τ	Range	Accuracy
Lat', Lon', Hei'	23	(−256, 255.999999881)	0.000000119
$Lat'_{off}, Lat'_{scale}, Lon'_{off}, Lon'_{scale}$	23	(−256, 255.999999881)	0.000000119
H'_{off}, H'_{scale}	23	(−256, 255.999999881)	0.000000119
$Line'_{off}, Line'_{scale}, Samp'_{off}, Samp'_{scale}$	18	(−8192, 8191.999996185)	0.000003815
a'_i, b'_i, c'_i, and d'_i $(i = 1-20)$	30	(−2, 1.999999999)	0.000000001

TABLE 10.6
Scale Factors for Parameters of Second Area

	τ	Range	Accuracy
Lat', Lon', Hei'	23	(−256, 255.999999881)	0.000000119
$Lat'_{off}, Lat'_{scale}, Lon'_{off}, Lon'_{scale}$	23	(−256, 255.999999881)	0.000000119
H'_{off}, H'_{scale}	21	(−1024, 1023.999999523)	0.000000477
$Line'_{off}, Line'_{scale}, Samp'_{off}, Samp'_{scale}$	16	(−32,768, 32,767.999984741)	0.000015259
a'_i, b'_i, c'_i, and d'_i $(i = 1-20)$	30	(−2, 1.999999999)	0.000000001

(a) (b) (c)

FIGURE 10.15 Orthoimage for the first study area (a) by FPGA, (b) by MATLAB, (c) the difference image between (a) and (b).

(a) (b) (c)

FIGURE 10.16 Orthoimage for the second study area (a) by FPGA, (b) by MATLAB, (c) the difference image between (a) and (b).

The orthorectification results obtained using the PC-based software are shown in Figures 10.15b and 10.16b. The contrast-enhanced difference images for two study areas are shown in Figures 10.15c and 10.16c, respectively. As shown in Figures 10.15c and 10.16c, the contrast-enhanced difference images show a few discrepancies that are present. The orthorectification images obtained by FPGA and PC are not visually different by inspection. The numerical differences between FPGA and PC become apparent when observing the difference images shown in Figures 10.15c and 10.16c. These pixel position differences are mainly caused by the bit width and scale factor of fixed-point data (French and Balster 2014). According to French and Balster (2014), the pixel position difference can be decreased with the increase in bit width and scale factor. Error analysis between the proposed method and the FP-RPC algorithm implemented on PC is provided in the next section.

10.4 DISCUSSION

10.4.1 Error Analysis

To quantitatively evaluate the accuracy of the proposed orthorectification method, the root-mean-square error (RMSE) (Shi and Shaker 2003; Reinartz et al. 2006) was utilized. Mathematically, the RMSEs of the image coordinates along the vertical axis (ΔI) and horizontal axis (ΔJ), and distance (ΔS) can be calculated using the following equations, respectively,

$$\Delta I = \sqrt{\frac{\sum_{h=1}^{n} (I'_h - I_h)^2}{n-1}} \quad \Delta J = \sqrt{\frac{\sum_{h=1}^{n} (J'_h - J_h)^2}{n-1}} \tag{10.15}$$

$$\Delta S = \sqrt{\frac{\sum_{h=1}^{n} (I'_h - I_h)^2 + (J'_h - J_h)^2}{n-1}} \tag{10.16}$$

where I'_h and J'_h are the image coordinates rectified by the proposed orthorectification method; I_h and J_h are the reference image coordinates; and n is the number of checkpoints.

To compute the RMSEs, 40 checkpoints for each study area were selected randomly (as shown in Figure 10.17). As shown in Figure 10.18, the differences in the values of image coordinates for the MATLAB-based and FPGA-based methods are given. Based on equations (10.15) and (10.16),

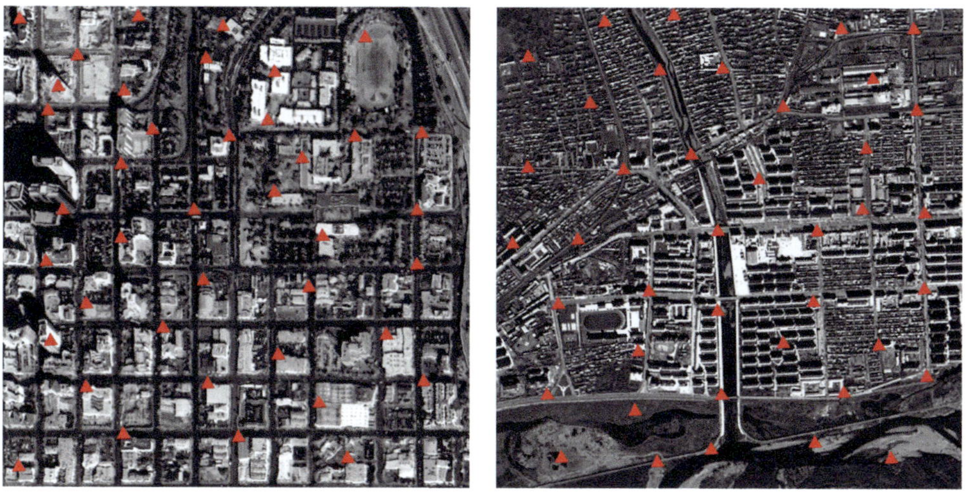

(a) (b)

FIGURE 10.17 Checkpoint distribution in (a) the first study area and (b) the second study area.

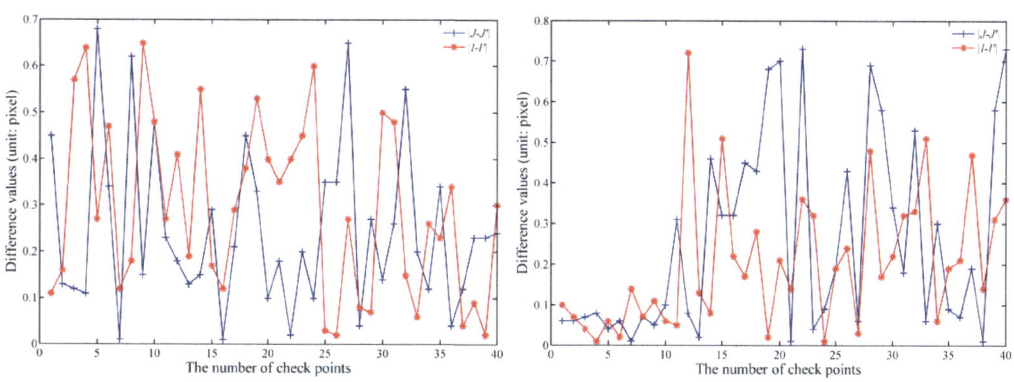

(a) (b)

FIGURE 10.18 Different statistical analyses for the FPGA-based method and MATLAB-based method for (a) the first study area and (b) the second study area.

the RMSEs (ΔI, ΔJ, and ΔS) are 0.35, 0.30, and 0.46 pixels, respectively, for the first study area; meanwhile, they are 0.27, 0.36, and 0.44 pixels, respectively, for the second study area. Moreover, other statistics were also calculated (as shown in Table 10.7).

According to the calculation results of equations (10.15) and (10.16), the orthorectification results obtained using the proposed method are considered acceptable because the RMSEs are less than one pixel (Schowengerdt 2007a, 2007b; Richards and Jia 1999). However, as shown in Figure 10.18, it should be noted that there still exist differences in the image coordinates acquired by the FPGA-based and MATLAB-based methods. These differences may be caused by the algorithms implemented by FPGA hardware, for example, the fixed-point computation, which propagate and accumulate.

10.4.2 Processing Speed Comparison

The aim of this section is to present the processing time as the increase in the size of the satellite image and evaluate the processing speed of the FPGA-based orthorectification method and the MATLAB-based method.

The processing time has been recorded as an average of 10 runs of the RPC orthorectification algorithm for each image. The average processing time for difference size of images is presented in Table 10.8. A plot of the image size vs. processing time is shown in Figure 10.19. The speed-up of the method can be defined as the MATLAB time taken divided by the time taken on the FPGA for the performance of the RPC algorithm (Senthilnath et al. 2014). In the image case considered, the maximum speed-up is acquired from a size of 1024 × 1024 pixels, where the speed-up is about 11,095.8709. From the results of Table 10.8, it can be demonstrated that the speed-up increases with the size of image.

The processing speed is one of the most important indicators for evaluating on-board processing. To evaluate and compare the speed of the proposed FPGA-based orthorectification method and the MATLAB-based method, the throughput, which is a normalized metric, is used and represents the capacity in terms of the number of pixels processed per second. For the proposed method,

TABLE 10.7
Statistical Analysis for the Different Image Coordinates Obtained by MATLAB and FPGA (Unit: pixel)

Study Area No.		Maximum	Minimum	Mean	STD
First	$\lvert I'-I \rvert$	0.65	0.02	0.29	0.19
	$\lvert J'-J \rvert$	0.68	0.01	0.25	0.18
Second	$\lvert I'-I \rvert$	0.72	0.01	0.20	0.16
	$\lvert J'-J \rvert$	0.73	0.01	0.26	0.24

TABLE 10.8
Average Processing Time for RPC Orthorectification Implementation on MATLAB and FPGA

No.	Image Size (pixels)	MATLAB Time (s)	FPGA Time (ms)	Speed-up
1	256 × 256	1.0515	0.09686	10,855.8745
2	512 × 512	4.2461	0.3875	11,008.8151
3	1024 × 1024	17.1986	1.5500	11,095.8709

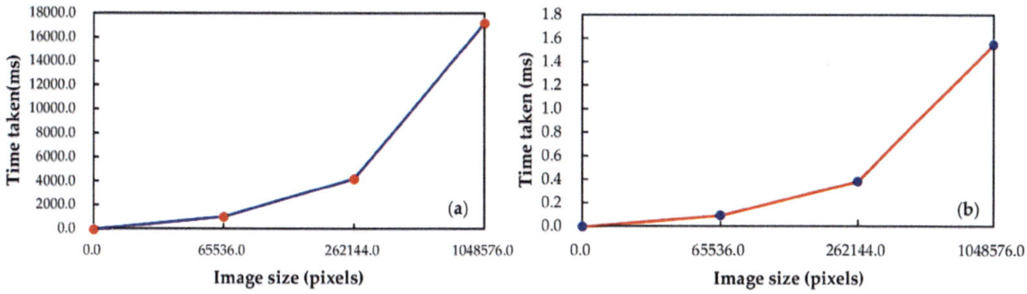

FIGURE 10.19 Image size vs. time taken to perform RPC orthorectification algorithm. (a) MATLAB-based platform and (b) FPGA-based platform.

TABLE 10.9
Utilization Ratio of Resources for the Proposed FPGA-Based Orthorectification

#		Utilization	Available	Utilization Ratio (%)
Slice logic	Slice LUTs	90,634	203,800	44.42
	Slice registers	22,798	407,600	5.59
IO		368	500	73.60

the average throughput is approximately 675.67 Mpixels/s. However, for the MATLAB-based method, the average throughput is approximately 61,677.49 pixels/s. This means that the proposed FPGA-based method has higher processing capacity than the MATLAB-based method.

10.4.3 Resource Consumption

Besides the speed of processing, the utilization ratio of each type of resource is also a key indicator when assessing the quality of a proposed method. As is well known, it can be determined whether a selected device meets the requirements of a design scheme by analyzing the utilization ratio of hardware resources. If the utilization ratio of a type of resource reaches 60%–80%, the selected device satisfies the requirement of the design scheme.

Thus, after implementing the proposed method, some important resources, such as lookup tables (LUTs), registers, and total pins, are analyzed. As shown in Table 10.9, the slice logics contain slice LUTs and slice registers. The utilization ratios of LUTs and registers are 44.42% and 5.59%, respectively. The utilization of input and output (IO) is 368, which is 73.60% of the total IOs.

In short, according to the above comprehensive utilization ratios for resources, it can be demonstrated that the resources of the selected FPGA can meet the design requirement of the proposed FPGA-based orthorectification method.

10.5 CONCLUSIONS

This chapter describes an orthorectification method, namely, the FPGA-based FP-RPC method for orthorectification on board spacecraft/satellite, in order to accelerate the orthorectification speed for remotely sensed images. The proposed FPGA-based FP-RPC method contains three main submodules: Read_parameter_mod, Coordinate_transform_mod, and Interpolation_mod based on the bilinear interpolation algorithm.

To validate the orthorectification accuracy, an orthophoto that was orthorectified by a PC-based platform (MATLAB 2012a) was used as a reference. Two datasets, IKONOS and SPOT-6 images, were used to validate the proposed FPGA-based FP-RPC method. The RMSE, which is associated with the maximum, minimum, standard deviation (STD), and mean of row and column coordinates' differences, was used. It can be concluded from these quantitative analyses that the proposed method can meet the demand of orthorectification in practice.

Moreover, a comparison of the processing speed was also performed for the proposed FPGA-based FP-RPC method and the PC-based RPC method. The throughput of the FPGA-based FP-RPC method and the PC-based RPC method are 675.67 megapixels per second and 61,070.24 pixels/s, respectively. Therefore, it can be shown that the processing speed of the FPGA-based FP-RPC method is faster (by approximately 11,000 times) than the processing speed of the MATLAB-based RPC method. In terms of resource consumptions, it can be found that the utilization ratios of ALUTs, registers, and IO are 44.42%, 5.59%, and 73.60%, respectively.

REFERENCES

Balster, E. J.; Fortener, B. T.; Turri, W. F., Integer computation of lossy JPEG2000 compression. *IEEE Transactions on Image Processing*, 20, 2011, pp. 2386–2391.

Ferrer, M. A.; Alonso, J. B.; Travieso, C. M., Offline geometric parameters for automatic signature verification using fixed-point arithmetic. *IEEE Transactions on Pattern Analysis and Machine Intelligence*, 27, 2005, pp. 993–997. https://doi.org/10.1109/TPAMI.2005.125

Fraser, C. S.; Hanley, H. B.; Yamakawa, T., Three-dimensional geopositioning accuracy of IKONOS imagery. *Photogrammetric Record*, 17, 2002, pp. 465–479. https://doi.org/10.1111/0031-868X.00199

French, J. C.; Balster, E. J., A fast and accurate orthorectification algorithm of aerial imagery using integer arithmetic. *IEEE Journal of Selected Topics in Applied Earth Observations and Remote Sensing*, 7, 2014, pp. 1826–1834. https://doi.org/10.1109/JSTARS.2013.2292009

Grodecki, J.; Dial, G., Block adjustment of high-resolution satellite images described by rational polynomials. *Photogrammetric Engineering & Remote Sensing*, 69, 2003, pp. 59–68. https://doi.org/10.14358/PERS.69.1.59

Hoja, D.; Schneider, M.; Müller, R.; Lehner, M.; Reinartz, P., Comparison of orthorectification methods suitable for rapid mapping using direct georeferencing and RPC for optical satellite data. In: Presented at the *ISPRS Conference 2008*, Peking (China), 3–11 July, 2008, pp. 1617–1624.

Reinartz, P.; Müller, R.; Lehner, M.; Schroeder, M., Accuracy analysis for DSM and orthoimages derived from SPOT HRS stereo data using direct georeferencing. *ISPRS Journal of Photogrammetry and Remote Sensing*, 60, 2006, pp. 160–169. https://doi.org/10.1016/j.isprsjprs.2005.12.003

Richards, J. A.; Jia, X., Remote sensing digital image analysis: an introduction. In *Remote Sensing Digital Image Analysis: An Introduction*, Richards, J. A.; Jia, X. Eds., Springer Berlin Heidelberg: Berlin, Heidelberg, 1999, pp. 39–74.

Schowengerdt, R. A., Chapter 7 – Correction and calibration. In *Remote Sensing*, Schowengerdt, R. A. Ed., 3rd ed.; Academic Press: Burlington, NJ, 2007a, pp. 285–XXIII.

Schowengerdt, R. A., Chapter 8 – Image registration and fusion. In *Remote Sensing*, Schowengerdt, R. A. Ed., 3rd ed.; Academic Press: Burlington, NJ, 2007b; pp. 355–XXVI.

Senthilnath, J.; Sindhu, S.; Omkar, S. N., GPU-based normalized cuts for road extraction using satellite imagery. *Journal of Earth System Science*, 123, 2014, pp. 1759–1769. https://doi.org/10.1007/s12040-014-0513-1

Shi, W.; Shaker, A., Analysis of terrain elevation effects on IKONOS imagery rectification accuracy by using non-rigorous models. *Photogrammetric Engineering & Remote* Sensing, 69, 2003, pp. 1359–1366. https://doi.org/10.14358/PERS.69.12.1359

Tao, C. V.; Hu, Y., A comprehensive study of the rational function model for photogrammetric processing. *Photogrammetric Engineering & Remote Sensing*. 67, 2001, pp. 1347–1357.

Wang, H.; Ellis, E. C., Spatial accuracy of orthorectified IKONOS imagery and historical aerial photographs across five sites in China. *International Journal of Remote Sensing*, 26, 2005, pp. 1893–1911. https://doi.org/10.1080/01431160512331326684

Yan, T.; Wang, Y.; Wang, P., Rational polynomial coefficient estimation via adaptive sparse PCA-based method. *Remote Sensing*, 16(16), 2024, p. 3018. https://doi.org/10.3390/rs16163018

Yang, G.; Xin, X.; Wu, Q., A study on ortho-rectification of SPOT6 image. In: Presented at the *2017 International Conference on Mechanical and Mechatronics Engineering (ICMME* 2017*)*, Kuala Lumpur, Malaysia, 2017.

Yang, G. D.; Zhu, X., Ortho-rectification of SPOT 6 satellite images based on RPC models. *Applied Mechanics and Materials*, 392, 2013, pp. 808–814. https://doi.org/10.4028/www.scientific.net/AMM.392.808

Zhang, G.; Qiang, Q.; Luo, Y.; Zhu, Y.; Gu, H.; Zhu, X., Application of RPC model in orthorectification of spaceborne SAR imagery. *Photogrammetric Record*, 27, 2012, pp. 94–110. https://doi.org/10.1111/j.1477-9730.2011.00667.x

Zhang, R.; Zhou, G.; Zhang, G.; Zhou, X., RPC-based orthorectification for satellite images using FPGA. *Sensors*, 18, 2018, p. 2511. https://doi.org/10.3390/s18082511

Zhou, G.; Jiang, L.; Huang, J.; Zhang, R.; Liu, D.; Zhou, X.; Baysal, O., FPGA-based on-board geometric calibration for linear CCD array sensors. *Sensors*. 18, 2018, pp. 1794–1812. https://doi.org/10.3390/s18061794

11 True² Orthoimage Map Generation

11.1 INTRODUCTION

Many investigations have demonstrated that the generation of high-resolution urban digital orthophotomaps (DOMs) using the existing procedures and algorithms, proposed in 1990 by the U.S. National Digital Orthophoto Program, has encountered many problems, such as incomplete orthorectification, occlusion, ghost image, and shadow. A comprehensive discussion regarding these problems can be found in Zhou et al. (2004, 2005). Thus, the generation of the so-called true orthophotomaps (TOMs) has become obligatory at the end of the 20th century and the beginning of the 21st century (Amhar et al. 1998; Schickier and Thorpe 1998; Di et al. 2019; Skarlatos 1999; Zhou et al. 2016, 2018b, 2020, 2022b; Greenfeld 2001; Haggag et al. 2018; Mayr 2002; Rau et al. 2002; Shoab et al. 2021; Jauregui et al. 2002; Siachalou 2004; Biasion et al. 2004; Shin and Lee 2021; Yao and Zhang 2006; Xie and Zhou 2008; Zhou and Jezek 2002; Zhou and Schickler 2004; Zhou 2009, 2020; Zhang et al. 2017; Huang and Zhang 2011; Yu et al. 2014; Gharibi and Habib 2018; Liu et al. 2021; Wang et al. 2022).

However, when a TOM superimposes onto a corresponding digital terrain model (DTM) in a flat and/or hilly area, the terrestrial texture is clearly visible (Figure 11.1a), but when a TOM superimposes onto a digital building model (DBM), the detailed façade textures of a building are not visible (Figure 11.1b). The impacts of these problems have significantly influenced the usefulness of TOMs in industries such as micro-climate monitoring, micro-environment analysis, and cellphone transmission station distribution, since incomplete building façade information cannot be tolerated for these applications.

In other words, traditional TOM only provides two-dimensional (2D) (*XY*) coordinates and building roofs' texture information, while the three-dimensional (3D) attributes and the textures of the building facades cannot be provided at all. For this reason, this chapter describes a new method for orthorectification of image, with which the TOM provides not only 2D (*XY*) coordinates and building roofs' texture information but also the 3D (*XYZ*) and the textures of the building's 360° facades. This is called true square orthophotomap, i.e., T²OM.

11.2 PRINCIPLES OF TRUE² ORTHOIMAGE MAP (T²OM) GENERATION

T²OM is defined as a DOM that can provide measurable 3D (*XYZ*) coordinates and textures for both the roof and the facade of a building. This means that the T²OM not only has traditional TOM characteristics but also provides the 3D geometric information (*X*, *Y*, *Z*) and the textures for the facade of a building. The method of generating a T²OM includes the four basic steps below:

1. DBM-based single-building T²OM generation, which consists of orthorectifying both the building roof and building facades: a concept named "superpixel" is proposed for the storage of building texture, building *ID*, etc.
2. DBM-based multiple-building T²OM generation: merging the DBM-based single-building T²OM, including organization of the building *ID*, building façade, building corner coordinates, etc.
3. DTM-based T²OM generation for the orthorectification of gently and continuously elevated hilly areas.
4. DTM- and DBM-based T²OM merging, which is for merging DTM- and DBM-based T²OM for the creation of an entire T²OM.

 DOI: 10.1201/9781003632108-11

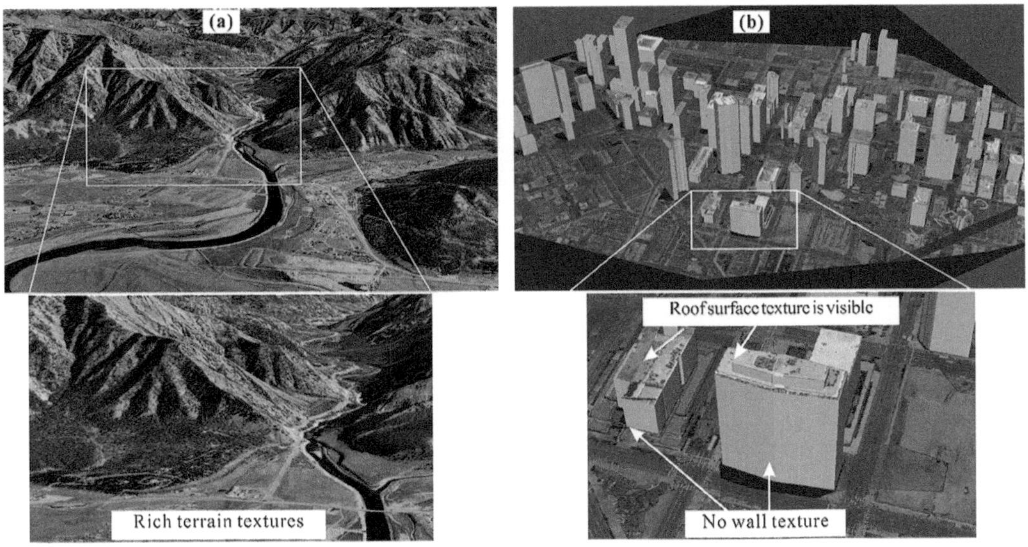

FIGURE 11.1 (a) Terrestrial textures visible in flat and hilly areas when TOM is superimposed on DTM (http://www.pcvr.com.cn/html/software/softwarei.html (accessed on August 5, 2022)), and (b) building façade texture visible when DOM is superimposed on DBM.

11.2.1 Generation of a DBM-Based Single-Building T²OM

To clearly describe the process for the generation of T²OM for buildings, a single building is first taken as an example (see Figure 11.2), presuming that the DBM for the single building and the exterior/interior orientational parameters (EOPs/IOPs) for an image are known. The steps for the generation of a T²OM for a single building are below.

11.2.1.1 Step 1: DBM-Based Building Roof Orthorectification, Which Consists Of

1. *Determining the size of the T²OM*: The resulting DBM-based single-building T²OM is expressed as a raster image with pixels arranged in rows and columns. Since the resulting orthoimage is orthorectified from raster image input (called original image) using the DBM data, the size of the output image is defined (Zhou et al. 2005) as

$$X_0 = \max\{\min\{X_D\},\min\{X_I\}\},\quad Y_0 = \max\{\min\{Y_D\},\min\{Y_I\}\}$$

$$X_1 = \min\{\max\{X_D\},\max\{X_I\}\},\quad Y_1 = \min\{\max\{Y_D\},\max\{Y_I\}\} \tag{11.1}$$

where X_0 and Y_0 are the coordinates of the lower-left corner of the output image; X_1 and Y_1 are the coordinates of the upper-right corner of the output image; X_D and Y_D are the X and Y coordinates of the DBM; X_I and Y_I are the X and Y coordinates of the original image; and max and min denote the maximum and minimum of the elements in the blanket. All of the coordinates here refer to the geodetic coordinate system required in the resulting T²OM.

2. *Computing the X, Y coordinates of each pixel*: In Figure 11.2, P (I, J) is a given point pixel on the roof of the T²OM building, and their raster rows and columns can be transformed to the coordinates of the output T²OM, i.e.,

$$X = X_0 + I \times P_x$$

$$Y = Y_0 + J \times P_y \tag{11.2}$$

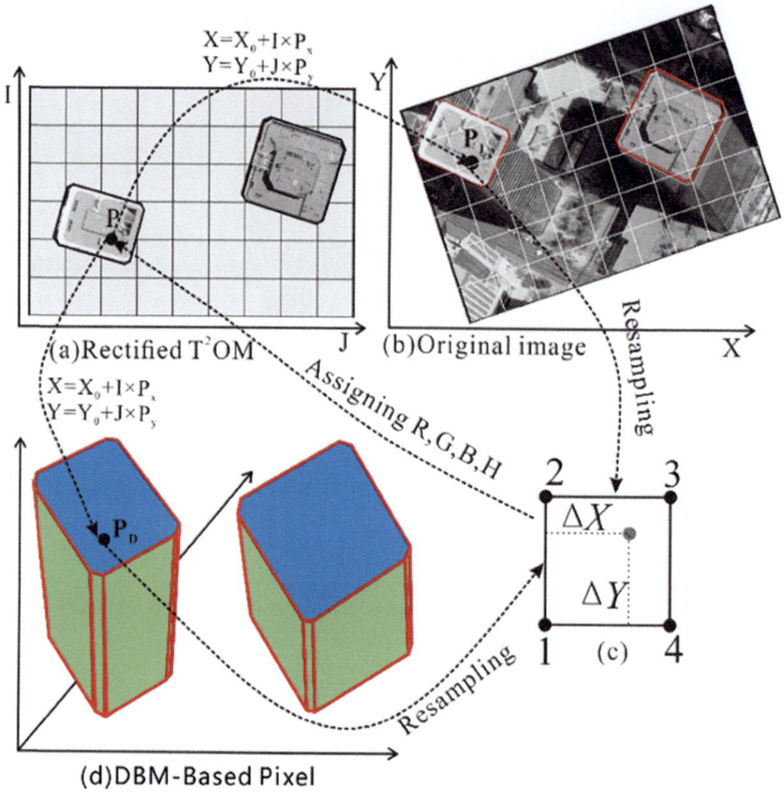

FIGURE 11.2 The procedure of DBM-based building roof orthorectification. (a) Rectified T²OM, (b) original image, (c) resampling, and (d) DBM-based pixel. (Courtesy of Zhou et al. 2022b.)

where X and Y represent the coordinates of pixels; X_0 and Y_0 are the coordinates of the lower-left corner of the output roof T²OM; P_x and P_y are the sizes of the pixels in the X and Y directions, respectively; and I and J are the rows and columns of P points, respectively.

3. *Computing the Z coordinate of P (I, J)*: To perform orthorectification, we also need to know the Z coordinates of the pixel P (I, J) in the output roof T²OM and this is obtained from DBM. However, DBM data only have vector coordinates at corner points. Therefore, it is necessary to interpolate an elevation to the roof pixel of the building. As shown in Figure 11.3, the elevation (height) is obtained only for pixels with corner points (blue pixels in Figure 11.3), while the other pixels (orange in Figure 11.3) are calculated by

$$h = a_0 j - a_1 i - a_2 \tag{11.3}$$

where h is the raster height value; i, j is the row number of the raster; a_0, a_1, and a_2 are the equation weight values defined by

$$Aj + Bi + Ch + D = 0 (C \neq 0) \tag{11.4}$$

$$h = -\frac{A}{C} j - \frac{B}{C} i - \frac{D}{C} \tag{11.5}$$

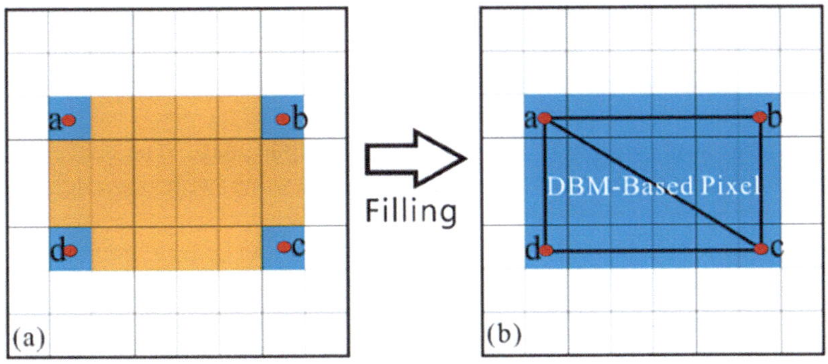

FIGURE 11.3 Assigning an elevation value to the building's roof. (a) Before the elevation value is filled, (b) after the elevation value is filled, where *a*, *b*, *c*, and *d* represent the corners of DBM.

$$a_0 = -\frac{A}{C}, \quad a_1 = -\frac{B}{C}, \quad a_2 = -\frac{D}{C} \tag{11.6}$$

where each triangular surface weight can be calculated from equation (11.2) in the case of three known vertices of the triangular surface and *A*, *B*, and *C* are the equation weight values.

When the pixels on the roof of the building have an elevation, we have to convert the pixel's geodetic coordinates back into the column and row by

$$R_D = (X - X_{0D}) / P_D, \quad C_D = (Y - Y_{0D}) / P_D \tag{11.7}$$

where P_D denotes the pixel size of the DBM image. X_{0D} and Y_{0D} are the lower-left corner coordinates of the DBM image. K_D and L_D will not generally be exact integers. Thus, an interpolation must be performed to determine Z. Usually, a bilinear interpolation method of the following form is employed (see Figure 11.2c):

$$Z = \left\{ \begin{array}{l} \left[\left[Z_1 \Delta X + (1 - \Delta X) Z_4 \right] + \left[Z_2 \Delta X + (1 - \Delta X) Z_3 \right] \right. \\ \left. + \left[Z_1 \Delta Y + (1 - \Delta Y) Z_4 \right] + \left[Z_2 \Delta Y + (1 - \Delta Y) Z_3 \right] \right] \end{array} \right\} / 4 \tag{11.8}$$

where $\Delta X = R_D - R_m$ and $\Delta Y = C_D - C_m$, in which R_m is the R_D rounded to its maximal integer and C_m is C_D rounded to its maximal integer. After this estimation, we then know the coordinate (*X*, *Y*, *Z*) of the pixel.

4. *Computing the corresponding coordinate in the original image*: To orthorectify the source image, the corresponding coordinate of the source image pixel in the output image is calculated by

$$x_I = x_{0I} - f \frac{a_1 (X - X_s) + b_1 (Y - Y_s) + c_1 (Z - Z_s)}{a_3 (X - X_s) + b_3 (Y - Y_s) + c_3 (Z - Z_s)}$$

$$y_I = y_{0I} - f \frac{a_2 (X - X_s) + b_2 (Y - Y_s) + c_2 (Z - Z_s)}{a_3 (X - X_s) + b_3 (Y - Y_s) + c_3 (Z - Z_s)} \tag{11.9}$$

where x_I and y_I are the corresponding coordinates of the pixel *P*(*X*, *Y*) in the source image; X_s, Y_s, and Z_s are the exposure stations; *f* is the focal length; $a_i = \{a_1, a_2, a_3\}$, $b_i = \{b_1, b_2, b_3\}$, and $c_i = \{c_1, c_2, c_3\}$ are the elements of the rotation matrix and are the functions of the three exterior orientation angles (ϕ, ω, κ). These elements have to be computed using at least three ground control points (GCPs).

5. *Assigning the gray value to pixels*: Since the grid of pixels in the source image rarely matches the grid of the output orthoimage, a resampling of the pixels has to be performed in order to assign gray value to the pixels in the output image. The nearest neighbor is employed because it directly transfers the original data values without averaging them. The computational procedure is illustrated in Figure 11.2.
6. *Storing the data for DBM-based building roof T^2OM*: As can be seen from the above, T^2OM needs to store more information than the traditional TOM does, such as building roof texture, facade texture, and façade *Z* coordinates. For this reason, "superpixel" is presented and has the following characteristics (see Figure 11.4): (1) it inherits the original image gray information; (2) the gray value, elevation, building *ID*, and facade texture index *ID* are stored; (3) each pixel coordinate is directly interconnected with the building *ID* and façade texture *ID*.

The detailed descriptions of *G*, *S*, *H*, and *ID* in the superpixel are as follows:

1. *G* (*I*, *J*) stands for the storage of the gray at *i*th row and *j*th column in the image coordinate systems whose value is 0–255.
2. *S* (*I*, *J*) stands for the storage of the building corner coordinate subdivision grid identification value, which occupies 8 bits. That is, by dividing a single pixel into 256 subdivision sequences, the accuracy of vector to grid data conversion is improved. For a given point P (X_P, Y_P), this can be expressed as (i_p, j_p) after the conversion of the vector to the grid. The lost information is $(X_P - X_0) - i_p \cdot \Delta X$. With this information, *Sx* can be calculated through equation (11.8).

$$S_x = \frac{\left(X_P - X_0 - \Delta x \cdot i_p\right) \cdot 16}{\Delta x} \tag{11.10}$$

where ΔX is the image resolution and . is the function of rounding down. (X_0, Y_0) is the top-left point of the image.

The calculation method used for S_y is the same as that used for S_x. At this time, S is expressed as two values. This makes the storage of *S* very difficult. Therefore, (S_x, S_y) is converted into a one-dimensional form by means of the Morton code (Zhou et al. 2018a).
3. *H* (*I*, *J*) stands for the storage of the building height or DTM height in a floating format.
4. *ID* stands for the storage of building *ID*. An *ID* can be used to call for the facade texture. A large city may have hundreds of buildings; therefore, 12 bits are designed to store 0–4095 buildings.

The T^2OM data for the roof of the building generated by the above steps are stored in the superpixel and shown in Table 11.1.

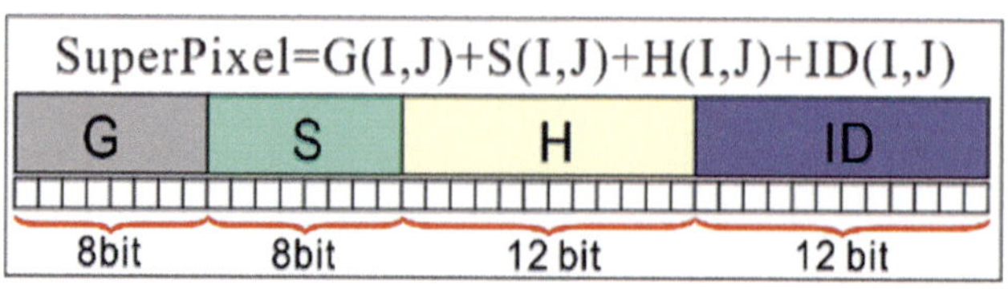

FIGURE 11.4 Superpixel occupies a total of 40 bits in the computer, where *I*, *J* represent the row and column, *G* (*I*, *J*) denotes gray, *S* (*I*, *J*) denotes the corner coordinate subdivision grid identification value, *H* (*I*, *J*) denotes elevation, and *ID* (*I*, *J*) represents the identification of a building.

TABLE 11.1
The Superpixel for the Roof Texture Data-Generated DBM-Based Building

DBM-Based Pixel	Row	Column	Gray Value	Subordinal	Height	Building ID
*BP*1	*I*(*BP*1)	*J*(*BP*1)	*G*1	$id_{(BP1)}(r, c)$	$h_{(BP1)}$	B_1
*BP*2	*I*(*BP*2)	*J*(*BP*2)	*G*2	0	$h_{(BP2)}$	B_1
…	…	…	…	…	…	…
BPi	*I*(*BPi*)	*J*(*BPi*)	*Gi*	$id_{(BPi)}(r, c)$	$h_{(BPi)}$	B_1
BPi + 1	*I*(*BPi* + 1)	*J*(*BPi* + 1)	*Gi* + 1	0	$h_{(BPi+1)}$	B_1
…	…	…	…	…	…	…

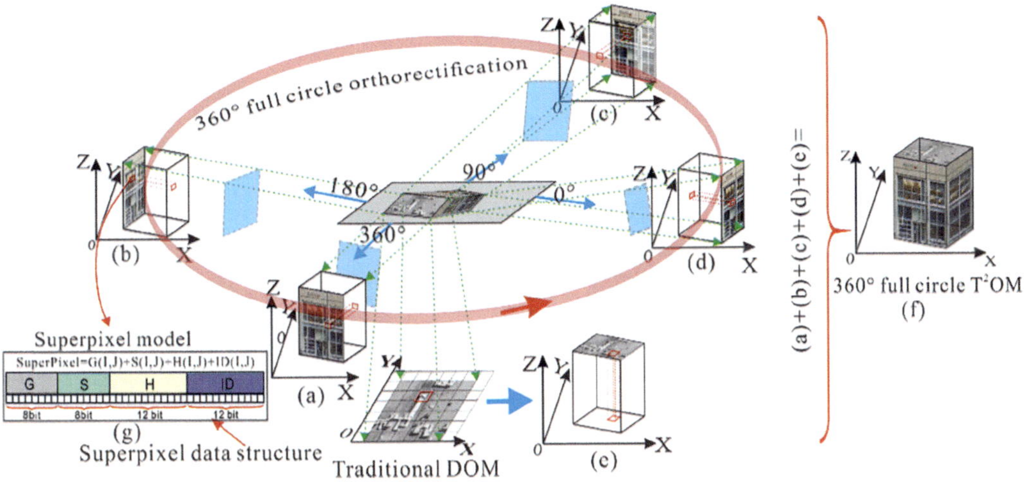

FIGURE 11.5 True three-dimensional (360°) full-circle T²OM generation. (a) Orthorectification for a building facade in 360° direction, (b) orthorectification for a building facade in 180° direction, (c) orthorectification for a building facade in 90° direction, (d) orthorectification for a building facade in 0° direction, (e) orthorectification for the building's roof, (f) the generation of T²OM for a building after a full circle (360°) orthorectification, and (g) explanation of a superpixel data structure. (Courtesy of Zhou et al. 2022b.)

11.2.1.2 Step 2: DBM-Based Orthorectification for a Single-Building Facade

To obtain the façade texture and the 3D coordinates of a building, the orthorectification for building facade in four directions is also performed in T²OM, and the four directions are determined according to the minimum bounding box (a detailed description has been given by Chan and Tan (2001)). The basic idea for the orthorectification of a building façade is: four directions at 0° for the front façade (Figure 11.5a), 90° for the left façade (Figure 11.5b), 180° for the back façade (Figure 11.5c), and 270° for the right façade (Figure 11.5d) are orthorectified, respectively. For example, the collinear equation for the 0° directional facade texture is adopted (see Figure 11.6a), i.e.,

$$y_g = -f\frac{a_1(Y_G - Y_S) + b_1(Z_G - Z_S) + c_1(X_G - X_S)}{a_3(Y_G - Y_S) + b_3(Z_G - Z_S) + c_3(X_G - X_S)} + y_0$$

$$z_g = -f\frac{a_2(Y_G - Y_S) + b_2(Z_G - Z_S) + c_2(X_G - X_S)}{a_3(Y_G - Y_S) + b_3(Z_G - Z_S) + c_3(X_G - X_S)} + z_0 \qquad (11.11)$$

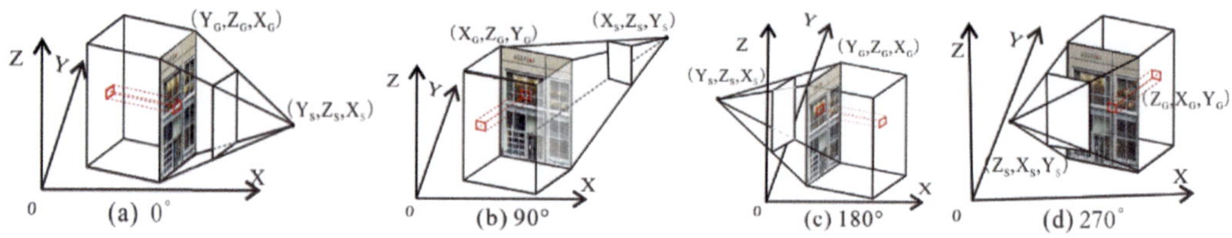

FIGURE 11.6 Orthorectification methods for the building facade through four different directions. (a) Orthorectification for a building facade in 0° direction, (b) orthorectification for a building facade in 90° direction, (c) orthorectification for a building facade in 180° direction, and (d) orthorectification for a building facade in 270° direction. (Courtesy of Zhou et al. 2022b.)

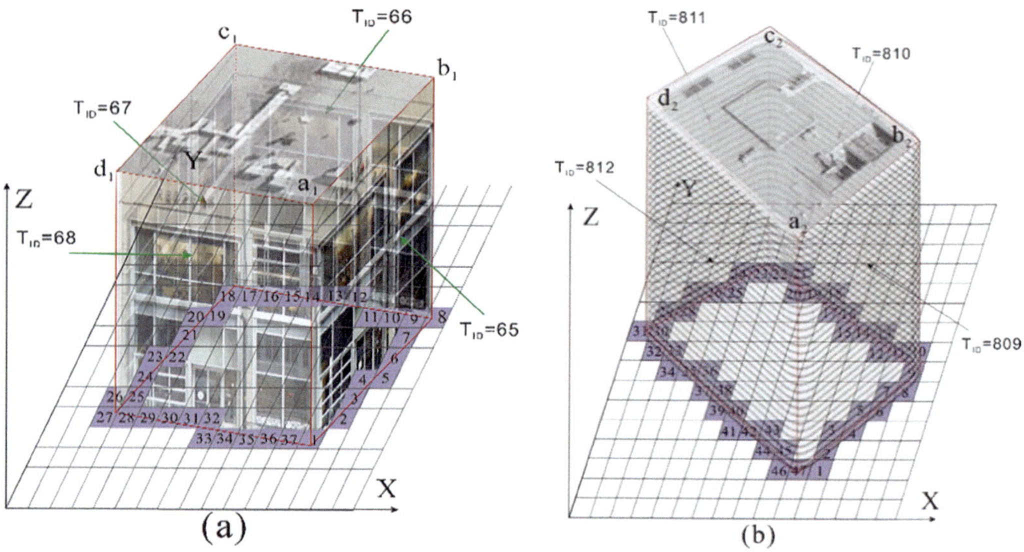

FIGURE 11.7 Building 3D spaghetti data structure, where *TID* represents the wall texture *ID* (e.g., $T = 809$); a_1, b_1, c_1, d_1, a_2, b_2, c_2, and d_2 represent building's roof corners. (a) 3D data structure of a building with a height of 25.8 m; (b) 3D data structure of a building with a height of 46.4 m. (Courtesy of Zhou et al. 2022b.)

The facade texture in 0° direction is orthorectified into the *ZOY* plane (see Figure 11.6a). Similarly, the collinear equation for the orthorectification of the other three directions at 90°, 180°, and 270° can be orthorectified into the *ZOX* (see Figure 11.6b), *YOZ* (see Figure 11.6c), and *YOX* (see Figure 11.6d) planes, respectively.

The details of the orthorectification of a single building can be further described as follows: As shown in Figure 11.7, the buildings in Figure 11.7a and b are 25.8 and 46.4 m, respectively, with an elevation resolution of 0.2 m. There are four planes (a_1b_1, b_1c_1, c_1d_1, and d_1a_1) for the building facade in Figure 11.7a, and the corresponding texture index data are 65, 66, 67, and 68, respectively. There are four planes (a_2b_2, b_2c_2, c_2d_2, and d_2a_2) for the building facade in Figure 11.7b, and the corresponding texture index data are 809, 810, 811, and 812, respectively (see Tables 11.2 and 11.3).

The buildings in Figure 11.7a and b have 37 and 47 façade superpixels, respectively, and the superpixels are stored for each pixel as shown in Tables 11.2 and 11.3. The corresponding pixels, textures, and elevations of each facade of the building in Figure 11.7 can be obtained, and the detailed storage contents are shown in Tables 11.4 and 11.5.

TABLE 11.2
The Building Wall Texture Superpixel Data in Figure 11.7a

Pixel	Gray (8 bit)	Height (*H*, 12 bit)	Texture ID (*TID*, 12 bit)	Notes	
1	11000110	000100000010	000001000001	Gray = 198	Wall 1
2	11010011	000100000010	000001000001	Gray = 211	*H* = 25.8
3	11010000	000100000010	000001000001	Gray = 208	*TID* = 65
…	…	…	…	…	
8	11010011	000100000010	000001000001	Gray = 211	
9	11001000	000100000010	000001000010	Gray = 200	Wall 2
10	11101000	000100000010	000001000010	Gray = 232	*H* = 25.8
…	…	…	…	…	*TID* = 66
18	11011101	000100000010	000001000010	Gray = 221	
19	11011101	000100000010	000001000011	Gray = 221	Wall 3
20	11011000	000100000010	000001000011	Gray = 216	*H* = 25.8
…	…	…	…	…	*TID* = 67
27	11011000	000100000010	000001000011	Gray = 216	
28	11010001	000100000010	000001000100	Gray = 209	Wall 4
29	11011010	000100000010	000001000100	Gray = 218	*H* = 25.8
…	…	…	…	…	*TID* = 68
37	11011010	000100000010	000001000100	Gray = 218	

TABLE 11.3
The Building Façade Texture Superpixel Data in Figure 11.7b

Pixel	Gray (8 bit)	Height (*H*, 12 bit)	Texture ID (*TID*, 12 bit)	Notes	
1	11010101	000111010000	001100101001	Gray = 213	Wall 1
2	11111101	000111010000	001100101001	Gray = 253	*H* = 46.4
3	11110001	000111010000	001100101001	Gray = 241	*TID* = 809
…	…	…	…	…	
9	11010100	000111010000	001100101001	Gray = 212	
10	11010111	000111010000	001100101010	Gray = 215	Wall 2
11	11010101	000111010000	001100101010	Gray = 213	*H* = 46.4
…	…	…	…	…	*TID* = 810
22	11010101	000111010000	001100101010	Gray = 213	
23	11011111	000111010000	001100101011	Gray = 223	Wall 3
24	11010111	000111010000	001100101011	Gray = 215	*H* = 46.4
…	…	…	…	…	*TID* = 811
30	11010111	000111010000	001100101011	Gray = 215	
31	11010101	000111010000	001100101100	Gray = 213	Wall 4
32	11010111	000111010000	001100101100	Gray = 215	*H* = 46.4
…	…	…	…	…	*TID* = 812
47	11010101	000111010000	001100101100	Gray = 213	

11.2.2 Generation for DBM-Based Multiple-Building T²OMs

There are usually many buildings in a city. Therefore, the next step is to generate the multiple-building T²OM on the basis of the generation of a single-building T²OM. Part 2 in Figure 11.8 shows the process of generation for multiple-building T²OMs, in which each building is assigned a unique identification (BuindingID) and the building *ID* is used to control the display and hiding of each

TABLE 11.4
The Building Wall Texture Index in Figure 11.7a

Wall ID	Wall Pixel Index	Texture Index	Height
a_1b_1	1, 2, 3, 4, 5, 6, 7, 8	65	25.8
b_1c_1	9, 10, 11, 12, 13, 14, 15, 16, 17, 18	66	25.8
c_1d_1	19, 20, 21, 22, 23, 24, 25, 26, 27	67	25.8
d_1a_1	28, 29, 30, 31, 32, 33, 34, 35, 36, 37	68	25.8

TABLE 11.5
The Building Wall Texture Index in Figure 11.7b

Wall ID	Wall Pixel Index	Texture Index	Height
a_2b_2	1, 2, 3, 4, 5, 6, 7, 8, 9	809	46.4
b_2c_2	10, 11, 12, 13, 14, 15, 16, 17, 18, 19, 20, 21, 22	810	46.4
c_2d_2	23, 24, 25, 26, 27, 28, 29, 30	811	46.4
d_2a_2	31, 32, 33, 34, 35, 36, 37, 38, 39, 40, 41, 42, 43, 44, 45, 46, 47	812	46.4

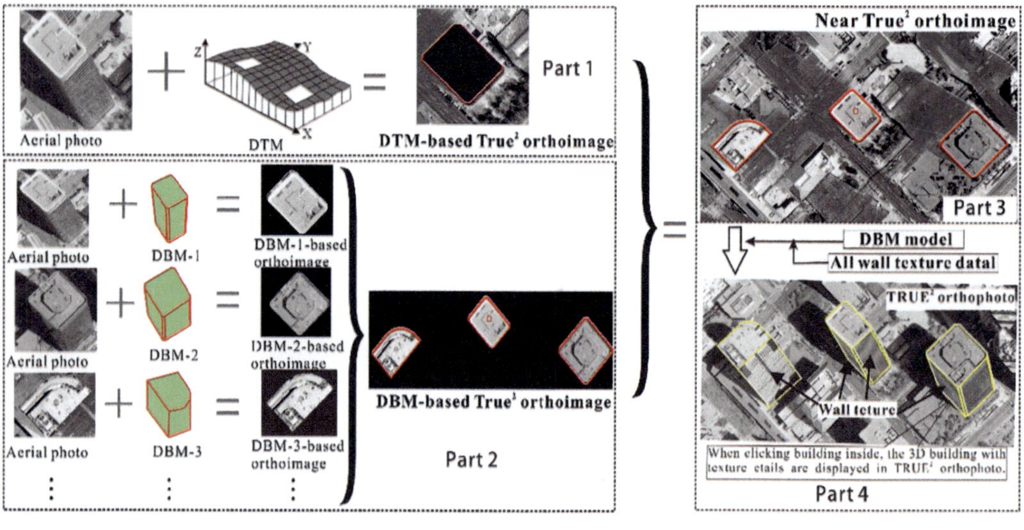

FIGURE 11.8 DBM-based generation for multiple-building T²OM and merged DTM-based and DBM-based T²OM. (Courtesy of Zhou et al. 2022b.)

building (see Table 11.6). In addition, the information of a single-building model is divided into top surface, elevation, and bottom surface. Additionally, each roof and facade is assigned a separate identity (RoofID, WallID) (see Tables 11.7 and 11.8). The face *ID*s are associated with the building top surface table and building wall table, and each face point *ID* (Points) is recorded in the roof table and wall table, and the point *ID*s are associated with the building corner point information table (see Table 11.9). The corner point information is expressed using multiple horizontal projection polygons, and each corner point is provided with 3D vector information in the corner point information table. Building textures are divided into top surface textures and wall textures in the table. In the building, the top surface table does not need to identify the top surface texture identity; its information can be associated with the 2D T²OM multiple-building data table in having the building identity (BuindingID) from which the top surface texture value and 3D coordinate information

TABLE 11.6
Building Relationships

Building ID	Type	Roof ID	Wall ID	Properties	Others
B_1	Volume	R_{11}	$W_{11}, W_{12}, W_{13}\ldots$	Brick structure	...
B_2	Volume	R_{12}	$W_{21}, W_{22}, W_{23}\ldots$	Reinforced concrete structure	...
B_3	Volume	R_{13}	$W_{31}, W_{32}, W_{33}\ldots$	Steel structure	...
...	...	...	...	...	...

TABLE 11.7
The Relationship of Building Roof Textures

Roof ID	Type	Texture ID	Point ID	Others
R_{11}	Polygon	$T^R{}_{11}$	$P^R{}_{11}, P^R{}_{12}, P^R{}_{13}\ldots$	...
R_{12}	Polygon	$T^R{}_{12}$	$P^R{}_{21}, P^R{}_{22}, P^R{}_{23}\ldots$	...
R_{13}	Polygon	$T^R{}_{13}$	$P^R{}_{31}, P^R{}_{32}, P^R{}_{33}\ldots$	...
...	...	...	...	...

TABLE 11.8
The Relationship of Building Wall Textures

Wall ID	Type	Texture ID	Point ID	Others
W_{11}	Polygon	$T^W{}_{11}$	$P^w{}_{11}, P^w{}_{12}, P^w{}_{13}\ldots$	...
W_{12}	Polygon	$T^W{}_{12}$	$P^w{}_{21}, P^w{}_{22}, P^w{}_{23}\ldots$	...
W_{13}	Polygon	$T^W{}_{13}$	$P^w{}_{31}, P^w{}_{32}, P^w{}_{33}\ldots$	...
...	...	...	...	...

TABLE 11.9
Building Vertex Relationships

Point ID	*X, Y, Z* Coord.			Pixel Coord.		Others
$P^w{}_{11}$	$X^W{}_{11}$	$Y^W{}_{11}$	$Z^W{}_{11}$	$I^W{}_{11}$	$J^W{}_{11}$	...
$P^w{}_{12}$	$X^W{}_{12}$	$Y^W{}_{12}$	$Z^W{}_{12}$	$I^W{}_{12}$	$J^W{}_{12}$	...
$P^w{}_{13}$	$X^W{}_{13}$	$Y^W{}_{13}$	$Z^W{}_{13}$	$I^W{}_{13}$	$J^W{}_{13}$	...
...	...	...	...	...	...	...
$P^R{}_{11}$	$X^R{}_{11}$	$Y^R{}_{11}$	$Z^R{}_{11}$	$I^R{}_{11}$	$J^R{}_{11}$	...
$P^R{}_{12}$	$X^R{}_{12}$	$Y^R{}_{12}$	$Z^R{}_{12}$	$I^R{}_{12}$	$J^R{}_{12}$	...
$P^R{}_{13}$	$X^R{}_{13}$	$Y^R{}_{13}$	$Z^R{}_{13}$	$I^R{}_{13}$	$J^R{}_{13}$	...
...	...	...	...	...	...	...

value of the building is obtained, and the real building top surface texture is obtained by rendering a single pixel at a single vector point in turn. For the wall texture of the building, the texture ID of the wall is associated with the texture data in the database. The texture name (TextureName), the address of the file uploaded in the computer (FileAddress), the data of the texture saved in binary (Binary), the format of the texture (Format), the size of the texture (Size) recorded in bytes (Byte) in the database, and the date when the texture was saved (Date) are recorded (see Table 11.10). Parts 1, 3, and 4 in Figure 11.8 are described in detail in Sections 11.3.3 and 11.3.4.

TABLE 11.10
Data Texture Table

Texture ID	Texture Name	File Address	Date	Format	Others
T^{W}_{11}	WTN11	WFA11	WD11	WF11	...
T^{W}_{12}	WTN12	WFA12	WD12	WF12	...
T^{W}_{13}	WTN13	WFA13	WD13	WF13	...
...	...	...	...	...	...
T^{R}_{11}	PTN11	PFA11	PD11	PF11	...
T^{R}_{12}	PTN12	PFA12	PD12	PF12	...
T^{R}_{13}	PTN13	PFA13	PD13	PF13	...
...	...	...	...	...	...

11.2.3 Generation of DTM-Based T²OM

Part 1 in Figure 11.8 is DTM-based T²OM generation, which orthorectifies the displacement caused by terrestrial elevation, i.e., orthorectifies the terrain into an upright position in a given map coordinate. Therefore, the digital differential orthorectification method is applied for this purpose. The details of this method can be found in Zhou et al. (2004) and Zhou (2020). Similarly, the given DTM data, data structure, and data storage for each pixel in the DTM-based T²OM are similar to that of the DBM-based T²OM, but the building ID is assigned as "none".

11.2.4 Merging DTM- and DBM-Based T²OMS

In view of the different structures of the DBM- and DTM-based T²OMs, an entire T²OM generation needs merging algorithms (see Figure 11.8). To do this, the logic operation <or> is performed with the superpixel *ID* of 0 or non-zero. To eliminate possible boundary confusion in the merging process, the following judgment conditions are executed: with the DTM-based T²OM as the base map (see Part 1 in Figure 11.8), when the same grid number appears in the DTM- and DBM-based T²OMs at the same time, only the DBM-based T²OM is retained (see Part 2 in Figure 11.8). This is because the building area determined by the horizontal projection polygon is not a regular rectangle or divided along the grid direction. Therefore, there are actually more grid elements located at the building boundary than at the real building boundary. Retaining the grid of the DBM-based T²OM will ensure the accuracy of the building location to the greatest extent. This is helpful for the 3D T²OM display. Merge the DTM-based and DBM-based true-squared orthophotos to obtain a near-true-squared image (see Part 3 in Figure 11.8). Linking the DBM model and wall textures to the building data enables the ability to display building façade textures and 3D coordinates (XYZ) measurable at any point (see Part 4 in Figure 11.8).

11.3 EXPERIMENTS AND ANALYSIS

Figure 11.9 is a flowchart for T²OM generation, divided into five parts. Part 1 shows two experimental datasets (high-resolution images, control points, orientation parameters, DBM, and DTM) from Denver, CO, USA, and Nanning, China, which are used as input data. Part 2 shows DBM-based T²OM generation, which consists of orthorectifying both building roofs and building facades, with which "superpixel" is used for the storage of building texture, building ID, etc. Part 3 shows DTM-based T²OM generation, which is for orthorectification of gentle and continuously elevated hilly areas. Part 4 shows the merging of DBM-based and DTM-based T²OMs as the output data, i.e., T²OM. Part 5 shows the accuracy evaluation for the generated T²OM using GCPs.

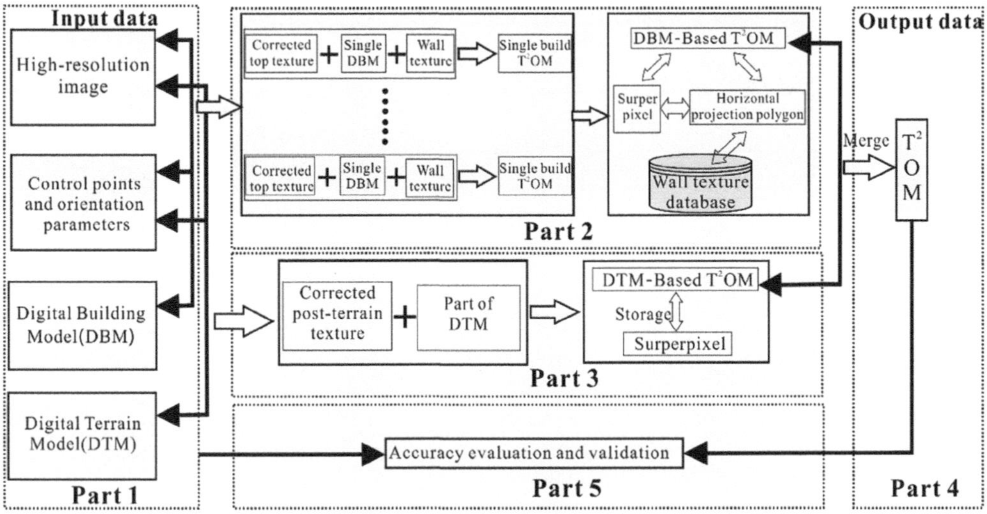

FIGURE 11.9 The proposed flowchart for T²OM generation.

TABLE 11.11
The File Format of the T²OM *Fus*

File Section	Properties	Description
File flag block	m_FileProperty	Identifier "*fus*" (char type)
	m_Version	Version number (int type)
Image header information	m_UpleftCoordinateX	Image coordinate lower right *X* value (double type, units: meters)
	m_UpleftCoordinateY	Image coordinate lower right *Y* value (double type, units: meters)
	m_TMaxZ	The highest point in the DTM file (double type, in meters)
	m_TMinZ	The lowest point in the DTM file (double type, in meters)
	m_BMaxZ	Maximum building height in DBM (double type in meters)
	m_BMinZ	Minimum building height in DBM (double type in meters)
	m_IntervalX	Unit interval in *X*-axis direction (double type in meters)
	m_IntervalY	Unit interval in *Y*-axis direction (double type, units: meters)
	m_FileHigh	Image height (int type)
	m_FileWidth	Image width (int type)
	Z_Tresolution	Topographic data unit elevation level (double type, units: meters)
	Z_Bresolution	Building data unit elevation level (type double, in meters)
	Build_Num	Number of building object elements (type int)
Image pixels information	T²OM_Gray	Pixel gray component (unsigned char type)
	T²OM_Ordinal	Subdivision grid order (unsigned char type)
	T²OM_Height	Elevation level high 8 bits (unsigned char type)
	T²OM_HI	Elevation level low 4 bits, logo high 4 bits (unsigned char type)
	T²OM_Index	Marker data low 8 bits (unsigned char type)

11.3.1 Metadata of T²OM

This experiment was implemented using the programming language C++. To store the T²OM in binary form, a file format *fus* needed to be designed to store superpixels due to the restriction of traditional bit-storing in computers. Table 11.11 shows the entire file format, which consists of a file flag block, image header file information, and image pixel information. These three parts were written to the *fus* file in binary form.

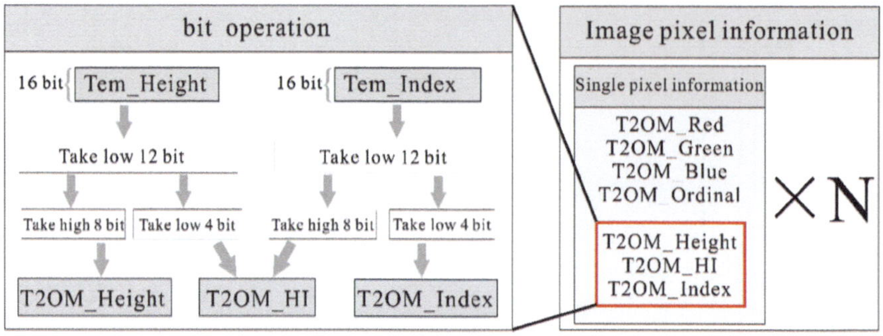

FIGURE 11.10 Saving pixel information by bit manipulation. (Courtesy of Zhou et al. 2022b.)

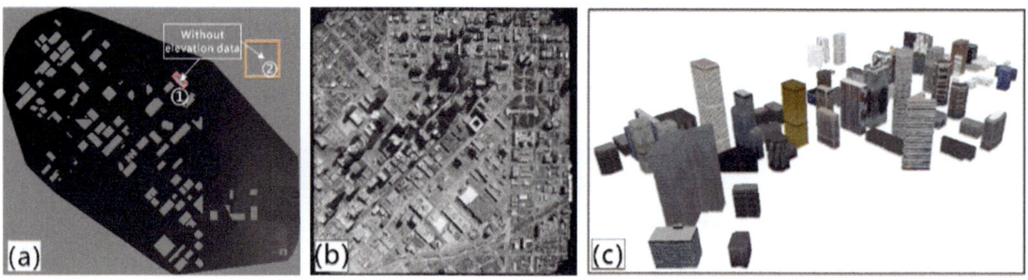

FIGURE 11.11 The experimental dataset 1. (a) DTM data, where areas ① and ② represent the areas without elevation data, (b) aerial image data, and (c) DBM data.

In the superpixel data structure, information about the elevation level *H* and building identification *ID* need 12 bits of memory. However, there is no data type of this size in the computer. Therefore, 12 bits of information are saved by the bit operation. As shown in Figure 11.10, this method first opens up two unsigned short types of data (16 bit), TEM_Height and TEM_Index, and uses them to record *H* and *ID*, respectively. Then, the data are put into variables (T^2OM_height, T^2OM_hi, and T^2OM_index) through the bit operation. Finally, the information in the superpixel is combined, as shown on the right side of the figure.

11.3.2 T^2OM Generation

This section uses Dataset 1 and Dataset 2 to describe the generation process of generating T^2OM and to verify the feasibility of the method in this chapter.

11.3.2.1 Experimental Result with Dataset 1

The experimental Dataset 1 includes digital surface model data, aerial imagery data, and DBM data. A brief description is given as follows:

1. *DTM data*: Figure 11.11a shows DTM data from Denver, CO, USA, which is represented as a height–depth map, where the darker the color is, the lower the height, and vice versa, because the topography of the city is relatively flat. Thus, the elevations shown on the ground are relatively similar (the colors shown are similar). The accuracy of plane surface coordinates and vertical coordinates is about 0.1 and 0.2 m, respectively. The horizontal datum is GRS 1980, and the vertical datum is NAD83.

2. *Aerial image data*: Figure 11.11b shows the original aerial image acquired using the RC30 aerial camera lens in Denver, CO. The flight altitude in Denver is 1650 m higher than the mean ground elevation of the imaged area. Aerial photographs were initially recorded on film and then scanned into digital format at a pixel resolution of 25 μm.
3. *DBM data:* Figure 11.11c shows Denver DBM data, and these buildings with a ground resolution of about 25.4 cm per pixel were identified. Each building model contains building corner point information and elevation texture information.

11.3.2.1.1 Step 1. Generation of DBM-Based T²OM

1. *DBM-based building roof orthorectification*: DBM-based roof orthorectification corrects only the displacements caused by the buildings and does not take into account the displacements caused by the terrain. Therefore, the generated DBM-based T²OM (see Figure 11.12) only corrects the building texture and not the texture of the terrain area, so the texture at the terrain is black (background value). Where (a) and (c) are two buildings in the T²OM, it can be seen that the roof textures of the buildings are obtained accurately.
2. *DBM-based building façade orthorectification*: To obtain the facade textures and 3D coordinates of the buildings, the orthorectification of the building facade textures was also performed in T²OM. The building facade textures for Dataset 1 were selected from the existing texture library and the same textures were used in all four directions, and the results of facade texture correction are shown in Figure 11.13. Figure 11.13a and b shows the results of the façade texture correction for (a) and (c) in Figure 11.12, respectively, and the angles marked in Figure 11.13 are consistent with those in Figure 11.12.

11.3.2.1.2 Step 2. Generation of DTM-Based T²OM

The correction of buildings needs to be followed by the orthorectification of nonbuilding areas. The DTM-based differential correction of nonbuildings is performed to obtain the corrected texture of the image. At the same time, superpixels are generated by overlaying each data information, and finally the DTM-based T²OM is obtained (see Figure 11.14). Because this part only corrects the terrain texture and not the building area texture, the texture at the building is black (background value). The DTM-based T²OM also has ghosts and shadows in the texture area, because Dataset 1 lacks complementary images, so occlusion detection, compensation for textures, and shadow detection and compensation operations are not performed in this step.

11.3.2.1.3 Step 3. Merging DTM- and DBM-Based T²OMs

Finally, the merging of the DBM with the DTM's T²OM yields the result shown in Figure 11.15b, which fills the superpixels at the roof texture exactly to the area where the original background value is black. Figure 11.15a and c shows the enlarged views of two of the buildings.

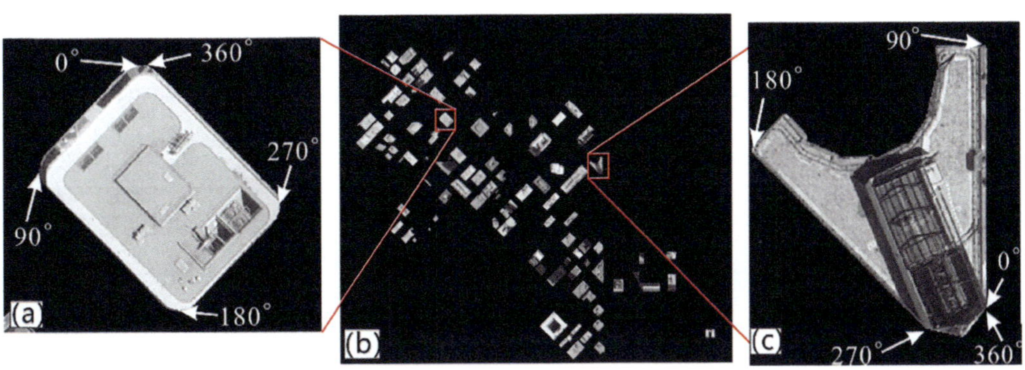

FIGURE 11.12 Generation results of DBM-based T²OM on dataset 1. (b) DBM-based T²OM and (a) and (c) are enlarged windows of the two regions.

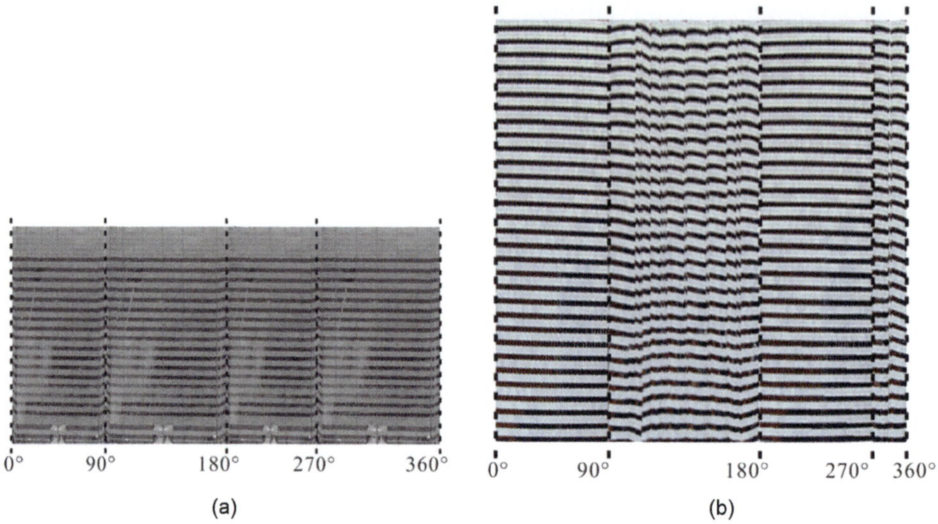

FIGURE 11.13 Façade texture orthorectification results (a) and (b) are the orthorectification results of the building Façade texture in (a) and (c) in Figure 11.12.

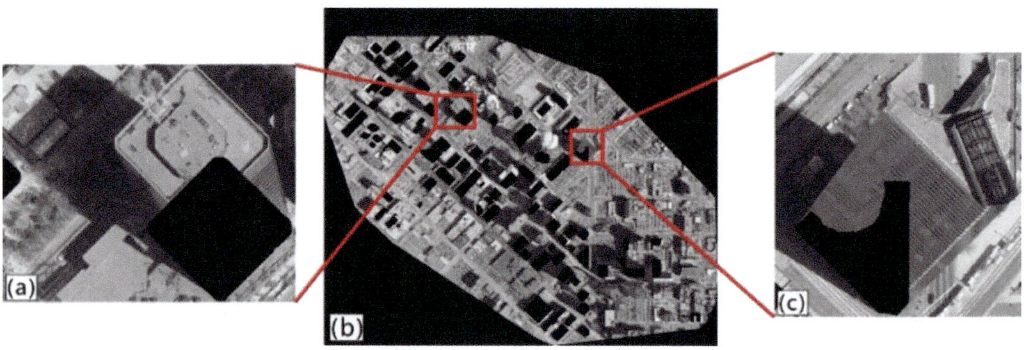

FIGURE 11.14 Generation results of DTM-based T^{2}OM on dataset 1. (b) DTM-based T^{2}OM and (a) and (c) are enlarged windows of the two regions.

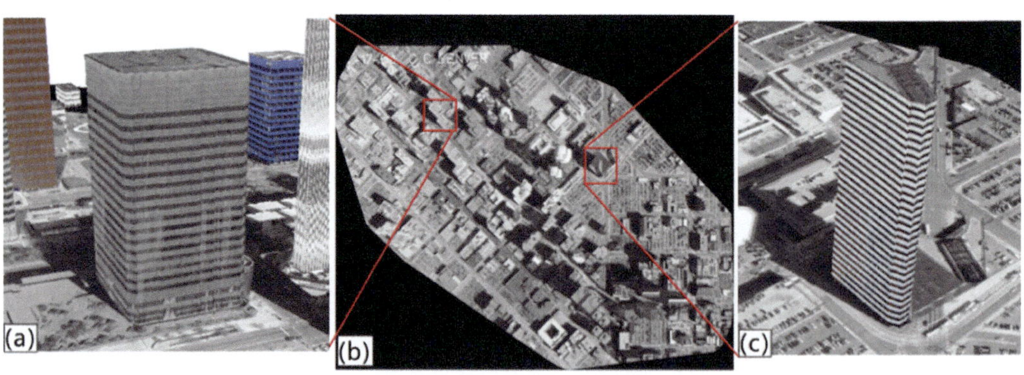

FIGURE 11.15 Merging DTM- and DBM-based T^{2}OM on dataset 1. (b) T^{2}OM and (a) and (c) are enlarged windows of the two regions.

11.3.2.2 Experimental Result with Dataset 2

The experimental Dataset 2 includes digital surface model data, aerial imagery data, and DBM data. A brief description is given as follows.

1. *DTM data*: Figure 11.16a shows DTM data from Nanning, China, represented as a height–depth map, where the darker the color is, the lower the height, and vice versa, because the topography of the city is relatively flat. Thus, the elevations shown on the ground are relatively similar (the colors shown are similar). The accuracy of the plane surface coordinates and vertical coordinates is about 0.1 and 0.2 m, respectively. The horizontal datum is GRS 1980, and the vertical datum is NAD83.
2. *Aerial image data*: Figure 11.16b shows the original aerial image acquired using the CMOS lens in Nanning. The flight altitude in Nanning is 200 m higher than the average ground elevation of the imaging area.
3. *DBM data*: Figure 11.16c shows Nanning DBM data, and buildings with a ground resolution of about 25.4 cm per pixel were identified. Each building model contains building corner point information and elevation texture information.

11.3.2.2.1 Step 1. Generation of DBM-Based T^{2}OM

1. *DBM-based building roof orthorectification*: The DBM-based roof orthorectification corrects only the displacements caused by the buildings and does not take into account the displacements caused by the terrain. Therefore, the generated DBM-based T^{2}OM (see Figure 11.17) only corrects the building texture and not the texture of the terrain area, so the texture of the terrain is black (background value). Four buildings of the T^{2}OM are denoted as (a–d), and the roof textures of the buildings are obtained accurately.
2. *DBM-based building façade orthorectification*: In order to obtain the building facade texture and 3D coordinates, the building facade texture is also orthorectified in T^{2}OM. The orthorectification results are shown in Figure 11.18. Figure 11.18a–d shows the results of the façade texture correction for (a–d) in Figure 11.17, respectively, and the angles marked in Figure 11.18 are consistent with those in Figure 11.17.

11.3.2.2.2 Step 2. Generation of DTM-Based T^{2}OM

The correction of buildings needs to be followed by the orthorectification of nonbuilding areas. The DTM-based differential correction of nonbuildings is performed to obtain the corrected texture of the image. At the same time, superpixels are generated by overlaying each data information, and finally the DTM-based T^{2}OM is obtained (see Figure 11.19). Because this part only corrects the terrain texture and not the building area texture, the texture of the building is black (background value). The DTM-based T^{2}OM also has ghosting and shadows in the texture region, and the

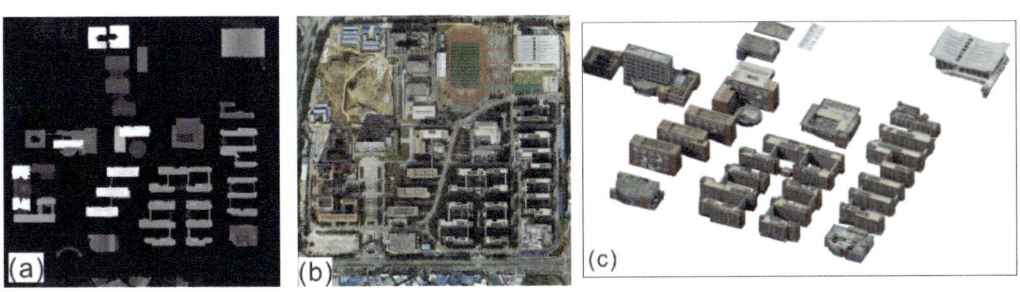

FIGURE 11.16 The experimental dataset 2. (a) DTM data, (b) aerial image data, and (c) DBM data.

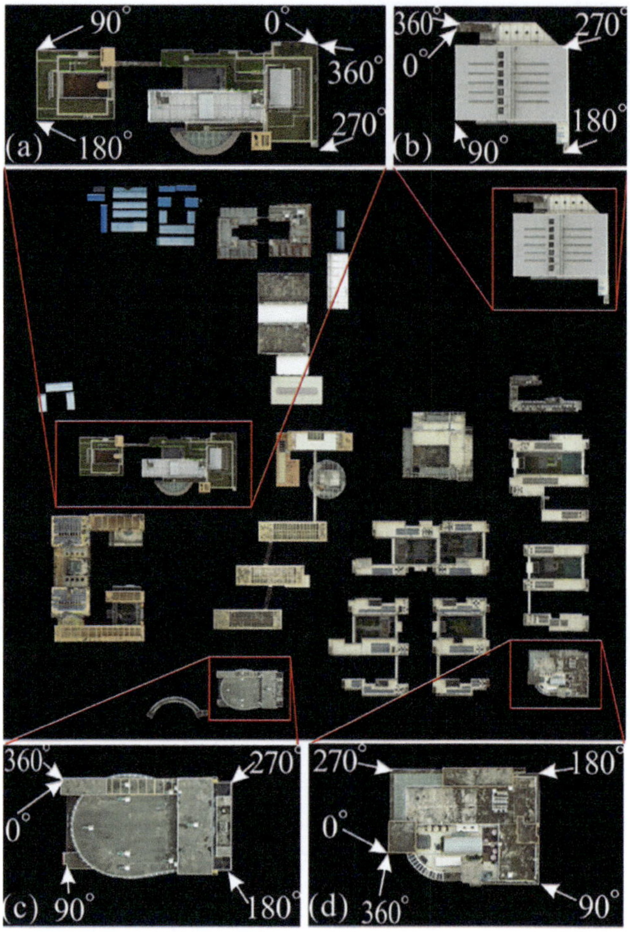

FIGURE 11.17 Generation results of DBM-based T²OM on dataset 2. (a)–(d) are enlarged windows of the four regions.

occlusion detection and compensation of textures and shadow detection and compensation operations in this chapter are adopted from Zhou et al. (2005).

11.3.2.2.3 Step 3. Merging DTM- and DBM-Based T²OMs

Finally, the merging of the DBM with the DTM's T²OM yields the result shown in Figure 11.20, which fills the superpixels at the roof texture exactly to the area where the original background value is black. Figure 11.20a–d shows the enlarged views of four of the buildings.

11.3.2.3 T²OM 3D Measurement

With the 3D measurement function, the elevation of any point in the scene can be obtained. Figure 11.21 shows elevation information from a point on the selected facade. Figure 11.22 shows the color information, true 3D coordinates, and attribute information of each pixel in the acquired point. *Selective Hide* can also be used to display individual buildings, as shown in Figure 11.23. In addition, based on the fact that the horizontal projection polygon of a building can completely record building information, we can obtain complete information about corner points and facade through the *3D Building Information Display* function, as shown in Figure 11.24. The *Building Distance Measurement* function calculates the minimum and maximum distances between two buildings by using the building corner information, as shown in Figure 11.25.

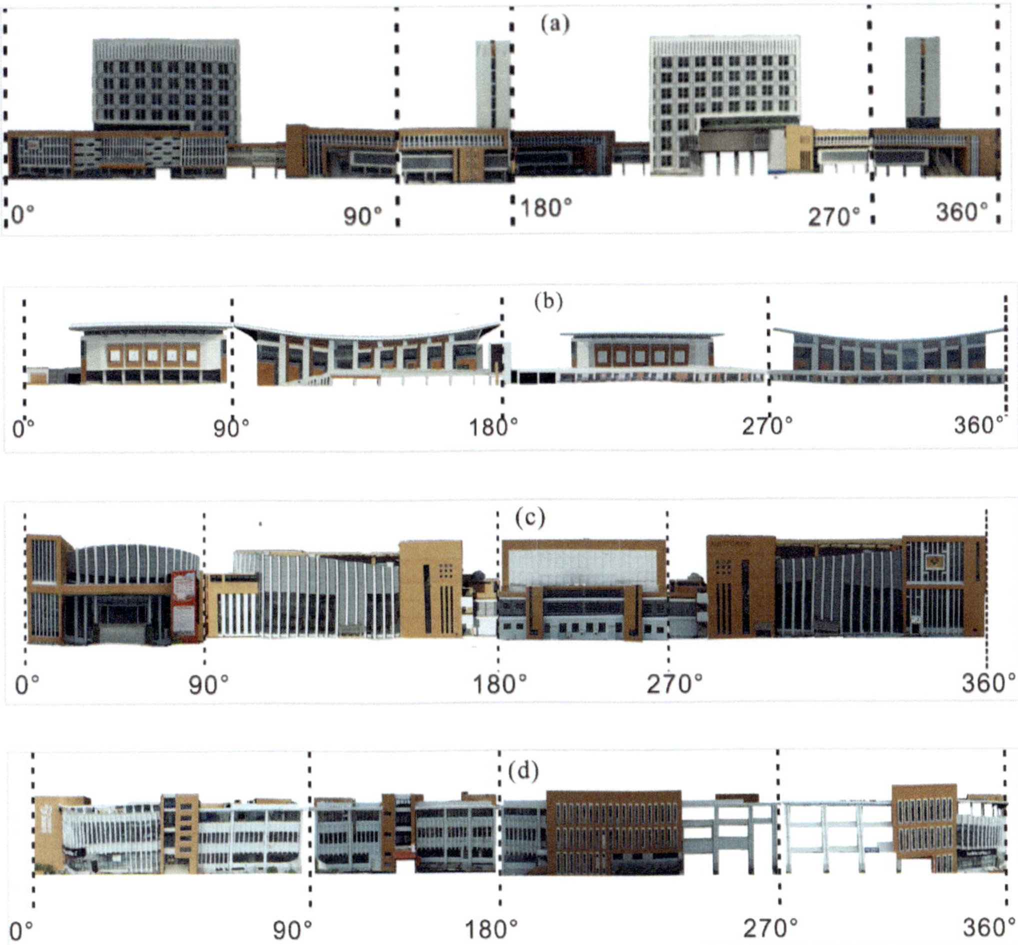

FIGURE 11.18 Façade texture orthorectification results, (a)–(d) are the orthorectification results of the building Façade texture in (a)–(d) in Figure 11.17.

11.3.3 Accuracy Evaluation and Analysis

In our method, T^2OM generation is obtained by merging the DTM- and DBM-based T^2OMs. There can be errors in multiple steps of the process. In addition, because the superpixel uses an elevation series instead of the double-form true elevation value, errors exist in the recorded height of the superpixel. Based on the generated T^2OM, this section evaluates the matching accuracy between the building horizontal projection polygons.

To ensure the accuracy of building location information in T^2OM, it is necessary to evaluate the construction accuracy of the horizontal projection polygon. In Figure 11.26, the grid detected as the building edge is represented in blue. Building corners are represented in green. The horizontally projected polygon is displayed in red. As can be seen from the enlarged Figure 11.26e–g, the extracted blue edge is completely consistent with the green building corner. This proves that the accuracy of the building horizontal projection polygon construction is sufficient.

Four buildings identified in the T^2OM data were randomly selected to evaluate the accuracy of information recording in T^2OM, as shown in Figure 11.27. First, roof corner points of each building above were extracted from the DBM. Then, the coordinates of the corner points in the DBM were compared with the grid number, subdivision, and elevation series recorded in the superpixels.

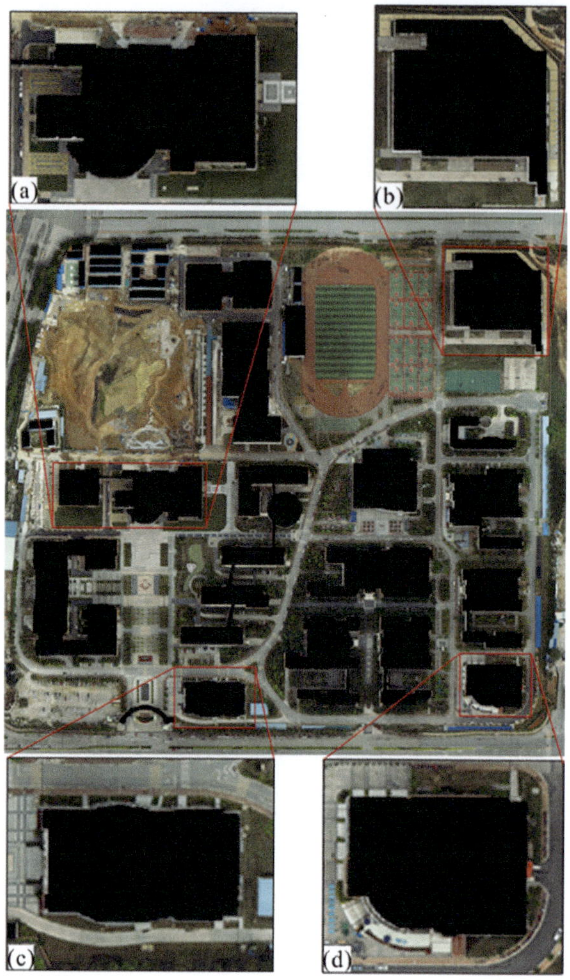

FIGURE 11.19 Generation results of DTM-based T²OM. (a)–(d) are enlarged windows of the four regions.

Table 11.12 describes the 3D coordinates before and after coding, where x_{ori}, y_{ori}, and z_{ori} are the 3D coordinates before coding; $S(r, c)$ and H are the subdivision and elevation series, respectively; and X_{bc}, Y_{bc}, and Z_{bc} are the 3D coordinates calculated using superpixels. Through calculations, the average errors of the X, Y, and Z coordinate components were determined to be 0.017, 0.025, and 0.09 m, respectively. Compared with the resolution of the original data in plane coordinates of 0.1 m and the elevation of 0.2 m, it can be concluded that the coding method recorded by superpixels can greatly reduce the level of error generated in the process of converting vector data into grid data.

11.3.4 Discussion

From the above two sets of experimental results, it is concluded that our proposed T²OM can realize the switching display of two-dimensional flat TOMs and three-dimensional buildings and use superpixels to save three-dimensional information so that the accuracy of three-dimensional measurement can be controlled within 0.0625 m.

In addition, it is feasible to expand the traditional pixel storage method to increase the amount of information expressed by a grid image. In our method, first, a large number of heterogeneous data are unified to achieve centralized management. This makes the reconstruction and display of

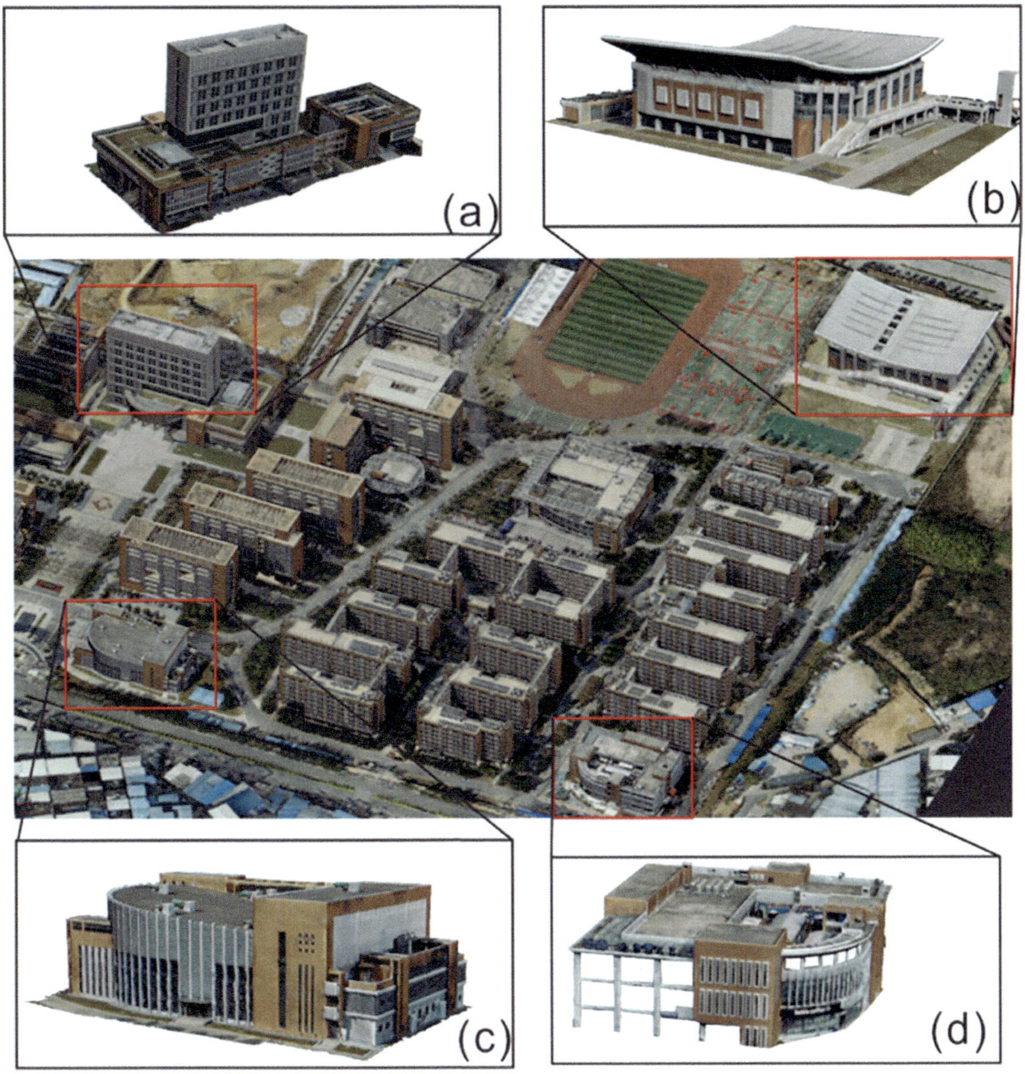

FIGURE 11.20 Merging DTM- and DBM-based T^{2}Oms on dataset 2. (a)–(d) are enlarged windows of the four regions.

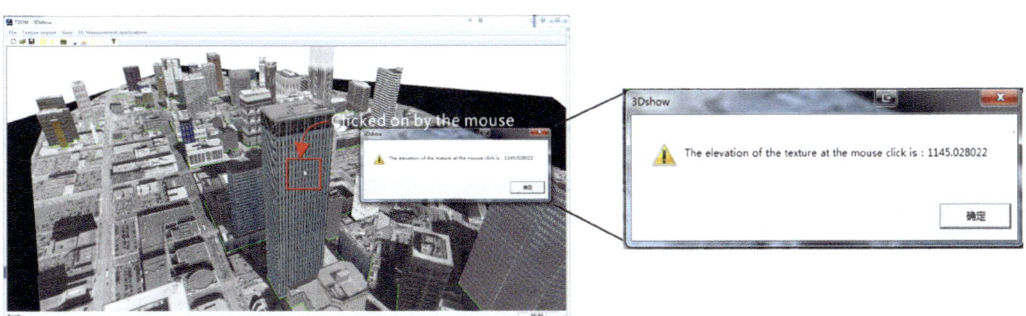

FIGURE 11.21 Measuring the elevation of any point on the façade. (Courtesy of Zhou et al. 2022b.)

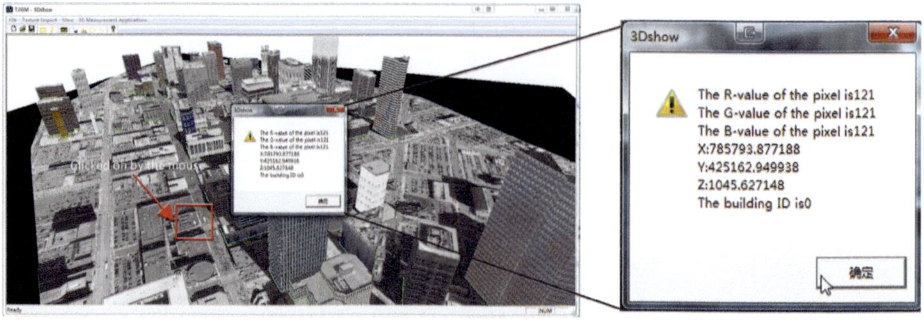

FIGURE 11.22 View of the superpixel information.

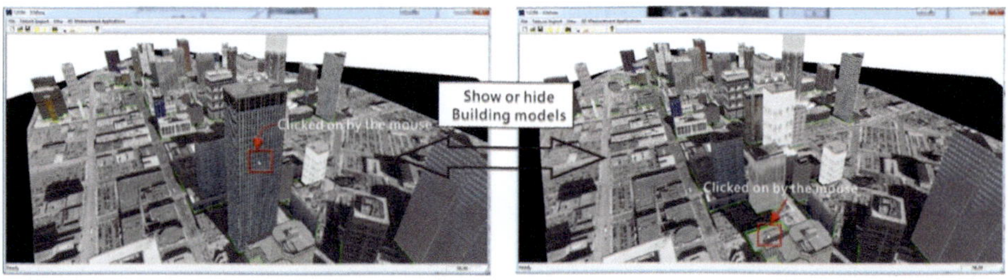

FIGURE 11.23 Selectively showing or hiding information.

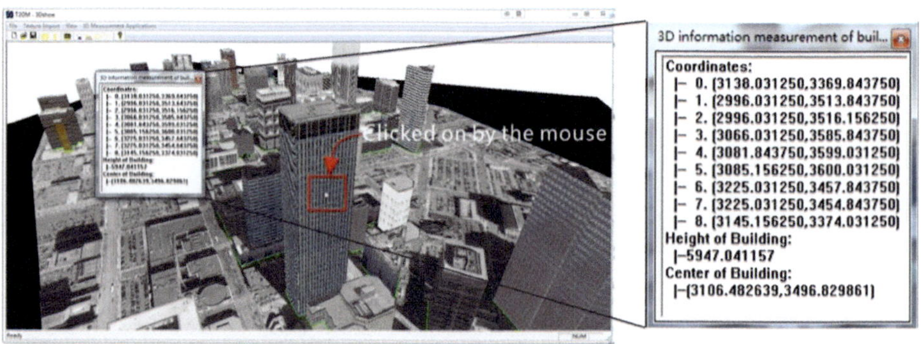

FIGURE 11.24 View information about the building.

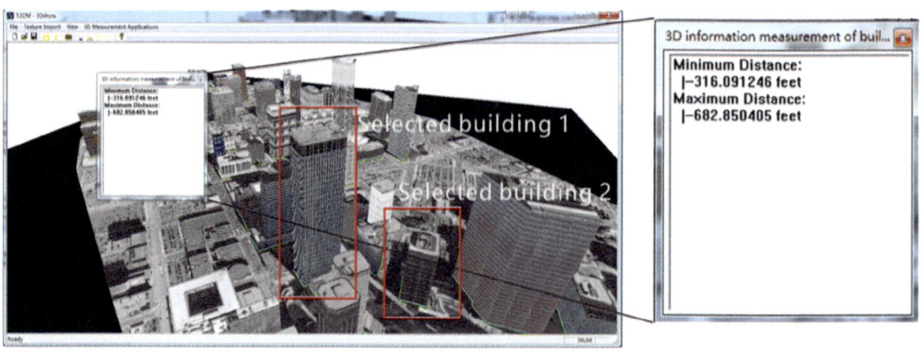

FIGURE 11.25 Measuring the maximum and minimum distances between two buildings.

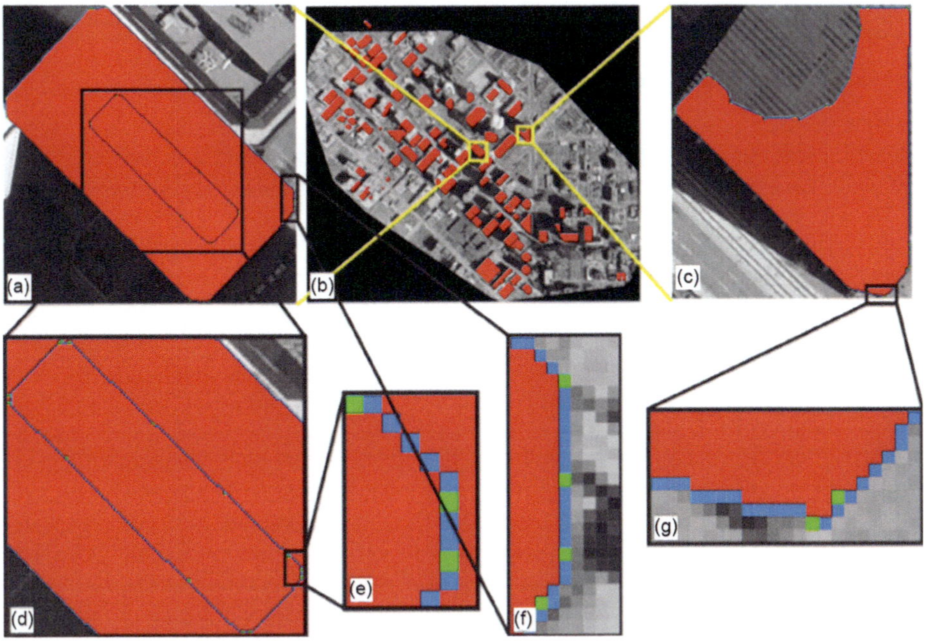

FIGURE 11.26 Acquisition of horizontal projection polygon corner points: (a) building model consisting of two voxels, (b) graph of the corner point detection results, (c) building model consisting of a single voxel, (d)–(f) local enlargement of (a), and (g) local enlargement of (c).

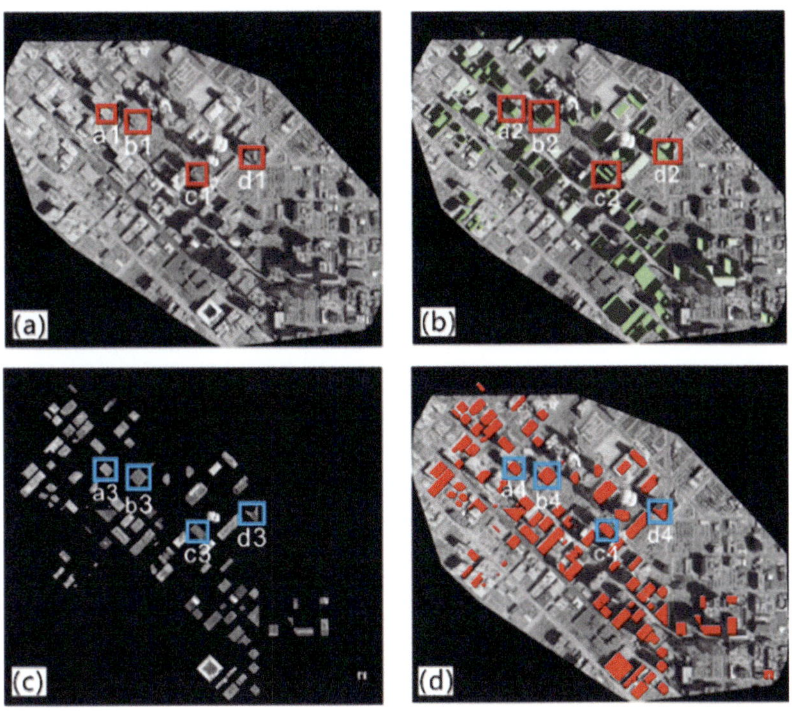

FIGURE 11.27 Analysis of T^2OM generation results. (a) Output orthophoto, (b) orthophoto superimposed on the DBM, (c) extracted roof texture, (d) red area used to represent the superpixels generated with the DBM. Where a1–a4 are four regions with the same geographical position in (a)–(d); b1–b4 are four regions with the same geographical position in (a)–(d); c1–c4 are four regions with the same geographical position in (a)–(d); d1–d4 are four regions with the same geographical position in (a)–(d).

TABLE 11.12
3D Coordinate Values Before and After Encoding

Building	X_{ori}	Y_{ori}	Z_{ori}	*id(r, c)*	*h*	X_{bc}	Y_{bc}	Z_{bc}
a	1286.850	1306.000	5551.700	14	1973	1286.840	1306.030	5551.600
	1346.850	1245.000	5551.700	14	1973	1346.840	1245.030	5551.600
	1355.000	1237.975	5551.700	241	1973	1355.030	1237.970	5551.600
	1356.000	1237.850	5551.700	209	1973	1356.030	1237.840	5551.600
b	1931.937	1424.000	5533.400	15	1876	1931.910	1424.030	5533.330
	1932.150	1440.000	5533.400	3	1876	1932.160	1440.030	5533.330
	1910.850	1462.000	5533.400	14	1876	1910.840	1462.030	5533.330
	1906.000	1464.150	5533.400	33	1876	1906.030	1464.160	5533.330
c	2990.850	1244.000	5462.500	14	1499	2990.840	1244.030	5462.330
	2912.850	1151.000	5462.500	14	1499	2912.840	1151.030	5462.330
	2911.850	1148.000	5462.500	14	1499	2911.840	1148.030	5462.330
	2964.850	1093.000	5462.500	14	1499	2964.840	1093.030	5462.330
d	3339.000	1882.850	5755.800	209	3057	3339.030	1882.840	5755.760
	3341.096	1883.000	5755.800	2	3057	3341.090	1883.030	5755.760
	3344.850	1881.000	5755.800	14	3057	3344.840	1881.030	5755.760
	3356.054	1864.000	5755.800	1	3057	3356.030	1864.030	5755.760

a 3D model easier, because superpixels accurately store each location with elevation information. However, the proposed method still has defects. The main problem is that storage of the T²OM requires twice as much memory space as that of the TOM. This makes it difficult to store and transfer data, especially across large-scale urban areas. A solution is to compress the bit width of storage components, such as *S*, *H*, and *ID*, by a statistical method. For example, in the compression of *H*, the height change curve for the whole study area can be counted, and the elevation series can be reduced from 4096 to 256. In addition, in areas without buildings, the removal of *ID* space to reduce memory consumption can be considered. Another defect that cannot be ignored is that the description of the geometric structure of building facades is not refined enough. On facades, there are balconies protruding from walls and windows recessed into walls. The geometric correction and line extraction of these facade structures is still difficult at present (Fan et al. 2021).

11.4 CONCLUSIONS

In light of the problem that traditional DOMs/TOMs only provide building roofs' 2D (*X*, *Y*) attributes and gray information and cannot provide 3D information or building facade textures at all, this chapter describes the generation of T²OM, which is radically different from the traditional generation of DOMs/TOMs, since the T²OM is able to provide 3D and detailed textures of building roofs and facades.

Two sets of experimental results demonstrate that the T²OM can maintain the traditional DOM/TOM characteristics, i.e., provide 2D *XY* coordinates and displaced building texture, but also provide the 3D *XYZ* coordinates of buildings' roofs and facades. The accuracy of 3D measurement on a T²OM can achieve 0.025 m (0.3 pixel).

Nevertheless, the proposed method needs to be improved; for example, when the number of buildings in a city is large, if all of them are loaded into the memory according to the original texture data, the 3D display may occupy a large amount of memory and the refresh speed will be reduced. Therefore, memory loading according to the visible area and the compression of loaded memory are needed to reduce the memory occupancy and improve the refresh speed.

REFERENCES

Amhar, F.; Jansa, J.; Ries, C., The generation of true orthophotos using a 3D building model in conjunction with a conventional DTM. *International Archives of Photogrammetry and Remote Sensing*, 32, 1998, pp. 16–22.

Biasion, A.; Dequal, S.; Lingua, A., A new procedure for the automatic production of true orthophotos. *International Archives of the Photogrammetry, Remote Sensing and Spatial Information Sciences*, 35, 2004, pp. 1682–1777.

Chan, C.; Tan, S., Determination of the minimum bounding box of an arbitrary solid: An iterative approach. *Computers & Structures*, 79, 2001, pp. 1433–1449. https://doi.org/10.1016/s0045-7949(01)00046-3

Di, K.; Jia, M.; Xin, X.; Wang, J.; Liu, B.; Li, J.; Xie, J.; Liu, Z.; Peng, M.; Yue, Z.; Liu, J.; Chen, R.; Zhang, C., High-resolution large-area digital orthophoto map generation using LROC NAC images. *Photogrammetric Engineering & Remote Sensing*, 85, 2019, pp. 481–491. https://doi.org/10.14358/pers.85.7.481

Fan, H.; Wang, Y.; Gong, J., Layout graph model for semantic façade reconstruction using laser point clouds. *Geo-spatial Information Science*, 24, 2021, pp. 403–421. https://doi.org/10.1080/10095020.2021.1922316

Gharibi, H.; Habib, A., True orthophoto generation from aerial frame images and LiDAR data: An update. *Remote Sensing*, 10, 2018, p. 581. https://doi.org/10.3390/rs10040581

Greenfeld, J., Evaluating the accuracy of digital orthophoto quadrangles (DOQ) in the context of parcel-based GIS. *Photogrammetric Engineering & Remote Sensing*, 67, 2001, pp. 199–206.

Haggag, M.; Zahran, M.; Salah, M., Towards automated generation of true orthoimages for urban areas. *American Journal of Geographic Information System*, 7, 2018, pp. 67–74. http://article.sapub.org/10.5923.j.ajgis.20180702.03.html

Huang, X.; Zhang, L., Morphological building/shadow index for building extraction from high-resolution imagery over urban areas. *IEEE Journal of Selected Topics in Applied Earth Observations and Remote Sensing*, 5, 2011, pp. 161–172. https://doi.org/10.1109/jstars.2011.2168195.

Jauregui, M.; Vílchez, J.; Chacón, L. A procedure for map updating using digital mono-plotting. *Computers & Geosciences*, 28, 2002, pp. 513–523. https://doi.org/10.1016/s0098-3004(01)00068-1

Liu, X.; Zhou, G.; Zhang, W.; Luo, S., Study on local to global radiometric balance for remotely sensed imagery. *Remote Sensing*, 13, 2021, p. 2068. https://doi.org/10.3390/rs13112068

Mayr, W., True orthoimages. *GIM International*, 37, 2002, pp. 37–39.

Rau, J.-Y.; Chen, N.-Y.; Chen, L.-C., True orthophoto generation of built-up areas using multi-view images. *Photogrammetric Engineering & Remote Sensing*, 68, 2002, pp. 581–588.

Schickier, W.; Thorpe, A., Operational procedure for automatic true orthophoto generation. *International Archives of Photogrammetry and Remote Sensing*, 32, 1998, pp. 527–532.

Shin, Y. H.; Lee, D.-C., True orthoimage generation using airborne LiDAR data with generative adversarial network-based deep learning model. *Journal of Sensors,* 2021, 2021, p. 4304548. https://doi.org/10.1155/2021/4304548

Shoab, M.; Singh, V. K.; Ravibabu, M. V., High-precise true digital orthoimage generation and accuracy assessment based on UAV images. *Journal of the Indian Society of Remote Sensing*, 50, 2021, pp. 613–622. https://doi.org/10.1007/s12524-021-01364-z

Siachalou, S. Urban orthoimage analysis generated from IKONOS data. *International Archives of the Photogrammetry, Remote Sensing and Spatial Information Sciences*, 35, 2004, pp. 12–23.

Skarlatos, D., Orthophotograph production in urban areas. *Photogrammetric Record*, 16, 1999, pp. 643–650. https://doi.org/10.1111/0031-868x.00144

Wang, Q.; Zhou, G.; Song, R.; Xie, Y.; Luo, M.; Yue, T., Continuous space ant colony algorithm for automatic selection of orthophoto mosaic seamline network. *ISPRS Journal of Photogrammetry and Remote Sensing*, 186, 2022, pp. 201–217. https://doi.org/10.1016/j.isprsjprs.2022.02.011

Xie, W.; Zhou, G., Experimental realization of urban large-scale true orthoimage generation. In: *Proceedings of the ISPRS Congress*, Beijing, China, 3–11 July 2008, pp. 3–11.

Yao, J.; Zhang, Z. M., Hierarchical shadow detection for color aerial images. *Computer Vision and Image Understanding*, 102, 2006, pp. 60–69. https://doi.org/10.1016/j.cviu.2005.09.003

Yu, B.; Wang, L.; Niu, Z., A novel algorithm in buildings/shadow detection based on Harris detector. *Optik*, 125, 2014, pp. 741–744. https://doi.org/10.1016/j.ijleo.2013.07.043

Zhang, R.; Liu, N.; Huang, J.; Zhou, X., On-board ortho-rectification for images based on an FPGA. *Remote Sensing*, 9, 2017, p. 874. https://doi.org/10.3390/rs9090874

Zhou, G., Near real-time orthorectification and mosaic of small UAV video flow for time-critical event response. *IEEE Transactions on Geoscience and Remote Sensing*, 47, 2009, pp. 739–747. https://doi.org/10.1109/tgrs.2008.2006505

Zhou, G., *Urban High-Resolution Remote Sensing: Algorithms and Modeling*; CRC Press: Boca Raton, FL, 2020. https://doi.org/10.1201/9781003082439

Zhou, G.; Bao, X.; Ye, S.; Wang, H.; Yan, H., Selection of optimal building facade texture images from UAV-based multiple oblique image flows. *IEEE Transactions on Geoscience and Remote Sensing*, 59, 2020, pp. 1534–1552. https://doi.org/10.1109/tgrs.2020.3023135.

Zhou, G.; Chen, W.; Kelmelis, J. A.; Zhang, D., A comprehensive study on urban true orthorectification. *IEEE Transactions on Geoscience and Remote Sensing*, 43, 2005, pp. 2138–2147. https://doi.org/10.1109/TGRS.2005.848417

Zhou, G.; Jezek, K. C., Satellite photograph mosaics of Greenland from the 1960s era. *International Journal of Remote Sensing*, 23, 2002, pp. 1143–1159. https://doi.org/10.1080/01431160110060907

Zhou, G.; Li, H.; Song, R.; Wang, Q.; Xu, J.; Song, B., Orthorectification of fisheye image under equidistant projection model. *Remote Sensing*, 14, 2022a, p. 4175. https://doi.org/10.3390/rs14174175

Zhou, G.; Pan, Q.; Yue, T.; Wang, Q.; Sha, H.; Huang, S.; Liu, X., Vector and raster data storage based on Morton code. *ISPRS International Archives of the Photogrammetry, Remote Sensing and Spatial Information Sciences*, XLII-3, 2018a, pp. 2523–2526. https://doi.org/10.5194/isprs-archives-xlii-3-2523-2018

Zhou, G.; Schickler, W., True orthoimage generation in extremely tall building urban areas. *International Journal of Remote Sensing*, 25, 2004, pp. 5161–5178.

Zhou, G.; Schickler, W.; Thorpe, A.; Song, P.; Chen, W.; Song, C., True orthoimage generation in urban areas with very tall buildings. *International Journal of Remote Sensing*, 25, 2004, pp. 5163–5180. https://doi.org/10.1080/01431160410001705079

Zhou, G.; Wang, Q.; Huang, Y.; Tian, J.; Li, H.; Wang, Y., True^2 orthoimage map generation. *Remote Sensing*, 14, 2022b, p. 4396. https://doi.org/10.3390/rs14174396

Zhou, G.; Wang, Y.; Yue, T.; Ye, S.; Wang, W., Building occlusion detection from ghost images. *IEEE Transactions on Geoscience and Remote Sensing*, 55, 2016, pp. 1074–1084. https://doi.org/10.1109/TGRS.2016.2619184

Zhou, G.; Zhang, R.; Zhang, D.; Huang, J.; Baysal, O., Real-time ortho-rectification for remote-sensing images. *International Journal of Remote Sensing*, 40, 2018b, pp. 2451–2465. https://doi.org/10.1080/01431161.2018.1488296

12 Orthophotomap Mosaic

12.1 INTRODUCTION

A single digital orthophotomap (DOM) cannot usually cover a large area to be investigated since its dimension is limited; it is therefore necessary to mosaic many DOMs together for a large field. On the other hand, a DOM is produced based on the digital surface model (DSM) or digital terrain model (DTM) through pixel-by-pixel digital differential orthorectification (Zhou et al. 2016, 2017b, 2019; Zheng et al. 2017a; Li et al. 2019). However, because the ground objects, typically houses, are incompletely orthorectified in adjacent DOMs due to the lack of object elevation in DTM (Zheng et al. 2017a; Li et al. 2019), incomplete image distortion correction, etc., the geometric malposition of the same object along the seamline of the two adjacent images usually happen when mosaicking them. This means a low quality of the mosaicked DOM (Zhou et al. 2005, 2017a, 2020). The effective way to overcome this problem is to choose an optimal seamline to avoid geometric misalignments of the same object in two neighbor DOMs.

A good method for automatic seamline selection generally satisfies the following requirements (Pan et al. 2015; He et al. 2019):

1. the seamlines should be selected to avoid the areas where obvious ground objects appear;
2. multiple image mosaic seamline networks should be simultaneously selected; and
3. the vertices of the mosaic seamlines should be avoided on obvious objects as much as possible.

Despite the accomplishments above, almost a few of the existing methods can achieve a good performance in meeting these three requirements simultaneously. Therefore, this chapter describes a method to automatically select a mosaic seamline network, named "Continuous Space Ant Colony Algorithm for Automatic Selection of Orthophoto Mosaic Seamline Network".

12.2 DEVELOPMENT OF DOM MOSAIC

Most methods have been investigated to ensure the mosaic seamlines avoid obvious objects in the past decades (e.g., Williams and Shah 1992; Davis 1998; Kerschner 2001; Chon and Kim 2006), in which the frame-to-frame method is usually used to select multiple orthophotos mosaic seamlines. Many studies have been done in the area of selecting mosaic seamlines in the past decades. They are mainly divided into four categories below.

1. *Methods based on the relative geographic location of the images*: The representatives include, such as Hsu and Kumar (2002), who used the traditional Voronoi map algorithm to generate an initial mosaic seamline network, which divides the image into multiple spatial regions using the mid-perpendiculars of adjacent image centers and forms a simple mosaic seamline network based on the geographic location demarcation. However, there may be cases where the mosaic seamlines do not pass through the overlapping regions, resulting in some regions not being covered by the image, leading to their loss. Pan et al. (2009, 2013, 2018) developed an algorithm for area Voronoi diagrams with overlap (AVDO), but the initial mosaic seamlines generated by these methods do not consider the texture and color differences between the images in the overlapping regions and therefore need to be optimized. However, the methods based on the relative geographic location of the images are simple but do not take into account the differences between images.

DOI: 10.1201/9781003632108-12

2. *Methods based on auxiliary data*: Ma and Sun (2011) proposed a novel *A** algorithm that can automatically keep the seamlines away from obstacle areas and improve the quality of orthophotos. However, this method lacks an automatic algorithm that can filter non-ground signals, particularly when the terrain is complex. In this method, the extraction of obstacle regions is affected by the point cloud filtering accuracy, which sometimes requires manual assistance. Wang et al. (2012) proposed a new method to generate mosaic seamlines using vector roads to extract skeletons to create weighted maps and then used the Dijkstra algorithm to search for the least cost path. Although the seamlines generated by this method usually follow the centerline of the road, the method relies heavily on a vector road map corresponding to the orthophoto. Wan et al. (2013) proposed the construction of a weighted graph G (V, E) using only vector roads as auxiliary data, extracting the skeletons of overlapping regions of adjacent orthophotos, and then applying the Floyd–Warshall algorithm to find the least cost path as the optimal mosaic seamline. This method synthesizes regular seamlines with a small number of vertices, which can be easily corrected manually with higher efficiency, but cannot be applied under very complex conditions (e.g., with many bridges). Chen et al. (2014) proposed a new algorithm for automatic mosaic seamline selection using a DSM. The algorithm introduced a new model called the Ortho-Elevation Synchronization Model (OESM), given that the DSM elevation is not perfectly synchronized with the DOM. Although the algorithm can determine the effective mosaic polygons for each orthophoto for seamless stitching, it requires both DSM and digital elevation model. Zheng et al. (2017a) proposed an orthophoto mosaic method based on an edge-tracking algorithm to avoid vertex positions falling on features. If the DSM does not reflect the real surface, the final network of mosaic seamlines will be inaccurate, and the edge tracking algorithm will be affected by the incompleteness of the edges through features such as houses, thus affecting the quality of the mosaic. Zheng et al. (2017b) introduced an orthophoto mosaic method based on the weighted A^* algorithm. The initial seamline network is generated by the standard Voronoi map algorithm, which is generated based on DSM data, vertex repositioning based on the edge maps and new vertex positions, and edge maps for optimized seamlines using the weighted A^* algorithm. The method is applicable to UAV imagery in urban and mountainous areas and automatically generates acceptably large orthophotos. Both vertex repositioning and seamline optimization were processed based on a single edge map generated by the DSM. Therefore, this method is heavily influenced by the quality of the DSM. Wang et al. (2017) proposed a novel seam detection method based on vector building diagrams. The method tracks the centerline between the vector buildings to generate candidate joints that avoid passing through the buildings. The candidate seams are then optimized by considering the surrounding pixels to minimize the visual transition between the input images. However, the method depends on the degree of correlation or match between the vector buildings and the image to be spliced, and generating centerlines between buildings is time-consuming. Li et al. (2017, 2018) proposed combining semantic segmentation information to calculate the energy cost per pixel. First, a convolutional neural network was used to distinguish overlaps into six categories (buildings, cars, trees, low vegetation, water, and impervious surfaces). Then, the semantic segmentation results were used to calculate the energy cost, and the graph cut was applied to find the optimal seamline. In their method, the superpixel segmentation was introduced to greatly improve the optimization efficiency by decreasing the number of elements in graph cuts. Zheng et al. (2018) proposed an orthophoto mosaic method based on the weighted A^* algorithm for UAV images of most terrain types (flat or mountainous), which also performs vertex repositioning but is highly correlated with the accuracy of the DSM. Zhang et al. (2018) proposed constructing an energy cost function using a combination of chromatic aberration, gradient, and DSM of adjacent images and then applied a GC algorithm to find the optimal seamline. As this method is currently implemented on

the original image, the graph node is dense. Therefore, a multiscale approach or the use of superpixels as graph nodes is needed to reduce the computation time. However, the methods based on auxiliary data are effective but limited by the auxiliary data.

3. *Methods based on image gray difference*: Williams and Shah (1992) proposed a greedy algorithm for solving the snake model, which makes the dynamic planning algorithm stable and reliable and reduces the computational effort, but does not yield an optimal solution. Davis (1998) used the Dijkstra algorithm to solve the stitching problem caused by lens distortion, camera positioning errors, or shooting motion, but the exhaustive search strategy is essentially a blind search, inefficient and time-consuming, and requires the selection of the start and end points of mosaic seamlines. Kerschner (2001) proposed that the twin snakes algorithm is mainly for forested areas and is prone to stop at the local maximum, so it is not possible to obtain a globally optimal solution. Zhou and Jezek (2002) ensured the visual quality of the stitched image by eliminating the chromatic aberration on both sides of the stitching seamline. However, they did not consider the misalignment caused by stitching and focused on satellite images. Chon and Kim (2006) used dynamic planning to detect optimal mosaic seamlines, but the starting and ending points of mosaic seamlines had to be specified. Many scholars (Zhang et al. 2009; Ma et al. 2013; Zhou et al. 2018) applied the ant colony algorithm to select two adjacent image mosaic seamlines. They used the initial pheromones (i.e., a functional relationship between the gray difference and the distance away from the initial mosaic seamline) to add traction to guide the ants toward the globally optimal search. However, their methods make the ants enter traps leading to search stagnation and are not for mosaicking multiple orthophotos. Yuan et al. (2015) proposed an automatic detection algorithm for orthophoto mosaic seamlines based on differential orthophotos, but the algorithm only used a simple building area threshold segmentation. Li et al. (2016, 2019) proposed a novel multiframe joint optimization strategy to find seamlines in multiple overlapping orthophotos simultaneously. This method generates an orthophoto parallax that is hidden because of the lower orthophoto similarity that overlapping regions have and can detect optimal seamlines for stitching aerial and street view panoramic orthophotos in a graphical cutting energy minimization framework. Yu et al. (2017) proposed a novel adaptive global color balancing method aimed at eliminating the effect of the color difference between adjacent optical images to achieve seamless image stitching. This method combines global and local optimization strategies to remove chromatic aberrations between different target images adaptively without the need to assign reference images. However, the misalignment caused by the mosaic seamlines passing through the features cannot be avoided. The methods based on the image gray difference do not need auxiliary data and are widely applicable.
4. *Methods based on objects*: Pan et al. (2015) proposed a novel method of determining the optimal mosaic seamline using the region change rate. The method uses the Mean Shift segmentation algorithm to segment the overlapping area; then detects the change index of the segmented area; divides the overlapping area into three levels: obstacle regions, preferred regions, and general regions; and finally uses the difference cost to optimize the optimal mosaic seamline. However, this method is not sensitive to the threshold for distinguishing between changed and unchanged pixels. Pang et al. (2016) proposed a new semi-global matching-based method to guide seamline determination for urban orthophoto mosaics, which can greatly eliminate geometric discontinuities. However, this method focuses on geometric discontinuities and rarely considers radiative discontinuities. Pan et al. (2017) proposed the use of the morphological building index and normalized difference vegetation index to detect building areas and vegetation, where each type of area is assigned different weights to optimize the obtained pixel cost, using the optimal pixel cost as the final seamline to greatly improve the quality of the inlay. However, the methods based on the object are effective but difficult to obtain object-level information.

12.3 MODEL FOR OPTIMAL SELECTION OF MOSAIC SEAMLINES

To describe the algorithm, the selection of mosaic seamlines for two orthophotos is first considered and then for multiple orthophotos.

12.3.1 Selection of Mosaic Seamlines for Two Orthophotos

Assume that Figure 12.1a represents an overlapping area of two orthophotos, where the boxes represent pixels. Each pixel is considered a potential path point for ant colony walking. Further, assume that X and Y are the numbers of rows and columns of pixels (where $X = 8$ and $Y = 5$ in Figure 12.1a), S represents the ant nest (also called *starting point*), and E represents the food source (also called *ending point*). The SE polyline was assumed to be the initial route of the ant colony path, m is the number of ants involved in the search per iteration, and n is the number of iterations of the ant colony's circular search. The values of m and n depend on the number of rows and columns of path points, which are generally represented as $m = \sqrt{\frac{Y}{2}}$ and $n = \sqrt{\frac{X}{2}}$ (Stützle and Hoos 2000).

Based on these assumptions, Figure 12.1a shows that if the colony (m ants) is at (i, j), then their next path points should be selected from the candidates of pixels, $(i, j-1)$, $(i+1, j-1)$, and $(i+1, j)$. The criterion of the next path point is a probability function (Wells et al. 2012).

$$P_t^{ij}(i+1, j) = \frac{\tau_t(i+1, j)}{\tau_t(i, j-1)+\tau_t(i, j+1)+\tau_t(i+1, j-1)+\tau_t(i+1, j)+\tau_t(i+1, j+1)} \quad (t = 1, 2, \ldots, n) \tag{12.1}$$

where $(t = 1, 2, \ldots, n)$ denotes the tth iterative search, $P_t^{ij}(i+1, j)$ denotes the probability of selecting the next path point as $(i+1, j)$ when the tth iteration ant colony is at the current path point (i, j), and $\tau_t(i, j)$ denotes the value of the pheromone on the current path point (i, j) at the tth iterative search, which is calculated as

$$\tau_1(i, j) = \begin{cases} \frac{\tau_{\min} + \tau_{\max}}{2} * \left[1 - \frac{f(i, j)}{N}\right] & f(i, j) < N \\ 0 & f(i, j) > N \end{cases} \tag{12.2}$$

where $\tau_1(i, j)$ denotes the initial pheromone at the path point (i, j), $\tau_{\max}$ and $\tau_{\min}$ denote the upper and lower bounds of the path point pheromone, denoted by $\tau_{\max} = \frac{n}{2N}$ and $\tau_{\min} = \frac{\tau_{\max}}{2n}$, respectively. If $\tau_1(i, j)$ is the largest of the three candidate path points, the corresponding path point is the most attractive to ant colonies, and N is the threshold of the cost of the path point. $f(i, j)$ is the cost at the path point (i, j) (see the lower-left corner in 0(*a*)) and can be calculated using 0.

$$f(i, j) = \max\{g(i, j), G(i, j)\} \tag{12.3}$$

where $g(i, j)$ and $G(i, j)$ represent the gray difference and gray gradient of pixels (i, j) in the overlap area (OA), respectively. The gray difference can be calculated as

$$g(i, j) = \left|g_l(i, j) - g_r(i, j)\right| \tag{12.4}$$

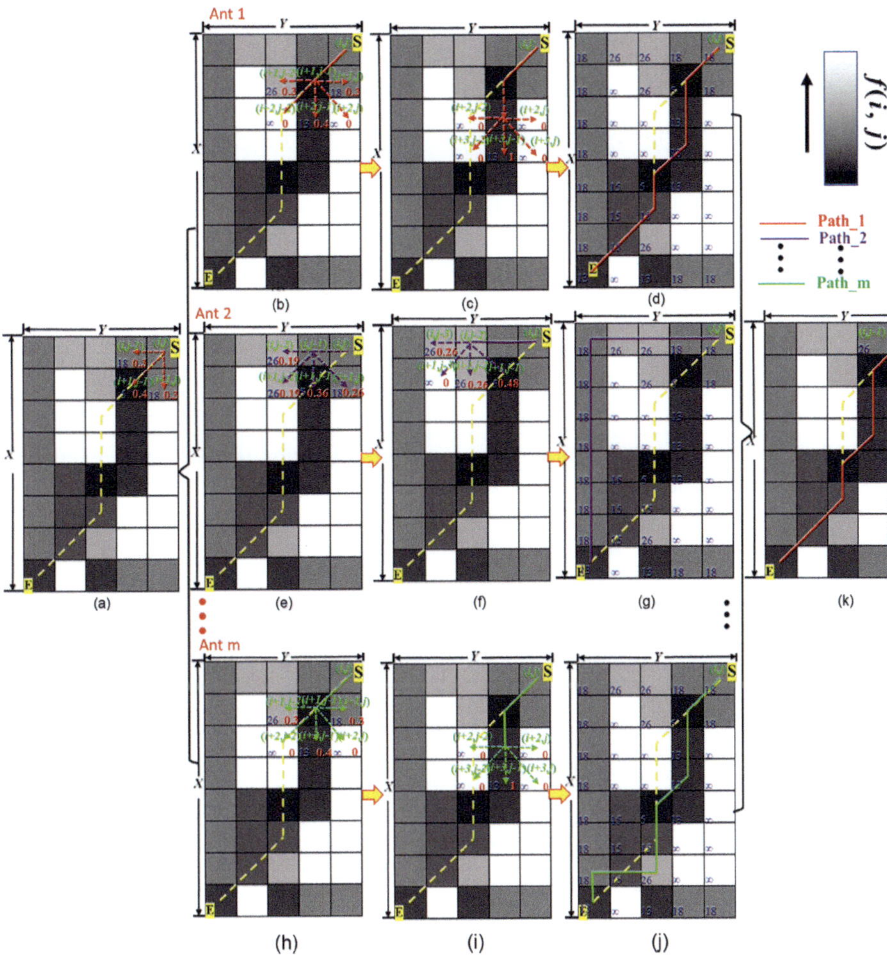

FIGURE 12.1 Process of selecting the optimal path at the 1st iteration of the continuous space ant colony algorithm. (Courtesy of Wang et al., 2022): (a) the basic principle of Ant algorithm, (b) the first step starting the point (i,j) in Ant 1, (c) the second step in Ant 1, (d) the completion of entire steps in Ant 1; (e) the first step starting the point (i,j-1) in Ant 2, (f) the second step in Ant 2, (g) the completion of entire steps in Ant 2, (h) the first step starting the point (i,j) in Ant m, (i) the second step in Ant m, (j) the completion of entire steps in Ant m; (k) is all completion of the steps through the above Ant algorithm.

where $g_l(i,j)$ and $g_r(i,j)$ denote the gray of pixels (i,j) at the left and right orthophotos, respectively. The gradient difference $G(i,j)$ is the most gray gradient out of the left image gray gradient, $G_l(i,j)$ and the right image gray gradients,$G_r(i,j)$ and can be calculated as

$$G(i,j) = \max\left(G_l(i,j), G_r(i,j)\right) \tag{12.5}$$

where $G_l(i,j)$ and $G_r(i,j)$ are the gray gradients at the left and right orthophotos at pixel (i,j) in the overlap region, respectively, which are calculated as

$$\begin{cases} G_l(i,j) = \max\{d_{l1}, d_{l2}, d_{l3}, d_{l4}, d_{l5}, d_{l6}, d_{l7}, d_{l8}\} \\ G_r(i,j) = \max\{d_{r1}, d_{r2}, d_{r3}, d_{r4}, d_{r5}, d_{r6}, d_{r7}, d_{r8}\} \end{cases} \tag{12.6}$$

where $d_{lk}\left(k=1,2,...,8\right)$ and $d_{rk}\left(k=1,2,...,8\right)$ are the gray gradients at the left and right orthophoto pixels (i,j) (see Figure 12.2), respectively, and can be calculated as

$$\begin{cases} d_{m1}=\left|g_l\left(i-1,j-1\right)-g_l\left(i,j\right)\right|, & d_{m2}=\left|g_l\left(i-1,j\right)-g_l\left(i,j\right)\right| \\ d_{m3}=\left|g_l\left(i-1,j+1\right)-g_l\left(i,j\right)\right|, & d_{m4}=\left|g_l(i,j-1)-g_l\left(i,j\right)\right| \\ d_{m5}=g_l\left(i,j+1\right)-g_l\left(i,j\right)|, & d_{m6}=\left|g_l\left(i+1,j-1\right)-g_l\left(i,j\right)\right| \\ d_{m7}=\left|g_l\left(i+1,j\right)-g_l\left(i,j\right)\right|, & d_{m8}=\left|g_l\left(i+1,j+1\right)-g_l\left(i,j\right)\right| \end{cases} \tag{12.7}$$

where $m=l,r$.

Within the model, the m ants' search process in the first iteration is taken as an example. With equation (12.2), the initial pheromones can be calculated as

$$\tau_1\left(i,j-1\right)=\frac{\dfrac{4}{2\times 50}+\dfrac{\dfrac{4}{2\times 50}}{2\times 4}}{2}*\left[1-\frac{18}{50}\right]\approx 0.01440 \quad \text{(where } N=50 \text{ is assumed)}$$

Similarly, $\tau_1\left(i+1,j-1\right)\approx 0.02025$, $\tau_1\left(i+1j\right)=0.01440$.

So, $p_1^{ij}\left(i,j-1\right)=\dfrac{0.01440}{0.01440+0.02025+0.01440}\approx 0.30$, $p_1^{ij}\left(i+1,j-1\right)\approx 0.40$, $p_1^{ij}\left(i+1,j\right)\approx 0.30$.

With these calculated probabilities, approximately m ants will select the path point $\left(i,j\right)$ as the next path point (see Figure 12.1b) with a probability of 0.3, will select $\left(i+1,j-1\right)$ with a probability of 0.4, and will select $\left(i+1,j\right)$ with a probability of 0.3. Assuming that one of the ants (marked as ant 1) selects a path point $\left(i+1,j-1\right)$ at the path point $\left(i,j\right)$, the path points can be placed in the tabu list (see Table. 12.1), and its accumulated cost can be calculated using $\sum_{x=0}^{k} f_1\left(i_x,j_x\right)$. The tabu list is used to store the path points that the ant has already traveled, indicating that these path points will not be admitted to ants traveling repeatedly. Then, ant 1's current tabu list would be $Tabu1$ = {(i, j), (i + 1, j − 1)}, and the accumulated cost would be $\sum_{x=0}^{k} f_1\left(i_x,j_x\right)=\sum_{x=0}^{k-1} f_1\left(i_x,j_x\right)+f\left(i+1,j-1\right)=18+5=23$. When ant 1 is at the path point $\left(i+1,j\right)$, using equation (12.1), the probability of ant 1 searching for the next path point $\left(i+1,j-2\right)$, $\left(i+2,j-2\right)$, $\left(i+2,j-1\right)$, $\left(i+2,j\right)$, and $\left(i+1,j\right)$ at path points $\left(i+1,j-1\right)$ was calculated as 0.3, 0, 0.4, 0, and 0.3, respectively. Assuming that ant 1 has chosen $\left(i+2,j-1\right)$ as the next path point, its tabu list would be updated to $Tabu1$ = {(i, j), (i + 1, j − 1), (i + 2, j − 1)} (see Table 12.1), and its accumulated cost would be updated to $\sum_{x=0}^{k} f_1\left(i_x,j_x\right)=\sum_{x=0}^{k-1} f_1\left(i_x,j_x\right)+f\left(i+2,j-1\right)=23+13=36$. Then, the next point path is searched as above until path point E is selected, and a feasible path (Path_1) is obtained (see Figure 12.1d).

As the search of the continuous space ant colony algorithm selects the next path point based on probability, other (m − 1) ants will choose (m − 1) paths different from ant 1 (see Figure 12.1a, g, j). For example, if another ant (marked as ant 2) chooses path point $\left(i,j-1\right)$ (see Figure 12.1e) as the

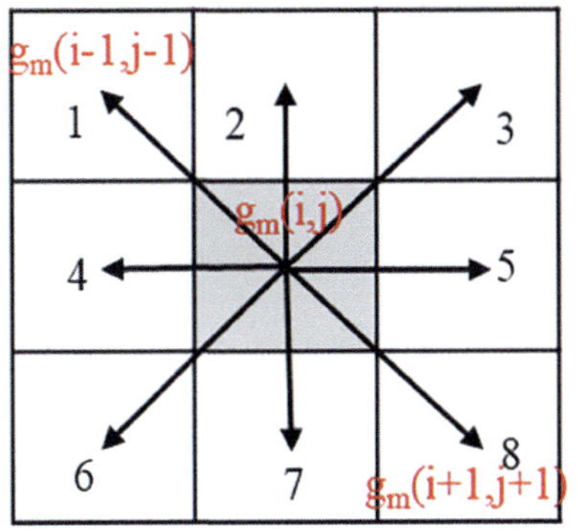

FIGURE 12.2 Eight neighborhoods.

TABLE 12.1
Tabu List of Ant Colony

Tabu List	Step 0	Step 1	Step 2	Step 3	...	Step *k*
$Tabu_1$	**(*i*,*j*)**	(*i*, *j*), **(*i* + 1, *j* − 1)**	(*i*, *j*), (*i* + 1, *j* − 1), **(*i* + 2,*j*-1)**	(*i*, *j*), (*i* + 1, *j* − 1), (*i* + 2, *j* − 1), **(*i* + 3,*j* − 1)**	...	(*i*, *j*), (*i* + 1, *j*), ..., **(*i* + 7,*j* − 4)**
$Tabu_2$	**(*i*,*j*)**	(*i*, *j*), **(*i*, *j* − 1)**	(*i*, *j*), (*i*, *j* − 1), **(*i*, *j* − 2)**	(*i*, *j*), (*i*, *j* − 1), (*i*, *j* − 2), **(*i*, *j* − 3)**	...	(*i*, *j*), (*i*, *j* − 1), ..., **(*i* + 7,*j* − 4)**
⋮	...	...	...	...	...	...
$Tabu_m$	**(*i*,*j*)**	(*i*, *j*), **(*i* + 1, *j* − 1)**	(*i*, *j*), (*i* + 1, *j* − 1), **(*i* + 2,*j* − 1)**	(*i*, *j*), (*i* + 1, *j* − 1), (*i* + 2, *j* − 1), **(*i* + 3,*j* − 1)**	...	(*i*, *j*), (*i* + 1, *j* − 1), ..., **(*i* + 7,*j* − 4)**

next path point at the path point (i, j), then ant 2 will also search for the next path point like ant 1 until it reaches path point E, also obtaining another feasible path named Path_2 (see Figure 12.1g). As there are m ants per iterative search, *m* feasible paths are obtained (see Figure 12.1j).

The average cost of the path traveled by each ant is compared with the average cost of the current optimal path. The path with the smallest average cost is marked as the current optimal path (the initial average cost of the current optimal path is the threshold of the cost (N)).

The average cost of Path_1 can be calculated as $\overline{f_1} = \frac{\sum f_1}{8} = 12$, the mean cost of Path_2 as $\overline{f_2} = \frac{\sum f_2}{12} = 19$, and the mean cost of Path_m as $\overline{f_m} = \frac{\sum f_m}{15} = 14.1$. As $\overline{f_1} < \overline{f_2} < \overline{\ldots < f_m} < \ldots < N$, Path_1 is set as the current optimal path (see Figure 12.1k). Up until now, the current optimal and feasible path was obtained after this iterative search, and multiple iterations of the ant colony search are required to obtain the globally optimal path.

Before the next iteration of the search, pheromones on the path points passed by each ant colony must be updated, with the update rule expressed as

$$\tau_t(i,j) = \begin{cases} \tau_{t-1}(i,j) + \Delta\tau & (i,j) \in Path_{opt} \\ \tau_{t-1}(i,j) - \Delta\tau & \text{otherwise} \end{cases} \tag{12.8}$$

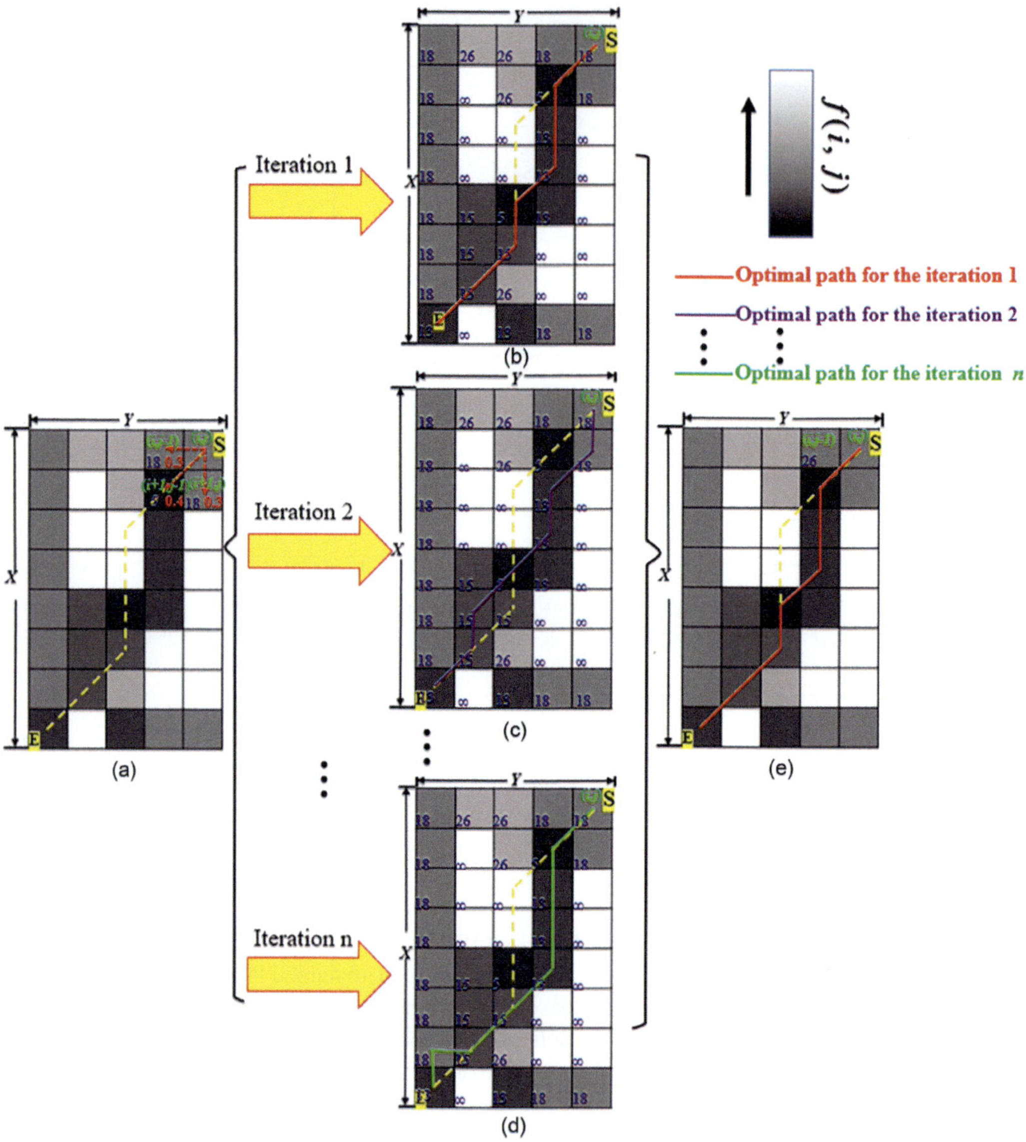

FIGURE 12.3 Process of selecting the optimal path for n iterations of the continuous space ant colony algorithm. (Courtesy of Wang et al., 2022): (a) the first iteration of Ant algorithm (Red line and yellow line represents the different path of an ant in iteration processing), (b) the second iteration of Ant algorithm, (c) the n iterations of Ant algorithm, and (d) the completion of n-time iterations.

where $\tau_t(i, j)$ and $\tau_{t-1}(i, j)$ denote the value of the pheromone on the path point (i, j) at the tth and $t-1$th iterative search, respectively, $Path_{opt}$ denotes the current optimal path, and $\Delta\tau$ denotes the amount of increased and volatile pheromones, with the magnitude of the increased and volatile amounts being inversely proportional to the maximum value of the cost of passing over the path, as shown below.

$$\Delta\tau = \frac{1}{\max\left\{f(i_1, j_1), f(i_2, j_2), \dots, f(i_k, j_k)\right\}} \tag{12.9}$$

where $f(i_1, j_1)$ denotes the gray at the path point (i_1, j_1) and k denotes the path length. To avoid the phenomenon of premature path maturity due to high pheromones at a path point, the pheromone values at each path point are restricted to a certain range $[\tau_{min}, \tau_{max}$. The pheromones at path points that exceed this range $[\tau_{min}, \tau_{max}$ can be updated as

$$\tau_t(i,j) = \begin{cases} \tau_{min}, & \text{if } \tau_t(i,j) < \tau_{min} \\ \tau_{max}, & \text{if } \tau_t(i,j) > \tau_{max} \end{cases} \tag{12.10}$$

where τ_{min} and τ_{max} indicate the lower and upper pheromone limits, respectively.

After the first iterative search of the ant colony, the above search process is repeated $(n - 1)$ times (see Figure 12.3b–d). The average cost of the optimal path obtained in each iteration is compared with the average cost of the current optimal path, and the path with the lowest average cost is set as the current optimal path until the number of iterations of the algorithm reaches n or the current optimal path converges in each iteration (the path deviation does not exceed three path points or the average cost of the current optimal path remains unchanged for ten consecutive times) (see Figure 12.3e).

12.3.2 Selection of Mosaic Seamline Network for Multiple Orthophotos

12.3.2.1 Construction of Initial Mosaic Seamline Nets for Multiple Orthophotos

During the mosaicking process, multiple orthophotos are often targeted over a large area. We therefore propose to first generate an initial mosaic seamline network by using AVDO to establish the relationships between the mosaic seamlines (Pan et al. 2009).

The generated initial mosaic seamlines only considered the relative position between the DOMs. Thus, the initial mosaic seamline network and vertices (the intersection of a mosaic seamline network) may be placed over the obvious features such as houses and need to be further optimized.

12.3.2.2 Optimal Selection of Multi-orthophoto Mosaic Seamline Networks

The proposed method for optimizing the selection of mosaic seamline networks using the continuous space ant colony algorithm in this chapter is implemented in the following process.

12.3.2.2.1 Adjusting the Position of the Vertex of the Mosaic Seamline

Figure 12.4a shows four adjacent orthophotos of two flight paths, and the green network seamlines are the initial network of mosaic seamlines generated using the AVDO algorithm. As shown in Figure 12.4b, several vertices were located at the OA of multiple orthophotos (see Figure 12.4b), and assuming the pixels are in the z-orthophoto OA, the pixel cost can be calculated according to equation (12.11).

$$\begin{pmatrix} f(i+1,j+1) & \cdots & f(i+l,j+l) \\ \vdots & \ddots & \vdots \\ f(i+k,j) & \cdots & f(i+k,j+l) \end{pmatrix} = \max_{x,y=1,\ldots z, x \neq y} \left[\begin{pmatrix} f_{xy}(i+1,j+1) & \cdots & f_{xy}(i+l,j+l) \\ \vdots & \ddots & \vdots \\ f_{xy}(i+k,j) & \cdots & f_{xy}(i+k,j+l) \end{pmatrix} \right] \quad (i,j) \in OA \tag{12.11}$$

where k and l represent the number of rows and columns of the OA, respectively, $f_{xy}(i, j)$ represents the gray substitute value between orthophoto x and orthophoto y at the pixel of the OA, which can be calculated from equations (12.2) to (12.6). After the gray value of the pixels was obtained, the initial pheromones of each pixel in the OA (see Figure 12.4a, fore-and-aft overlap area) between the orthophotos in the heading were calculated according to equation (12.12), and the probability of selecting the path point (i, j) was calculated according to equation (12.7).

$$\begin{pmatrix} \tau_1(i+1,j) & \tau_1(i+1,j+2) & \cdots & \tau_1(i+1,j+l) \\ \tau_1(i+2,j) & \tau_1(i+2,j+2) & \cdots & \tau_1(i+2,j+l) \\ \vdots & \vdots & \ddots & \\ \tau_1(i+k,j) & \tau_1(i+k,j+2) & \cdots & \tau_1(i+k,j+l) \end{pmatrix} \tag{12.12}$$

$$= a\begin{pmatrix} 1-\frac{f(i+1,j)}{N} & 1-\frac{f(i+1,j+2)}{N} & \cdots & 1-\frac{f(i+1,j+l)}{N} \\ 1-\frac{f(i+2,j)}{N} & 1-\frac{f(i+2,j+2)}{N} & \cdots & 1-\frac{f(i+2,j+l)}{N} \\ \vdots & \vdots & \ddots & \vdots \\ 1-\frac{f(i+k,j)}{N} & 1-\frac{f(i+k,j+2)}{N} & \cdots & 1-\frac{f(i+k,j+l)}{N} \end{pmatrix}$$

where $a = \frac{\tau_{\min} + \tau_{\max}}{2}$, $\tau_1(i, j)$ represents the initial pheromone at the path point (i, j), and the k and l expressions have the same meaning as in equation (12.11).

The ants were placed at the starting points, S_{01} and S_{11}, of each inter-voyage mosaic seamline, and their subsequent path points were selected in parallel downward following the steps of the continuous space ant algorithm that optimally selected the optimal mosaic seamline between the two orthophotos until all the ants completed one iteration of the continuous space ant algorithm to reach their respective end points, E_{01} and E_{11}. Assuming that the mosaic seamline selected by ant e in m ants was optimal in this iteration, an initial optimized selection of one mosaic seamline for each ant was obtained, as shown by the red curves ($S_{01}E_{01}$, $S_{11}E_{11}$) in Figure 12.4b. The pheromones of the path points that each ant passed were then updated based on equation (12.13), and the range of pheromones for each path point after the update was limited to $[\tau_{\min}, \tau_{\max}$, according to equation (12.10).

$$\left[\tau_t^1,\ldots,\tau_t^e,\ldots,\tau_t^m\right] = \left[\tau_{t-1}^1,\ldots,\tau_{t-1}^e,\ldots,\tau_{t-1}^m\right] + \begin{bmatrix} -\Delta\tau & \cdots & \Delta\tau & \cdots & -\Delta\tau \\ -\Delta\tau & \cdots & \Delta\tau & \cdots & -\Delta\tau \\ \vdots & & \ddots & \ddots & \vdots \\ -\Delta\tau & \cdots & \Delta\tau & \cdots & -\Delta\tau \end{bmatrix} \tag{12.13}$$

where $\Delta\tau$ is given by 0, $\tau_t^1 = \left[\tau_t^1(i_1, j_1), \tau_t^1(i_2, j_2), \ldots, \tau_t^1(i_n, j_n)\right]^T, \ldots, \tau_t^m = \left[\tau_t^m(i_1, j_1), \tau_t^m(i_2, j_2), \ldots, \tau_t^m(i_n, j_n)\right]^T$ is the size of the pheromone at the point of the path taken by ant 1, …, m in the colony at the tth search, respectively. The intersection of the adjusted red curves ($S_{01}E_{01}$, $S_{11}E_{11}$) with the green initial mosaic seamline (S_1E_1) between the airways provides the adjusted vertex positions shown in Figure 12.4b for S'_{01}, S'_{11}, S'_1 and E'_{01}, E'_{11}, E'_1, where the vertex between the airways is the starting point of the mosaic seamline and the end of the previous mosaic (except for the vertices at the ends), i.e., E'_{01} is both the end of the mosaic seamline $S_1 \sim E'_{01}$ and the starting point of the mosaic

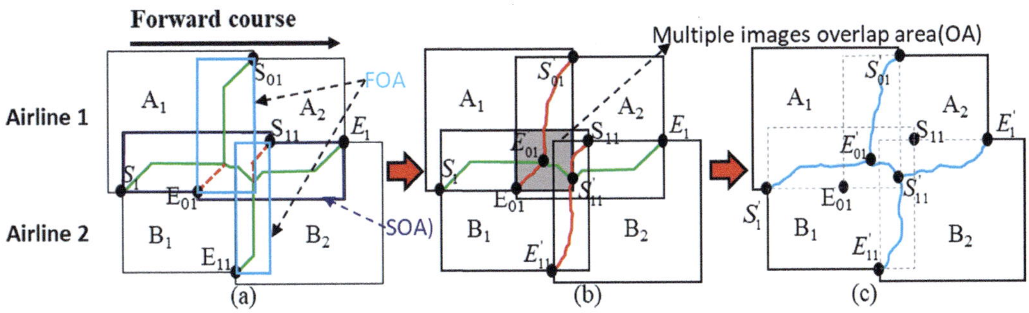

FIGURE 12.4 Multi-orthophoto mosaic network optimization selection. (Courtesy of Wang et al., 2022): (a) the mosaick processing when selecting different seamlines, (b) the selected seamline with four orthorectified images, and (c) the completed mosaic of four images.

seamline $E'_{01} \sim S'_{11}$. Since the vertices E'_{01} and S'_{11} are the intersection points of the AVDO mosaic line network and the mosaic line obtained by one iteration of the continuous space ant colony algorithm, i.e., these points are on the adjusted mosaic seamlines $S'_{01}E_{01}$ and $S_{11}E'_{11}$ obtained by one iteration of the continuous space ant colony algorithm, these vertices are also optimized. Although these vertices are not necessarily the optimal vertices, they also effectively avoid obvious ground features such as ground buildings.

12.3.2.2.2 Parallel Selection of Selective Mosaic Networks

During the simultaneous optimization of multiple orthophoto mosaic seamlines, in order to avoid the pheromones produced by the ants of the heading and inter-routing searches that interfere with each other during the search process, we propose dividing the pheromones into heading pheromones (distributed within the forward overlap area in Figure 12.4a) and side pheromones (distributed within the side overlap area in Figure 12.4a) where the heading and side overlap are common.

The ant colony for heading search only recognizes the heading pheromone, and the ant colony for the side search only recognizes the side pheromone. The manner in which their pheromones are set initial values and updated is the same and is shown by equations (12.12) and (12.13), respectively.

The initial pheromones on the pixels were calculated using equation (12.12). Ant colonies were then placed at each of the adjusted starting points (S'_{01}, S'_{11}, S'_1, E'_{01}, and S'_{11}), and their corresponding endpoints were E'_{01}, E'_{11}, E'_{01}, S'_{11}, and E'_1, respectively. Each mosaic seamline was obtained by following the steps of selecting the mosaic seamline between two orthophotos in parallel optimization with the continuous space ant colony algorithm, and the pheromone was updated after each iteration according to equation (12.13) until the number of iterations reached a predetermined value, or the current optimal paths obtained in each cycle converged (path deviation does not exceed three path points or the gray of the current optimal path remains unchanged for ten consecutive times).

$$Path_{opt} = \left\{ Path_{K_1}, Path_{K_2}, \ldots, Path_{K_K} \right\} \tag{12.14}$$

where K represents the number of segments into which the mosaic seamlines were divided and $Path_{K_1}, Path_{K_2}, \ldots, Path_{K_K}$ represent the optimal path for each segment of the mosaic seamlines. The basis for determining the optimal mosaic seamlines for each segment is given in equation (12.15), which ultimately minimizes the average gray value of each segment of the mosaic seamlines.

$$\begin{bmatrix} K_1 \\ K_2 \\ \vdots \\ K_K \end{bmatrix} = \begin{bmatrix} \underset{u=1,\ldots,n1}{\operatorname{argmin}}\left(\min\left(\sum_{v=1,\ldots,k_1} f_1(i_v, j_v),\ldots,\sum_{v=1,\ldots,k_{m1}} f_{m1}(i_v, j_v)\right)\right) \\ \underset{u=1,\ldots,n2}{\operatorname{argmin}}\left(\min\left(\sum_{v=1,\ldots,k_{m2}} f_1(i_v, j_v),\ldots,\sum_{v=1,\ldots,k_{m2}} f_{m2}(i_v, j_v)\right)\right) \\ \underset{u=1,\ldots,nK}{\operatorname{argmin}}\left(\min\left(\sum_{v=1,\ldots,k_{mK}} f_1(i_v, j_v),\ldots,\sum_{v=1,\ldots,k_{mK}} f_{mK}(i_v, j_v)\right)\right) \end{bmatrix} \tag{12.15}$$

where $\underset{u=1,\ldots,n1}{\operatorname{argmin}}\left(\min\left(\sum_{v=1,\ldots,k_1} f_1(i_v, j_v),\ldots,\sum_{v=1,\ldots k_{m1}} f_{m1}(i_v, j_v)\right)\right)$ denotes the index of the mosaic seamline that finds the minimum average cost obtained for that segment, and the combination of each optimally selected mosaic seamline, shown in Figure 12.4c, constitutes the final optimally selected mosaic seamline network.

$$Path_{opt} = \left\{Path_{S'_{01}\to E'_{01}}, Path_{S'_{11}\to E'_{11}}, Path_{S'_1\to E'_{01}}, Path_{E'_{01}\to S'_{11}}, Path_{S'_{11}\to E'_1}\right\} \tag{12.16}$$

In equation (12.16), $Path_{opt}$ denotes the global optimal path,$Path_{S'_{01}\to E'_{01}}$, $Path_{S'_{11}\to E'_{11}}$, $Path_{S'_1\to E'_{01}}$, $Path_{E'_{01}\to S'_{11}}$, and $Path_{S'_{11}\to E'_1}$chosen from S'_{01} to E'_{01}, S'_{11} to E'_{11}, S'_1 to E'_{01}, E'_{01} to S'_{11}, and S'_{11} to E'_1, respectively.

12.3.3 Ant Colony Response Strategies in Special Environments

During the search for the optimal path, the ants may encounter a special case of the concave region, shown in Figure 12.5a, where the white area is the obstacle and the black area is the feasible area. After the ants go through $a_3 \to b_3 \to c_3 \to d_3$, they find no downward optional nodes and some of the ants may go into a dead-end situation, which leads to slow or even failed algorithms. In this study, we adopted a retreat search strategy to avoid these problems and thus ensured a smooth search. The steps are as follows.

1. When ant K reaches the path point d_3, it finds that there are no downward path points available and has to select path points in both directions, but when it reaches path point d_2 or d_4, it finds that there are no more available path points. Then, it returns to the path point d_3 where the horizontal search began and marks the path point as infeasible. $Tabu(n(k)) = \{d_2, d_3, d_4\}$, where $Tabu(n(k))$ represents the obstacle list of all ants that search in the same direction as ant K and $n(k)$ represents the nth line that ant K has traversed.
2. Ant K withdraws from path point d_3 to path points c_3, $n(k) = n(k) - 1$, and $Tabu(n(k)) = c_3$ and continues to search horizontally according to step (1), i.e., goes to path point c_2 or path point c_4, finds no feasible path points, and marks the path point just traveled as infeasible, then $Tabu(n(k)) = \{c_2, c_3, c_4\}$.
3. Repeat step (2) until ant K withdraws from the path point c_3 to path points b_3, $n(k) = n(k) - 1$, and $Tabu(n(k)) = c_3$, at which point the ant searches laterally to reach either b_2 or b_4 and finds either $|Z_{b3}(K) - \{c_3\}|! = 0$, at which point ant K jumps out of the trap. $|Z_i(K)|$ represents the number of elements in the set of feasible nodes of ant K at this point. If the return to the original path point still does not allow the ant to jump out of the trap, the ant crosses the obstacle and finds a viable path point as the starting point of the adjusted mosaic.

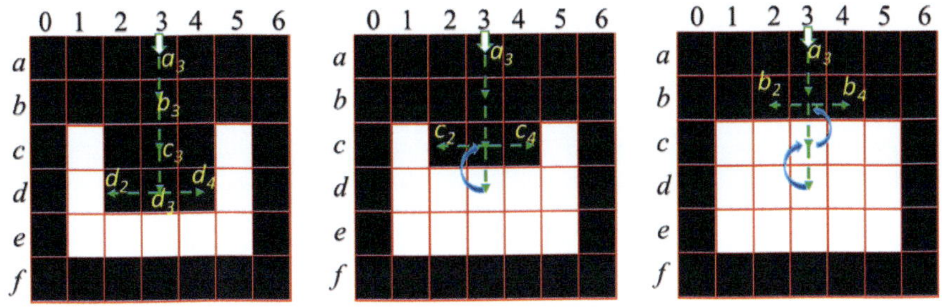

(a) (b) (c)

FIGURE 12.5 Schematic diagram of the evasive search strategy: (a) the search process, (b) a right search strategy, and (c) an completed search process.

TABLE 12.2
Details of the Two UAV Image Datasets

Descriptions \ Dataset	Dataset 1	Dataset 2
Location	Guilin, China	Nanning, China
Spatial resolution	0.1 m	0.1 m
The number of images	24	28
Image size	8048 × 5685 pixels	5472 × 3648 pixels
Mosaic type	Multiple images	Multiple images

12.4 EXPERIMENT AND ANALYSIS

12.4.1 Datasets

To verify the superiority of the method proposed in this chapter, experiments were conducted using two UAV image datasets from different regions. The first dataset contained 24 images with a size of 8048 × 5685 pixels^2 and was located in the City of Guilin, Guangxi, China. The image data were from three aerial strips with approximately 43% side overlap and 68% forward overlap. The second dataset contains 28 images with a size of 5472 × 3648 pixels and located in the City of Nanning, Guangxi, China. The image data were from three aerial strips with approximately 40% side overlap and 58% forward overlap. The details are listed in Table 12.2.

This experiment was implemented with the program developed using C++ based on OpenCV programming under Windows on a computer with a 3.4 GHz Intel core I7-4770 processor and 16 GB RAM.

12.4.2 Determining the Cost Threshold *N*

To ensure the running speed of the algorithm and reduce the workload of assessing the feasibility of a path point, the cost threshold of the feasible path point is first introduced to limit the scope of the ants' activities.

Figure 12.6a and d and b and e shows the left and right images of different regions, respectively. The gray scale cost map of the overlapping area of adjacent images can be calculated from equation (12.2). The results are shown in Figure 12.6c and f.

In practice, the threshold is determined by orthophoto quality. For orthophotos with poor quality, such as high noise, the threshold should be increased accordingly. Zhang et al. (2009) only consider

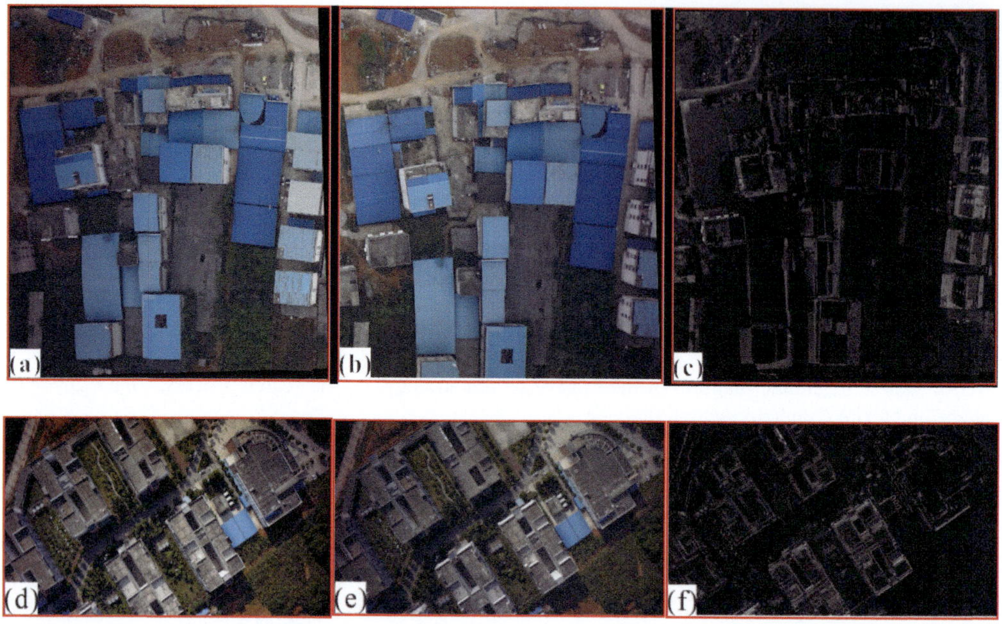

FIGURE12.6 Left orthophoto (a) and (d), right orthophoto (b) and (e), gray cost map (c) and (f).

the difference in gray and set the gray threshold to 30 per pixel. To better avoid obvious ground features on the ground, our method comprehensively considers the difference in gray and the difference in texture (gradient) and resets the cost threshold. In addition, in order to determine the cost threshold, a series of experiments was conducted. Figure 12.7a and d is a subset of Figure 12.6c and f, respectively. Figure 12.7b–d shows the feasible $(f \le N)$ and infeasible $(f > N)$ paths of ants whose gray threshold (N) was set to 15, 30, and 50, respectively. It can be seen that N set to 50 is reasonable, and most of the infeasible areas replaced the areas above the ground, such as buildings (see Figure 12.7). When$f(i,j) > N$, the path point is considered an obstacle point by the ant. The lower the threshold, N is generally set, the more areas are marked as obstacles. For example, when $N = 15$, there are too many areas marked as obstacles, i.e., it is impossible or difficult to distinguish between ground and buildings, which makes it difficult to find the optimal mosaic seamline. After the above tests, for high-resolution DOMs, when$N = 50$, it is easy to distinguish between ground and buildings, and other features, so $N = 50$ is used in the experiments of this chapter.

12.4.3 Automatic Selection of Seamline Network

This section uses Dataset 1 and Dataset 2 to verify the advantages and superiority of the method proposed in this chapter.

12.4.3.1 Experimental Result with Dataset 1

The selection of a mosaic network is finished through the following three steps.

Step 1: Construct the initial mosaic seamline network of multiple orthophotos, and the initial mosaic seamline network (see Figure 12.8a) generated by AVDO in Section 12.3.2.1. The mosaic line network generated by the AVDO method falls on buildings and other ground objects in many areas, such as area A, area B, area C, and area D in Figure 12.8.

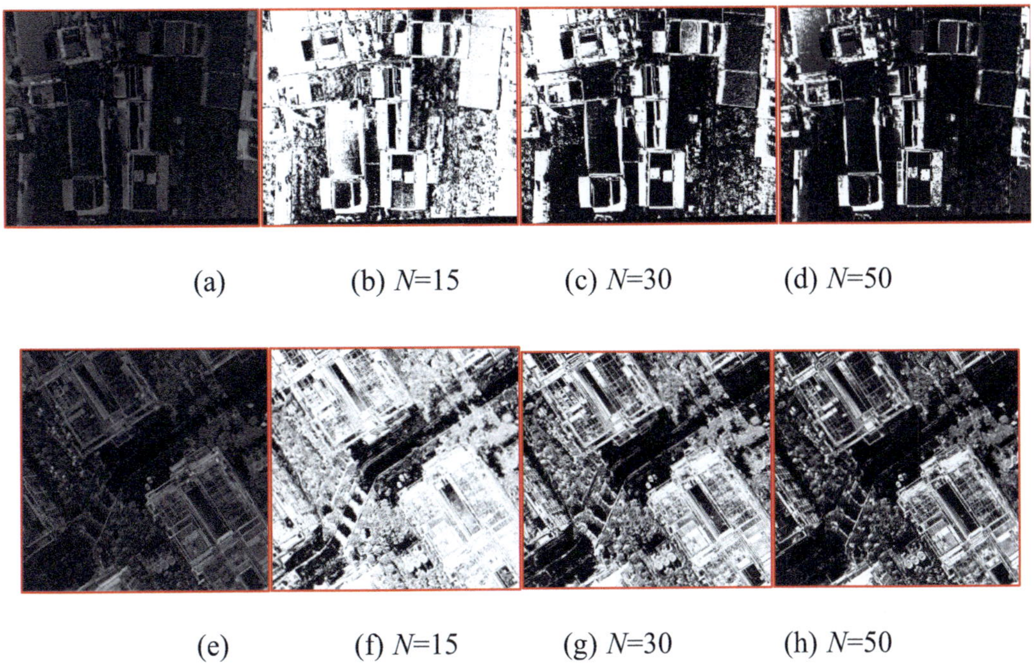

FIGURE 12.7 The threshold selection: orthophoto (a) and (e) are subsets of Figure 12.6c and f, respectively. The white areas (gray equal to 255) in (b), (c), (d), (f), (g), and (h) in the figure represent the area where the cost is higher than N.

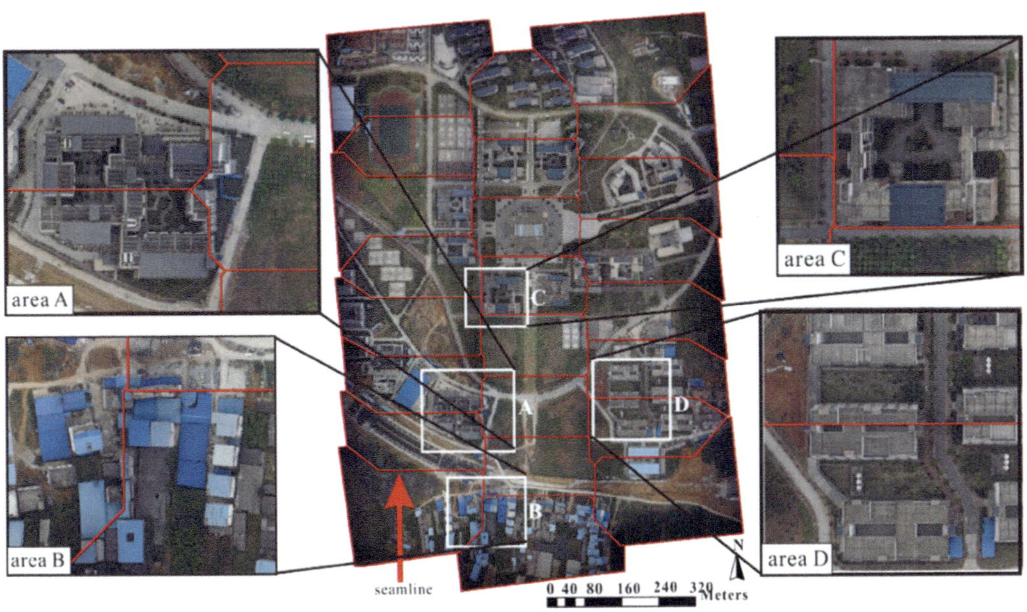

FIGURE 12.8 Initial mosaic seamline networks generated by AVDO.

Step 2: Adjust the position of the vertices of the mosaic seamline. Place the ant colony at the starting points of the initial mosaic seamline (S_{01}, S_{02}, S_{03}, etc. in Figure 12.9a) and use a once continuous space ant colony algorithm to search for the ending points (E_{01}, E_{02}, E_{03}, etc. in Figure 12.9a) to obtain the mosaic line of adjacent images. Then, the intersection formed by the mosaic lines of adjacent images and the initial mosaic seamlines formed before the flight belt is regarded as the adjusted starting point (S'_{01}, S'_{02}, S'_{03}, etc. in Figure 12.9b) or end point (E'_{01}, E'_{02}, E'_{03}, etc. in Figure 12.9b).

Step 3: Select the mosaic network in parallel. Place the ant colony at each adjusted starting point and search in parallel to the respective end points to obtain the optimal mosaic line of each segment and then connect these mosaic lines to form the global optimal mosaic seamline network (see Figure 12.10d) generated by our method.

Figure 12.10a–d shows the mosaic seamline network generated by the Qmosaic, OrthoVista, Dijkstra, and our method, respectively.

It can be seen from the mosaicked orthophotos that the seamline network generated by our method adjusts the vertex to avoid falling onto obvious objects and effectively avoids the buildings. The seamlines generated by the other two methods partially passed ground objects, such as buildings, resulting in visual discontinuity.

12.4.3.2 Experimental Result with Dataset 2

The selection of a mosaic network is finished through the following three steps.

Step 1: Construct the initial mosaic seamline network of multiple orthophotos and the initial mosaic seamline network (see Figure 12.11a) generated by AVDO in Section 12.3.2.1. The mosaic line network generated by the AVDO method falls on buildings and other ground objects in many areas, such as area *A*, area *B*, area *C*, and area *D* in Figure 12.11.

(a) (b)

FIGURE 12.9 (a) The starting and ending points of the mosaic line before adjustment and (b) the starting and ending points of the mosaic line after adjustment.

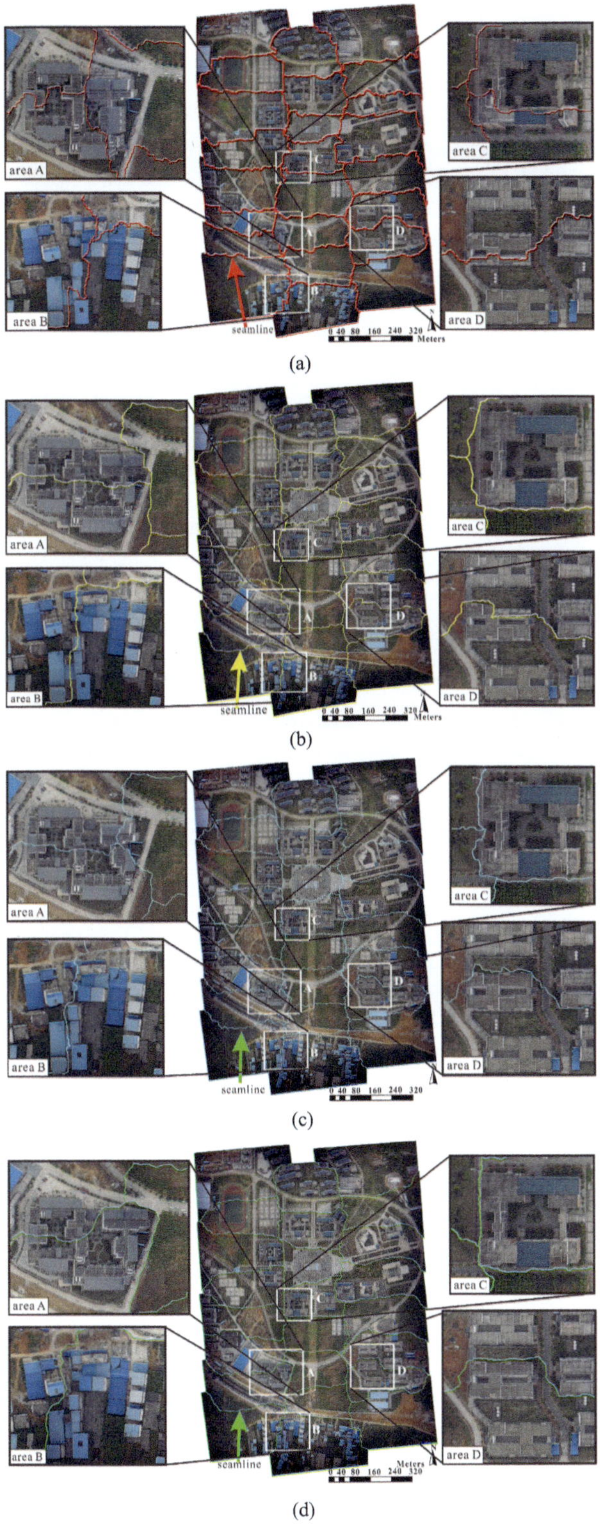

FIGURE 12.10 Mosaic seamline networks generated by different methods on Dataset 1: (a) mosaic seamline network generated using Qmosaic, (b) mosaic seamline network generated using OrthoVista, (c) Dijkstra, and (d) mosaic seamline network generated using our method. (Courtesy of Wang et al. 2022.)

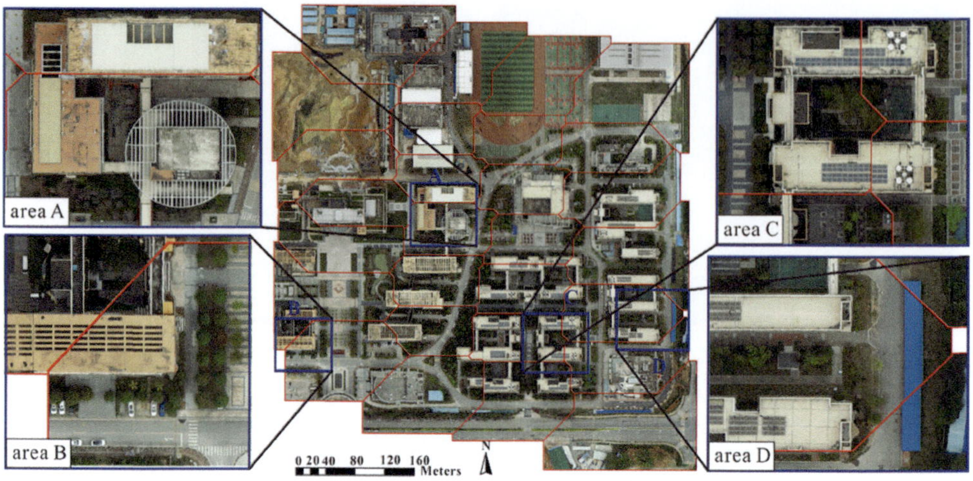

FIGURE 12.11 Initial mosaic seamline networks generated by AVDO. (Courtesy of Wang et al. 2022.)

(a) (b)

FIGURE 12.12 (a) The starting and ending points of the mosaic line before adjustment and (b) the starting and ending points of the mosaic line after adjustment. (Courtesy of Wang et al. 2022.)

Step 2: Adjust the position of the vertices of the mosaic seamline. Place the ant colony at the starting points of the initial mosaic seamline (S_{01}, S_{02}, S_{03}, etc. in Figure 12.12a), and use a once continuous space ant colony algorithm to search for the ending points (E_{01}, E_{02}, E_{03}, etc. in Figure 12.12a) to obtain the mosaic line of adjacent images. Then, the intersection formed by the mosaic lines of adjacent images and the initial mosaic seamlines formed before the flight belt is regarded as the adjusted starting point (S'_{01}, S'_{02}, S'_{03}, etc. in Figure 12.12b) or end point (E'_{01}, E'_{02}, E'_{03}, etc. in Figure 12.12b).

Step 3: Select the mosaic network in parallel. Place the ant colony at each adjusted starting point and search in parallel to ending points to obtain the optimal mosaic line of each segment, and then connect these mosaic lines to form the global optimal mosaic seamline network (see Figure 12.13d) generated by our method.

Figure 12.13a–d shows the mosaic seamline network generated by the Qmosaic, OrthoVista, Dijkstra, and our method, respectively.

It can be seen from the detailed image that the mosaic seamline network generated by our method adjusts the vertex to avoid falling onto obvious objects, and the mosaic seamline generated effectively avoids the buildings. The mosaic seamlines generated by the other two methods partially penetrated ground objects, such as buildings, resulting in visual discontinuity.

12.4.4 Discussion and Analysis

To compare the quality of the mosaic seamlines generated by different methods, we quantitatively count the number of mosaic seamlines passing through ground objects, the average cost, and the running time. The results are listed in Table 12.3.

As observed from Table 12.3, it can be concluded that our method saves 50 and 53 seconds for Dataset 1 and Dataset 2, respectively, when compared with OrthoVista. It can be concluded that our method saves 19 and 22 seconds for Dataset 1 and Dataset 2, respectively, when compared with Dijkstra. Our method runs longer than the Qmosaic, but the mosaic seamline network generated by our method passes through fewer ground objects. The average cost of the mosaic seamline network generated by our method in Dataset 1 is the lowest (15.56), which decreases 28.59% relative to that from OrthoVista (21.79), decreases 15.20% relative to that from Dijkstra (18.35), and decreases 16.70% relative to that from Qmosaic (18.68). The average cost of the mosaic seamline network generated by our method in Dataset 2 is the lowest (19.28), which decreases 26.77% relative to that from OrthoVista (26.33), decreases 14.69% relative to that from Dijkstra (22.60), and decreases 18.17% relative to that from Qmosaic (23.56).

To further verify the superiority of our method, this chapter uses the cost (see equation 12.3) to represent pixel differences, which include gray-scale differences and gradient differences. This is because the smaller the pixel difference that a mosaic line passes through, the better the quality of the selected mosaic seamlines (Pan et al. 2009). The detailed statistics are shown in Tables 12.4 and 12.5.

From Table 12.4, it can be obtained that

i. The average cost of the mosaic seamlines generated by our method in the four test areas (Figure 12.10) is 31.57, 19.79, 28.05, and 11.51, which decreases 9.72%, 19.49%, 25.08%, and 12.4% relative to that from OrthoVista; decreases 3.01%, 12.75%, 26.32%, and 12.80% relative to that from Dijkstra; and decreases 2.50%, 8.80%, 19.83%, and 17.96% relative to that from Qmosaic.
ii. The cost of mosaic seamlines from our method in the three test areas except area A is below 50. The length of mosaics seamlines with a cost of more than 50 in the four test areas selected by OrthoVista, Dijkstra, and Qmosaic accounts for about 5%.

From Table 12.5, it can be obtained that

i. The average cost of the mosaic seamlines generated by our method in the four test areas (Figure 12.13) is 10.98, 28.50, 21.31, and 18.86, which decreases 67.40%, 19.83%, 40.24%, and 25.63% relative to that from OrthoVista; decreases 64.42%, 3.91%, 42,34%, and 15.16% relative to that from Dijkstra; and decreases 64.82%, 5.66%, 34.71%, and 22.23% relative to that from Qmosaic.
ii. The length of mosaic seamlines with cost of more than 50 in the four test areas in Figure 12.13 selected by our method accounts for less than 5%, while the length of mosaic seamlines with cost (except Area *D*) of more than 50 selected by the other two methods accounts for more than 10%.

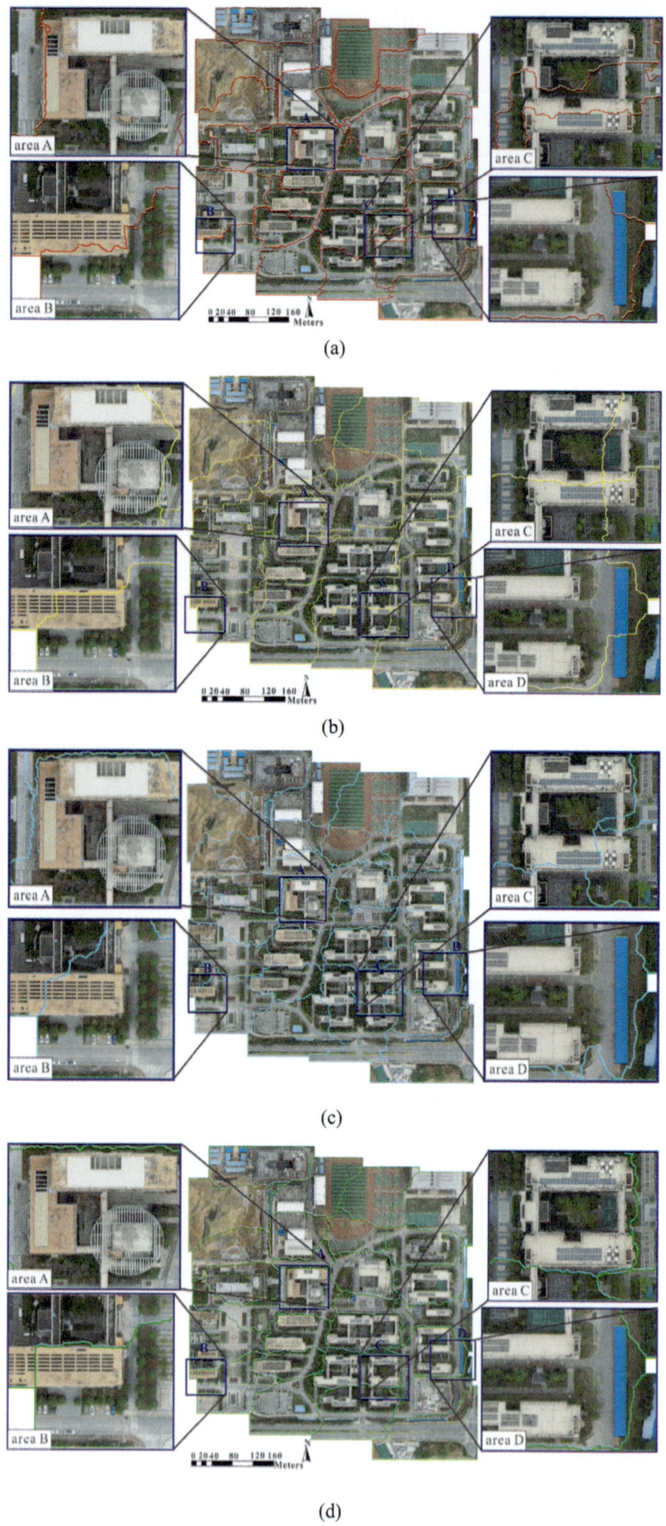

FIGURE 12.13 Mosaic seamline networks generated by different methods on Dataset 2: (a) mosaic seamline network generated using Qmosaic, (b) mosaic seamline network generated using OrthoVista, (c) Dijkstra, and (d) mosaic seamline network generated using our method. (Courtesy of Wang et al. 2022.)

TABLE 12.3
Statistical Analysis for Different Methods

	Number of Through Obvious Objects by Seamlines		Average Cost		Running Time (s)	
Methods	**Dataset 1**	**Dataset 2**	**Dataset 1**	**Dataset 2**	**Dataset 1**	**Dataset 2**
Qmosaic	10	11	18.68	23.56	47	102
OrthoVista	12	15	21.79	26.33	136	182
Dijkstra	11	12	18.35	22.60	105	151
Our method	6	8	15.56	19.28	86	129

TABLE 12.4
Statistics of Mosaic Seamlines in Test Area *A–D* in Dataset 1

Test Area	Methods	Maximum Cost	Minimum Cost	Average Cost	Cost ≤ 30	30 < Cost ≤ 50	50 < Cost ≤ 80	Cost > 80
Area *A*	Qmosaic	177	0	32.38	52.79%	35.56%	7.71%	3.94%
	OrthoVista	154	0	34.97	46.12%	35.31%	13.89%	4.68%
	Dijkstra	163	0	32.55	48.22%	34.99%	12.48%	4.31%
	Our method	140	0	31.57	52.69%	34.39%	8.21%	4.71%
Area *B*	Qmosaic	103	0	21.70	76.65%	9.46%	12.85%	1.04%
	OrthoVista	121	0	24.58	70.31%	17.01%	11.03%	1.65%
	Dijkstra	76	0	22.89	71.65%	19.89%	6.19%	2.27%
	Our method	48	0	19.97	79.09%	20.91%	0%	0%
Area *C*	Qmosaic	94	0	34.99	37.70%	49.02%	13.09%	0.2%
	OrthoVista	103	0	37.44	27.35%	57.94%	14.32%	0.39%
	Dijkstra	89	0	38.07	28.71%	58.84%	11.68%	0.77%
	Our method	49	0	28.05	55.34%	44.66%	0%	0%
Area *D*	Qmosaic	120	0	14.03	88.70%	8.04%	2.73%	0.53%
	OrthoVista	115	0	13.14	92.41%	5.04%	2.07%	0.58%
	Dijkstra	109	0	13.20	91.52%	5.01%	3.01%	0.46%
	Our method	48	0	11.51	97.78%	2.22%	0%	0%

TABLE 12.5
Statistics of Mosaic Seamlines in Test Area *A–D* in Dataset 2

Test Area	Methods	Maximum Cost	Minimum Cost	Average Cost	Cost ≤ 30	30 < Cost ≤ 50	50 < Cost ≤ 80	Cost > 80
Area *A*	Qmosaic	121	0	31.21	51.97%	36.26%	8.44%	3.33%
	OrthoVista	164	0	33.69	47.63%	38.21%	12.13%	2.03%
	Dijkstra	118	0	30.86	55.73%	36.333%	6.82%	1.12%
	Our method	47	0	10.98	98.85%	1.15%	0%	0%

(Continued)

TABLE 12.5 (*Continued*)
Statistics of Mosaic Seamlines in Test Area *A–D* in Dataset 2

Test Area	Methods	Maximum Cost	Minimum Cost	Average Cost	Cost ≤ 30	30 < Cost ≤ 50	50 < Cost ≤ 80	Cost > 80
Area *B*	Qmosaic	115	0	30.21	53.28%	36.73%	8.29%	1.70%
	OrthoVista	126	0	35.55	45.89%	36.57%	15.13%	2.41%
	Dijkstra	98	0	29.66	54.43%	37.32%	8.06%	0.19%
	Our method	102	0	28.50	58.56%	40.22%	0.90%	0.32%
Area *C*	Qmosaic	175	0	32.64	47.39%	41.99%	7.83%	2.79%
	OrthoVista	189	0	35.66	46.32%	37.46%	14.65%	1.37%
	Dijkstra	192	0	36.96	44.83%	38.49%	15.05%	1.63%
	Our method	87	0	21.31	75.68%	22.53%	1.61%	0.18%
Area *D*	Qmosaic	125	0	24.25	71.03%	27.55%	1.21%	0.21%
	OrthoVista	142	0	25.36	69.37%	27.66%	2.53%	0.44%
	Dijkstra	56	0	22.23	85.47%	14.53%	0%	0%
	Our method	45	0	18.86	85.39%	14.61%	0%	0%

12.5 CONCLUSIONS

To obtain a high-quality DOM mosaicked from multiple DOMs, this chapter describes a novel method for the selection of seamline network constructed by the multiple DOMs to be mosaicked. This method first establishes a cost function on the basis of gray differences and a gray gradient in the overlapping area. Then, a continuous space ant colony algorithm is established to iteratively select the least cost path for the optimal mosaic seamline to avoid the obstacles, such as the buildings, trees, and other obvious objects.

Although the method can search a high-quality mosaic line network intelligently well, the method proposed in this chapter still needs a few improvements. For example, this method can adjust the vertex adjustment of the image mosaic seamline network, but does not apply to the mosaic seamline vertices of the disordered routes.

REFERENCES

Chen, Q.; Sun, M.; Hu, X.; Zhang, Z., Automatic seamline network generation for urban orthophoto mosaicking with the use of a digital surface model. *Remote Sensing*, 6(12), 2014, pp. 12334–12359.

Chon, J.; Kim, H., Determination of the optimal seam-lines in image mosaicking with the dynamic programming (DP) on the converted cost space. *Lecture Notes in Computer Science*, 4029, 2006, pp. 750–757.

Davis, J., Mosaics of scenes with moving objects. In: *Proceedings of the IEEE Computer Society Conference on Computer Vision & Pattern Recognition*, Los Alamitos, CA, 1998, pp. 354–360.

He, J.; Sun, M.; Chen, Q.; Zhang, Z., An improved approach for generating globally consistent seamline networks for aerial image mosaicking. *International Journal of Remote Sensing*, 40(3), 2019, pp. 859–882.

Hsu, S.; Kumar, R., Automated mosaics via topology inference. *Computer Graphics & Applications IEEE*, 22(2), 2002, pp. 44–54.

Kerschner, M., Seamline detection in colour orthoimage mosaicking by use of twin snakes. *ISPRS Journal of Photogrammetry & Remote Sensing*, 56(1), 2001, pp. 53–64.

Li, L.; Tu, J.; Gong, Y.; Yao, J.; Li, J., Seamline network generation based on foreground segmentation for orthoimage mosaicking. *ISPRS Journal of Photogrammetry & Remote Sensing*, 148, 2019, pp. 41–53.

Li, L.; Yao, J.; Lu, X.; Tu, J.; Shan, J., Optimal seamline detection for multiple image mosaicking via graph cuts. *ISPRS Journal of Photogrammetry & Remote Sensing*, 113, 2016, pp. 1–16.

Li, L.; Yao, J.; Shi, S.; Yuan, S.; Zhang, Y.; Li, J., Superpixel-based optimal seamline detection in the gradient domain via graph cuts for orthoimage mosaicking. *International Journal of Remote Sensing*, 39(11–12), 2018, pp. 3908–3925.

Li, L.; Yao, J.; Xie, R.; Xia, M.; Xiang, B., Superpixel-based optimal seamline detection via graph cuts for panoramic images. In: *IEEE International Conference on Information & Automation*, Ningbo, China, 2017, pp. 1484–1489.

Ma, D.; Ding, N.; Cui, J.; Wang, J., Optimizing ortho-image mosaic seamline based on ant colony algorithm. *Science of Surveying and Mapping*, 38(2), 2013, pp. 117–108.

Ma, H.; Sun, J., Intelligent optimization of seam-line finding for orthophoto mosaicking with LiDAR point clouds. *Journal of Zhejiang University Science C: Computer & Electronics*, 12(5), 2011, pp. 417–429.

Pan, J.; Fang, Z.; Chen, S.; Ge, H.; Hu, F.; Wang, M., An improved seeded region growing-based seamline network generation method. *Remote Sensing*, 10(7), 2018, p. 1065.

Pan, J.; Wang, M.; Li, D.; Li, J., Automatic generation of seamline network using area Voronoi diagrams with overlap. *IEEE Transactions on Geoscience & Remote Sensing*, 47(6), 2009, pp. 1737–1744.

Pan, J.; Wang, M.; Li, J.; Yuan, S.; Hu, F., Region change rate-driven seamline determination method. *ISPRS Journal of Photogrammetry and Remote Sensing*, 105, 2015, pp. 141–154.

Pan, J.; Wang, M.; Ma, D.; Zhou, Q., Seamline network refinement based on area Voronoi diagrams with overlap. *IEEE Transactions on Geoscience & Remote Sensing*, 52(3), 2013, pp. 1658–1666.

Pan, J.; Yuan, S.; Li, J.; Yuan, S., Seamline optimization based on ground objects classes for orthoimage mosaicking. *Remote Sensing Letters*, 8(3), 2017, pp. 280–289.

Pang, S.; Sun, M.; Hu, X.; Zhang, Z., SGM-based seamline determination for urban orthophoto mosaicking. *ISPRS Journal of Photogrammetry & Remote Sensing*, 112, 2016, pp. 1–12.

Stützle, T.; Hoos, H., MAX-MIN ant system. *Future Generation Computer Systems*, 16(8), 2000, pp. 889–914.

Wan, Y.; Wang, D.; Xiao, J.; Lai, X.; Xu, J., Automatic determination of seamlines for aerial image mosaicking based on vector roads alone. *ISPRS Journal of Photogrammetry & Remote Sensing*, 76, 2013, pp. 1–10.

Wang, D.; Cao, W.; Xin, X.; Shao, Q.; Brolly, M.; Xiao, J.; Wan, Y.; Zhang, Y., Using vector building maps to aid in generating seams for low-attitude aerial orthoimage mosaicking: Advantages in avoiding the crossing of buildings. *ISPRS Journal of Photogrammetry and Remote Sensing*, 125(2017), 2017, pp. 207–224.

Wang, D.; Wan, Y.; Xiao, J.; Lai, X.; Huang, W.; Xu, J., Aerial image mosaicking with the aid of vector roads. *Photogrammetric Engineering & Remote Sensing*, 78(11), 2012, pp. 1141–1150.

Wang, Q.; Zhou, G.; Song, R.; Xie, Y.; Luo, M.; Yue, T., Continuous space ant colony algorithm for automatic selection of orthophoto mosaic seamline network. *ISPRS Journal of Photogrammetry and Remote Sensing*, 186, 2022, pp. 201–217. https://doi.org/10.1016/j.isprsjprs.2022.02.011

Wells, J. C. K.; Desilva, J. M.; Stock, J. T., The obstetric dilemma: an ancient game of Russian roulette, or a variable dilemma sensitive to ecology. *American Journal of Physical Anthropology*, 149(S55), 2012, pp. 40–71.

Williams, D. J.; Shah, M., A fast algorithm for active contours and curvature estimation. *CVGIP: Image Understanding*, 55(1), 1992, pp. 14–26.

Yu, L.; Zhang, Y.; Sun, M.; Zhou, X.; Liu, C., An auto-adapting global-to-local color balancing method for optical imagery mosaic. *ISPRS Journal of Photogrammetry & Remote Sensing*, 132(2017), 2017, pp. 1–19.

Yuan, X.; Duan, M.; Cao, J., A seam line detection algorithm for orthophoto mosaicking based on disparity image. *Acta Geodaetica et Cartographica Sinica*, 44(8), 2015, pp. 877–883.

Zhang, J.; Sun, M.; Zhang, Z., Automated seamline detection for orthophoto mosaicking based on ant colony algorithm. *Geomatics and Information Science of Wuhan University*, 36(6), 2009, pp. 675–678.

Zhang, Y.; Zhang, M.; Du, S.; Zou, Z.; Fan, C., Seamline optimisation for urban aerial ortho-image mosaicking using graph cuts. *The Photogrammetric Record*, 33(161), 2018, pp. 131–147.

Zheng, M.; Xiong, X.; Zhu, J., Automatic seam-line determination for orthoimage mosaics using edge-tracking based on a DSM. *Remote Sensing Letters*, 8(10), 2017a, pp. 977–986.

Zheng, M.; Xiong, X.; Zhu, J., A novel orthoimage mosaic method using a weighted A* algorithm – Implementation and evaluation. *ISPRS Journal of Photogrammetry and Remote Sensing*, 138, 2018, pp. 30–46.

Zheng, M.; Zhou, S.; Xiong, X.; Zhu, J., A novel orthoimage mosaic method using the weighted A* algorithm for UAV imagery. *Computers & Geosciences*, 109, 2017b, pp. 238–246.

Zhou, G., *Urban High-Resolution Remote Sensing: Algorithms and Modelling*, Taylor & Francis/CRC Press: New York, 2020.

Zhou, G.; Chen, W.; Kelmelis, J. A.; Zhang, D., A comprehensive study on urban true orthorectification. *IEEE Transactions on Geoscience and Remote Sensing*, 43(9), 2005, pp. 2138–2147.

Zhou, G.; Jezek, K. C., Satellite photograph mosaics of Greenland from the 1960s era. *International Journal of Remote Sensing*, 23(6), 2002, pp. 1143–1159.

Zhou, G.; Wang, Q.; Jia, B.; Pan, Q.; Wang, H., An improved adaptive ant colony algorithm for intelligent seamline detection of orthoimage mosaicking. In: *IGARSS* 2018–2018 *IEEE International Geoscience and Remote Sensing Symposium*, Wuhan, China, 2018, pp. 1320–1323.

Zhou, G.; Wang, Y.; Yue, T.; Ye, S.; Wang, W., Building occlusion detection from ghost images. *IEEE Transactions on Geoscience and Remote Sensing*, 55(2), 2017a, pp. 1074–1084.

Zhou, G.; Yue, T.; Shi, Y.; Zhang, R.; Huang, J., Second-order polynomial equation-based block adjustment for orthorectification of DISP imagery. *Remote Sensing*, 8(8), 2016, p. 680.

Zhou, G.; Zhang, R.; Liu, N.; Huang, J.; Zhou, X., On-board ortho-rectification for images based on an FPGA. *Remote Sensing*, 9(9), 2017b, p. 874.

Zhou, G.; Zhang, R.; Zhang, D.; Huang, J.; Baysal, O., Real-time ortho-rectification for remote-sensing images. *International Journal of Remote Sensing*, 40(5–6), 2019, pp. 2451–2465.

Appendix A
Partial Derivatives of Collinearity Equations

A.1 PARTIAL DERIVATIVES WITH RESPECT TO SENSOR'S EXTERIOR ORIENTATION ELEMENTS

If there is one ground control point (GCP), G, the satellite flying epoch is t.

Also, the position and attitude parameters of a sensor on-board spaceborne in the same strip can be expressed by a third-order polynomial (Li and Zhou 2002), i.e.,

$$X_S(t) = a_0 + a_1 t + a_2 t^2 + a_3 t^3$$

$$Y_S(t) = b_0 + b_1 t + b_2 t^2 + b_3 t^3$$

$$Z_S(t) = c_0 + c_1 t + c_2 t^2 + c_3 t^3$$

$$\phi_S(t) = d_0 + d_1 t + d_2 t^2 + d_3 t^3$$

$$\omega_S(t) = e_0 + e_1 t + e_2 t^2 + e_3 t^3$$

$$\kappa_S(t) = f_0 + f_1 t + f_2 t^2 + f_3 t^3$$

For the GCP, G, its collinearity equation is

$$x_g - x_0 = f\frac{r_{11}(X_G - X_S(t)) + r_{12}(Y_G - Y_S(t)) + r_{13}(Z_G - Z_S(t))}{r_{31}(X_G - X_S(t)) + r_{32}(Y_G - Y_S(t)) + r_{33}(Z_G - Z_S(t))} = f\frac{\overline{X}}{\overline{Z}}$$

$$y_g - y_0 = f\frac{r_{21}(X_G - X_S(t)) + r_{22}(Y_G - Y_S(t)) + r_{23}(Z_G - Z_S(t))}{r_{31}(X_G - X_S(t)) + r_{32}(Y_G - Y_S(t)) + r_{33}(Z_G - Z_S(t))} = f\frac{\overline{Y}}{\overline{Z}}$$

where

$$\overline{X} = r_{11}(X_G - X_S(t)) + r_{12}(Y_G - Y_S(t)) + r_{13}(Z_G - Z_S(t))$$

$$\overline{Y} = r_{21}(X_G - X_S(t)) + r_{22}(Y_G - Y_S(t)) + r_{23}(Z_G - Z_S(t))$$

$$\overline{Z} = r_{31}(X_G - X_S(t)) + r_{32}(Y_G - Y_S(t)) + r_{33}(Z_G - Z_S(t))$$

Partial differentials of (x_g, y_g) with respect to $(Xs(t), Y_S(t), Z_S(t))$ and $(\phi_S(t), \omega_S(t), \kappa_S(t))$ are

$$\frac{\partial x_R}{\partial X_S(t)} dX_S(t) = \frac{\partial x_R}{\partial X_S(t)}\left(da_0 + tda_1 + t^2 da_2 + t^3 da_3\right)$$

$$\frac{\partial x_R}{\partial Y_S(t)} dY_S(t) = \frac{\partial x_R}{\partial Y_S(t)}\left(db_0 + tdb_1 + t^2 db_2 + t^3 db_3\right)$$

$$\frac{\partial x_R}{\partial Z_S(t)} dZ_S(t) = \frac{\partial x_R}{\partial Z_S(t)}\left(dc_0 + tdc_1 + t^2 dc_2 + t^3 dc_3\right)$$

$$\frac{\partial x_R}{\partial \phi_S(t)} d\phi_S(t) = \frac{\partial x_R}{\partial \phi_S(t)}\left(dd_0 + tdd_1 + t^2 dd_2 + t^3 dd_3\right)$$

$$\frac{\partial x_R}{\partial \omega_S(t)} d\omega_S(t) = \frac{\partial x_R}{\partial \omega_S(t)}\left(de_0 + tde_1 + t^2 de_2 + t^3 de_3\right)$$

$$\frac{\partial x_R}{\partial \kappa_S(t)} d\kappa_S(t) = \frac{\partial x_R}{\partial \kappa_S(t)}\left(df_0 + tdf_1 + t^2 df_2 + t^3 df_3\right)$$

$$\frac{\partial y_R}{\partial X_S(t)} dX_S(t) = \frac{\partial y_R}{\partial X_S(t)}\left(da_0 + tda_1 + t^2 da_2 + t^3 da_3\right)$$

$$\frac{\partial y_R}{\partial Y_S(t)} dY_S(t) = \frac{\partial y_R}{\partial Y_S(t)}\left(db_0 + tdb_1 + t^2 db_2 + t^3 db_3\right)$$

$$\frac{\partial y_R}{\partial Z_S(t)} dZ_S(t) = \frac{\partial y_R}{\partial Z_S(t)}\left(dc_0 + tdc_1 + t^2 dc_2 + t^3 dc_3\right)$$

$$\frac{\partial y_R}{\partial \phi_S(t)} d\phi_S(t) = \frac{\partial y_R}{\partial \phi_S(t)}\left(dd_0 + tdd_1 + t^2 dd_2 + t^3 dd_3\right)$$

$$\frac{\partial y_R}{\partial \omega_S(t)} d\omega_S(t) = \frac{\partial y_R}{\partial \omega_S(t)}\left(de_0 + tde_1 + t^2 de_2 + t^3 de_3\right)$$

$$\frac{\partial y_R}{\partial \kappa_S(t)} d\kappa_S(t) = \frac{\partial y_R}{\partial \kappa_S(t)}\left(df_0 + tdf_1 + t^2 df_2 + t^3 df_3\right)$$

The partial derivatives are as follows:

$$\frac{\partial x_g}{\partial X_S(t)} = f\,\frac{\dfrac{\partial \bar{X}}{\partial X_S(t)}\bar{Z} - \dfrac{\partial \bar{Z}}{\partial X_S(t)}\bar{X}}{\bar{Z}\bar{Z}} = f\,\frac{-r_{11}\bar{Z} + r_{31}\bar{X}}{\bar{Z}\bar{Z}} = \frac{1}{\bar{Z}}\left(-fr_{11} + r_{31}x_R\right) \quad \text{(unit: mm/m)}$$

$$\frac{\partial x_f}{\partial Y_S(t)} = f\frac{\frac{\partial \bar{X}}{\partial Y_S(t)}\bar{Z} - \frac{\partial \bar{Z}}{\partial Y_S(t)}\bar{X}}{\bar{Z}\bar{Z}} = f\frac{-r_{12}\bar{Z} + r_{32}\bar{X}}{\bar{Z}\bar{Z}} = \frac{1}{\bar{Z}}\left(-fr_{12} + r_{32}x_g\right)$$

$$\frac{\partial x_R}{\partial Z_S(t)} = f\frac{\frac{\partial \bar{X}}{\partial Z_S(t)}\bar{Z} - \frac{\partial \bar{Z}}{\partial Z_S(t)}\bar{X}}{\bar{Z}\bar{Z}} = f\frac{-r_{13}\bar{Z} + r_{33}\bar{X}}{\bar{Z}\bar{Z}} = \frac{1}{\bar{Z}}\left(-fr_{13} + r_{33}x_g\right)$$

$$\frac{\partial x_g}{\partial \phi_S(t)} = f\frac{\frac{\partial \bar{X}}{\partial \phi_S(t)}\bar{Z} - \frac{\partial \bar{Z}}{\partial \phi_S(t)}\bar{X}}{\bar{Z}\bar{Z}} = \frac{f}{\bar{Z}\bar{Z}}\left\{\begin{array}{l}\bar{Z}\left[\frac{\partial r_{11}}{\partial \phi_S(t)}(X_G - X_S) + \frac{\partial r_{12}}{\partial \phi_S(t)}(Y_G - Y)_S + \frac{\partial r_{13}}{\partial \phi_S(t)}(Z_G - Z_S)\right] \\ -\bar{X}\left[\frac{\partial r_{31}}{\partial \phi_S(t)}(X_G - X_S) + \frac{\partial r_{32}}{\partial \phi_S(t)}(Y_G - Y_S) + \frac{\partial r_{33}}{\partial \phi_S(t)}(Z_G - Z_S)\right]\end{array}\right\}$$

$$= \frac{f}{\bar{Z}\bar{Z}}\left\{\begin{array}{l}\bar{Z}\left[-\sin\phi\cos\kappa(X_G - X_S) + \sin\omega\cos\phi\cos\kappa(Y_G - Y_S) - \cos\omega\cos\phi\cos\kappa(Z_G - Z_S)\right] \\ -\bar{X}\left[\cos\phi(X_G - X_S) + \sin\omega\sin\phi(Y_G - Y_S) - \cos\omega\sin\phi(Z_G - Z_S)\right]\end{array}\right\}$$

$$= \frac{f}{\bar{Z}\bar{Z}}\left\{\begin{array}{l}\bar{Z}\left[-\sin\phi\cos\kappa(X_G - X_S) + r_{11}\sin\omega(Y_G - Y_S) - r_{33}\cos\kappa(Z_G - Z_S)\right] \\ -\bar{X}\left[\cos\phi(X_G - X_S) + \sin\omega\sin\phi(Y_G - Y_S) - \cos\omega\sin\phi(Z_G - Z_S)\right]\end{array}\right\}$$

$$\frac{\partial x_g}{\partial \omega_S(t)} = f\frac{\frac{\partial \bar{X}}{\partial \omega_S(t)}\bar{Z} - \frac{\partial \bar{Z}}{\partial \omega_S(t)}\bar{X}}{\bar{Z}\bar{Z}}$$

$$= \frac{f}{\bar{Z}\bar{Z}}\left\{\begin{array}{l}\bar{Z}\left[\frac{\partial r_{11}}{\partial \omega_S(t)}(X_G - X_S) + \frac{\partial r_{12}}{\partial \omega_S(t)}(Y_G - Y_S) + \frac{\partial r_{13}}{\partial \omega_S(t)}(Z_G - Z_S)\right] \\ -\bar{X}\left[\frac{\partial r_{31}}{\partial \omega_S(t)}(X_G - X_S) + \frac{\partial r_{32}}{\partial \omega_S(t)}(Y_G - Y_S) + \frac{\partial r_{33}}{\partial \omega_S(t)}(Z_G - Z_S)\right]\end{array}\right\}$$

$$= \frac{f}{\bar{Z}\bar{Z}}\left\{\begin{array}{l}\bar{Z}\left[\begin{array}{l}0\cdot(X_G - X_S) + (-\sin\omega\sin\kappa + \cos\omega\sin\phi\cos\kappa)(Y_G - Y_S) \\ +(\cos\omega\sin\kappa + \sin\omega\sin\phi\cos\kappa(Z_G - Z_S)\end{array}\right] \\ -\bar{X}\left[0\cdot(X_G - X_S) - \cos\omega\cos\phi(Y_G - Y_S) - \sin\omega\cos\phi(Z_G - Z_S)\right]\end{array}\right\}$$

$$= \frac{f}{\bar{Z}\bar{Z}}\left\{\bar{Z}\left[-r_{13}(Y_G - Y_S) + r_{12}(Z_G - Z_S)\right] - \bar{X}\left[-r_{33}(Y_G - Y_S) + r_{32}(Z_G - Z_S)\right]\right\}$$

$$\frac{\partial x_g}{\partial \kappa_S(t)} = f\frac{\frac{\partial \bar{X}}{\partial \kappa_S(t)}\bar{Z} - \frac{\partial \bar{Z}}{\partial \kappa_S(t)}\bar{X}}{\bar{Z}\bar{Z}} = \frac{f}{\bar{Z}\bar{Z}}\left\{\begin{array}{l}\bar{Z}\left[\begin{array}{l}-\cos\phi\sin\kappa(X_G - X_S) + (\cos\omega\cos\kappa - \sin\omega\sin\phi\sin\kappa) \\ (Y_G - Y_S) + (\sin\omega\cos\kappa + \cos\omega\sin\phi\sin\kappa)(Z_G - Z_S)\end{array}\right] \\ -\bar{X}\left[0\cdot(X_G - X_S) + 0\cdot(Y_G - Y_S) + 0\cdot(Z_G - Z_S)\right]\end{array}\right\}$$

$$= \frac{f}{\bar{Z}\bar{Z}}\left\{\bar{Z}\left[r_{21}(X_G - X_S) + r_{22}(Y_G - Y_S) + r_{23}(Z_G - Z_S)\right]\right\}$$

$$= \frac{f}{\bar{Z}}(\bar{Y}) = y_g$$

$$\frac{\partial y_g}{\partial X_S(t)} = f\frac{\frac{\partial \overline{Y}}{\partial X_S(t)}\overline{Z} - \frac{\partial \overline{Z}}{\partial X_S(t)}\overline{Y}}{\overline{Z}\overline{Z}} = f\frac{-r_{21}\overline{Z} + r_{31}\overline{Y}}{\overline{Z}\overline{Z}} = \frac{1}{\overline{Z}}\left(-fr_{21} + r_{31}y_g\right)$$

$$\frac{\partial y_g}{\partial Y_S(t)} = f\frac{\frac{\partial \overline{Y}}{\partial Y_S(t)}\overline{Z} - \frac{\partial \overline{Z}}{\partial Y_S(t)}\overline{Y}}{\overline{Z}\overline{Z}} = z_R\frac{-r_{22}\overline{Z} + r_{32}\overline{Y}}{\overline{Z}\overline{Z}} = \frac{1}{\overline{Z}}\left(-fr_{22} + r_{32}y_g\right)$$

$$\frac{\partial y_g}{\partial Z_S(t)} = f\frac{\frac{\partial \overline{Y}}{\partial Z_S(t)}\overline{Z} - \frac{\partial \overline{Z}}{\partial Z_S(t)}\overline{Y}}{\overline{Z}\overline{Z}} = z_R\frac{-r_{23}\overline{Z} + r_{33}\overline{X}}{\overline{Z}\overline{Z}} = \frac{1}{\overline{Z}}\left(-fr_{23} + r_{33}y_g\right)$$

$$\frac{\partial y_g}{\partial \phi_S(t)} = f\frac{\frac{\partial \overline{Y}}{\partial \phi_S(t)}\overline{Z} - \frac{\partial \overline{Z}}{\partial \phi_S(t)}\overline{Y}}{\overline{Z}\overline{Z}} = \frac{f}{\overline{Z}\overline{Z}}\left\{\begin{array}{l}\overline{Z}\left[\frac{\partial r_{21}}{\partial \phi_S(t)}(X_G - X_S) + \frac{\partial r_{22}}{\partial \phi_S(t)}(Y_G - Y_S) + \frac{\partial r_{23}}{\partial \phi_S(t)}(Z_G - Z_S)\right] \\ -\overline{Y}\left[\frac{\partial r_{31}}{\partial \phi_S(t)}(X_G - X_S) + \frac{\partial r_{32}}{\partial \phi_S(t)}(Y_G - Y_S) + \frac{\partial r_{33}}{\partial \phi_S(t)}(Z_G - Z_S)\right]\end{array}\right\}$$

$$= \frac{f}{\overline{Z}\overline{Z}}\left\{\begin{array}{l}\overline{Z}\left[\sin\phi\sin\kappa(X_G - X_S) - \sin\omega\cos\phi\sin\kappa(Y_G - Y_S) + \cos\omega\cos\phi\sin\kappa(Z_G - Z_S)\right] \\ -\overline{Y}\left[\cos\phi(X_G - X_S) + \sin\omega\sin\phi(Y_G - Y_S) - \cos\omega\sin\phi(Z_G - Z_S)\right]\end{array}\right\}$$

$$= \frac{f}{\overline{Z}\overline{Z}}\left\{\begin{array}{l}\overline{Z}\left[\sin\phi\sin\kappa(X_G - X_S) + r_{21}\sin\omega(Y_G - Y_S) + r_{33}\sin\kappa(Z_G - Z_S)\right] \\ -\overline{Y}\left[\cos\phi(X_G - X_S) + \sin\omega\sin\phi(Y_G - Y_S) - \cos\omega\sin\phi(Z_G - Z_S)\right]\end{array}\right\}$$

$$\frac{\partial y_g}{\partial \omega_S(t)} = f\frac{\frac{\partial \overline{Y}}{\partial \omega_S(t)}\overline{Z} - \frac{\partial \overline{Z}}{\partial \omega_S(t)}\overline{Y}}{\overline{Z}\overline{Z}}$$

$$= \frac{f}{\overline{Z}\overline{Z}}\left\{\begin{array}{l}\overline{Z}\left[\frac{\partial r_{21}}{\partial \omega_S(t)}\cdot(X_G - X_S) + \frac{\partial r_{22}}{\partial \omega_S(t)}(Y_G - Y_S) + \frac{\partial r_{23}}{\partial \omega_S(t)}(Z_G - Z_S)\right] \\ -\overline{Y}\left[\frac{\partial r_{31}}{\partial \phi_S(t)}(X_G - X_S) + \frac{\partial r_{32}}{\partial \phi_S(t)}(Y_G - Y_S) + \frac{\partial r_{33}}{\partial \phi_S(t)}(Z_G - Z_S)\right]\end{array}\right\}$$

$$= \frac{f}{\overline{Z}\overline{Z}}\left\{\begin{array}{l}\overline{Z}\left[\begin{array}{l}0\cdot(X_G - X_S) + (-\sin\omega\cos\kappa - \cos\omega\sin\phi\sin\kappa)(Y_G - Y_S) \\ +(\cos\omega\cos\kappa - \sin\omega\sin\phi\sin\kappa)(Z_G - Z_S)\end{array}\right] \\ -\overline{Y}\left[0\cdot(X_G - X_S) - \cos\omega\cos\phi(Y_G - Y_S) - \sin\omega\cos\phi(Z_G - Z_S)\right]\end{array}\right\}$$

$$= \frac{f}{\overline{Z}\overline{Z}}\left\{\overline{Z}\left[-r_{23}(Y_G - Y_S) + r_{22}(Z_G - Z_S)\right] - \overline{Y}\left[-r_{33}(Y_G - Y_S) + r_{32}(Z_G - Z_S)\right]\right\}$$

$$\frac{\partial y_f}{\partial \kappa_S(t)} = f\frac{\frac{\partial \bar{Y}}{\partial \kappa_S(t)}\bar{Z} - \frac{\partial \bar{Z}}{\partial \kappa_S(t)}\bar{Y}}{\bar{Z}\bar{Z}}$$

$$= \frac{f}{\bar{Z}\bar{Z}}\left\{\begin{matrix}\bar{Z}\begin{bmatrix}-\cos\phi\cos\kappa(X_G - X_S) + (-\cos\omega\sin\kappa - \sin\omega\sin\phi\cos\kappa) \\ (Y_G - Y_S) + (-\sin\omega\sin\kappa + \cos\omega\sin\phi\cos\kappa)(Z_G - Z_S)\end{bmatrix} \\ -\bar{Y}\left[0\cdot(X_G - X_S) + 0\cdot(Y_G - Y_S) + 0\cdot(Z_G - Z_S)\right]\end{matrix}\right\}$$

$$= \frac{f}{\bar{Z}\bar{Z}}\left\{\bar{Z}\left[-r_{11}(X_G - X_S) - r_{12}(Y_G - Y_S) - r_{13}(Z_G - Z_S)\right]\right\}$$

$$= \frac{f}{\bar{Z}}(-\bar{X}) = -x_g$$

Finally, the coefficients of the observation equations are

$$\beta_{11} = \frac{\partial x_g}{\partial X_S(t)},\quad \beta_{12} = \frac{\partial x_g}{\partial Y_S(t)},\quad \beta_{13} = \frac{\partial x_g}{\partial Z_S(t)},\quad \beta_{14} = \frac{\partial x_g}{\partial \phi_S(t)},\quad \beta_{15} = \frac{\partial x_g}{\partial \omega_S(t)},$$

$$\beta_{16} = \frac{\partial x_g}{\partial k_S(t)},\quad \beta_{17} = t\cdot\beta_{11},\quad \beta_{18} = t\cdot\beta_{12},\quad \beta_{19} = t\cdot\beta_{13}$$

$$\beta_{110} = t\cdot\beta_{14},\quad \beta_{111} = t\cdot\beta_{15},\quad \beta_{112} = t\cdot\beta_{16},\quad \beta_{113} = t^2\cdot\beta_{11},\quad \beta_{114} = t^2\cdot\beta_{12},$$

$$\beta_{115} = t^2\cdot\beta_{13},\quad \beta_{116} = t^2\cdot\beta_{14},\quad \beta_{117} = t^2\cdot\beta_{15},\quad \beta_{118} = t^2\cdot\beta_{16},\quad \beta_{119} = t^3\cdot\beta_{11},$$

$$\beta_{120} = t^3\cdot\beta_{12},\quad \beta_{121} = t^3\cdot\beta_{13},\quad \beta_{122} = t^3\cdot\beta_{14},\quad \beta_{123} = t^3\cdot\beta_{15},\quad \beta_{124} = t^3\cdot\beta_{16}$$

$$\beta_{21} = \frac{\partial y_g}{\partial X_S(t)},\quad \beta_{22} = \frac{\partial y_g}{\partial Y_S(t)},\quad \beta_{23} = \frac{\partial y_g}{\partial Z_S(t)},\quad \beta_{24} = \frac{\partial y_g}{\partial \phi_S(t)},\quad \beta_{25} = \frac{\partial y_g}{\partial \omega_S(t)},$$

$$\beta_{26} = \frac{\partial y_g}{\partial k_S(t)},\quad \beta_{27} = t\cdot\beta_{21},\quad \beta_{28} = t\cdot\beta_{22},\quad \beta_{29} = t\cdot\beta_{23}$$

$$\beta_{210} = t\cdot\beta_{24},\quad \beta_{211} = t\cdot\beta_{25},\quad \beta_{212} = t\cdot\beta_{26},\quad \beta_{213} = t^2\cdot\beta_{21},\quad \beta_{214} = t^2\cdot\beta_{22},\quad \beta_{215} = t^2\cdot\beta_{23},$$

$$\beta_{216} = t^2\cdot\beta_{24},\quad \beta_{217} = t^2\cdot\beta_{25},\quad \beta_{218} = t^2\cdot\beta_{26},\quad \beta_{219} = t^3\cdot\beta_{21},\quad \beta_{220} = t^3\cdot\beta_{22},$$

$$\beta_{221} = t^3\cdot\beta_{23},\quad \beta_{222} = t^3\cdot\beta_{24},\quad \beta_{223} = t^3\cdot\beta_{25},\quad \beta_{224} = t^3\cdot\beta_{26}$$

A.2 PARTIAL DERIVATIVES WITH RESPECT TO SENSOR'S EXTERIOR ORIENTATION ELEMENTS AND (X_M, Y_M, Z_M)

In this adjustment system, if there is one tie point (TP), M, whose geodetic coordinates (X_M, Y_M, Z_M) are unknown, the satellite flying epoch is t. Also, the position and attitude parameters of a sensor on-board spaceborne in the same strip can be expressed by a third-order polynomial (Zhou et al. 2022), i.e.,

$$X_S(t) = a_0 + a_1 t + a_2 t^2 + a_3 t^3$$

$$Y_S(t) = b_0 + b_1 t + b_2 t^2 + b_3 t^3$$

$$Z_S(t) = c_0 + c_1 t + c_2 t^2 + c_3 t^3$$

$$\phi_S(t) = d_0 + d_1 t + d_2 t^2 + d_3 t^3$$

$$\omega_S(t) = e_0 + e_1 t + e_2 t^2 + e_3 t^3$$

$$\kappa_S(t) = f_0 + f_1 t + f_2 t^2 + f_3 t^3$$

For the tie point, *M*, its collinearity equation is

$$x_M - x_0 = f\frac{r_{11}(X_M - X_S(t)) + r_{12}(Y_M - Y_S(t)) + r_{13}(Z_M - Z_S(t))}{r_{31}(X_M - X_S(t)) + r_{32}(Y_M - Y_S(t)) + r_{33}(Z_M - Z_S(t))} = f\frac{\overline{X}}{\overline{Z}}$$

$$y_M - y_0 = f\frac{r_{21}(X_M - X_S(t)) + r_{22}(Y_M - Y_S(t)) + r_{23}(Z_M - Z_S(t))}{r_{31}(X_M - X_S(t)) + r_{32}(Y_M - Y_S(t)) + r_{33}(Z_M - Z_S(t))} = f\frac{\overline{Y}}{\overline{Z}}$$

where

$$\overline{X} = r_{11}(X_M - X_S(t)) + r_{12}(Y_M - Y_S(t)) + r_{13}(Z_M - Z_S(t))$$

$$\overline{Y} = r_{21}(X_M - X_S(t)) + r_{22}(Y_M - Y_S(t)) + r_{23}(Z_M - Z_S(t))$$

$$\overline{Z} = r_{31}(X_M - X_S(t)) + r_{32}(Y_M - Y_S(t)) + r_{33}(Z_M - Z_S(t))$$

Partial differentials of (x_m, y_m) with respect to $(X_S(t), Y_S(t), Z_S(t))$ and $(\phi_S(t), \omega_S(t), \kappa_S(t))$ within a third-order polynomial are

$$\frac{\partial x_m}{\partial X_S(t)} dX_S(t) = \frac{\partial x_m}{\partial X_S(t)}\left(da_0 + t da_1 + t^2 da_2 + t^3 da_3\right)$$

$$\frac{\partial x_m}{\partial Y_S(t)} dY_S(t) = \frac{\partial x_m}{\partial Y_S(t)}\left(db_0 + t db_1 + t^2 db_2 + t^3 db_3\right)$$

$$\frac{\partial x_m}{\partial Z_S(t)} dZ_S(t) = \frac{\partial x_m}{\partial Z_S(t)}\left(dc_0 + t dc_1 + t^2 dc_2 + t^3 dc_3\right)$$

$$\frac{\partial x_m}{\partial \phi_S(t)} d\phi_S(t) = \frac{\partial x_m}{\partial \phi_S(t)}\left(dd_0 + t dd_1 + t^2 dd_2 + t^3 dd_3\right)$$

$$\frac{\partial x_m}{\partial \omega_S(t)} d\omega_S(t) = \frac{\partial x_m}{\partial \omega_S(t)}\left(de_0 + t de_1 + t^2 de_2 + t^3 de_3\right)$$

$$\frac{\partial x_m}{\partial \kappa_S(t)} d\kappa_S(t) = \frac{\partial x_m}{\partial \kappa_S(t)} \left(df_0 + tdf_1 + t^2 df_2 + t^3 df_3 \right)$$

$$\frac{\partial y_m}{\partial X_S(t)} dX_S(t) = \frac{\partial y_m}{\partial X_S(t)} \left(da_0 + tda_1 + t^2 da_2 + t^3 da_3 \right)$$

$$\frac{\partial y_m}{\partial Y_S(t)} dY_S(t) = \frac{\partial y_m}{\partial Y_S(t)} \left(db_0 + tdb_1 + t^2 db_2 + t^3 db_3 \right)$$

$$\frac{\partial y_m}{\partial Z_S(t)} dZ_S(t) = \frac{\partial y_m}{\partial Z_S(t)} \left(dc_0 + tdc_1 + t^2 dc_2 + t^3 dc_3 \right)$$

$$\frac{\partial y_m}{\partial \phi_S(t)} d\phi_S(t) = \frac{\partial y_m}{\partial \phi_S(t)} \left(dd_0 + tdd_1 + t^2 dd_2 + t^3 dd_3 \right)$$

$$\frac{\partial y_m}{\partial \omega_S(t)} d\omega_S(t) = \frac{\partial y_m}{\partial \omega_S(t)} \left(de_0 + tde_1 + t^2 de_2 + t^3 de_3 \right)$$

$$\frac{\partial y_m}{\partial \kappa_S(t)} d\kappa_S(t) = \frac{\partial y_m}{\partial \kappa_S(t)} \left(df_0 + tdf_1 + t^2 df_2 + t^3 df_3 \right)$$

Partial differentials of (x_m, y_m) with respect to $(X_S(t), Y_S(t), Z_S(t))$ and $(\phi_S(t), \omega_S(t), \kappa_S(t))$ within collinearity are

$$\frac{\partial x_m}{\partial X_S(t)} = f \frac{\frac{\partial \bar{X}}{\partial X_S(t)} \bar{Z} - \frac{\partial \bar{Z}}{\partial X_S(t)} \bar{X}}{\bar{Z}\bar{Z}} = f \frac{-r_{11}\bar{Z} + r_{31}\bar{X}}{\bar{Z}\bar{Z}} = \frac{1}{\bar{Z}}(-fr_{11} + r_{31}x_m)$$

$$\frac{\partial x_m}{\partial Y_S(t)} = f \frac{\frac{\partial \bar{X}}{\partial Y_S(t)} \bar{Z} - \frac{\partial \bar{Z}}{\partial Y_S(t)} \bar{X}}{\bar{Z}\bar{Z}} = f \frac{-r_{12}\bar{Z} + r_{32}\bar{X}}{\bar{Z}\bar{Z}} = \frac{1}{\bar{Z}}(-fr_{12} + r_{32}x_m)$$

$$\frac{\partial x_m}{\partial Z_S(t)} = f \frac{\frac{\partial \bar{X}}{\partial Z_S(t)} \bar{Z} - \frac{\partial \bar{Z}}{\partial Z_S(t)} \bar{X}}{\bar{Z}\bar{Z}} = f \frac{-r_{13}\bar{Z} + r_{33}\bar{X}}{\bar{Z}\bar{Z}} = \frac{1}{\bar{Z}}(-fr_{13} + r_{33}x_m)$$

$$\frac{\partial x_m}{\partial \phi_S(t)} = f \frac{\frac{\partial \bar{X}}{\partial \phi_S(t)} \bar{Z} - \frac{\partial \bar{Z}}{\partial \phi_S(t)} \bar{X}}{\bar{Z}\bar{Z}} = \frac{z_R}{\bar{Z}\bar{Z}} \left\{ \begin{array}{l} \bar{Z}\left[\frac{\partial r_{11}}{\partial \phi_S(t)}(X_M - X_S) + \frac{\partial r_{12}}{\partial \phi_S(t)}(Y_M - Y_S) + \frac{\partial r_{13}}{\partial \phi_S(t)}(Z_M - Z_S) \right] \\ -\bar{X}\left[\frac{\partial r_{31}}{\partial \phi_S(t)}(X_M - X_S) + \frac{\partial r_{32}}{\partial \phi_S(t)}(Y_M - Y_S) + \frac{\partial r_{33}}{\partial \phi_S(t)}(Z_M - Z_S) \right] \end{array} \right\}$$

$$= \frac{f}{\bar{Z}\bar{Z}} \left\{ \begin{array}{l} \bar{Z}[-\sin\phi\cos\kappa(X_M - X_S) + \sin\omega\cos\phi\cos\kappa(Y_M - Y_S) - \cos\omega\cos\phi\cos\kappa(Z_M - Z_S)] \\ -\bar{X}[\cos\phi(X_M - X_S) + \sin\omega\sin\phi(Y_M - Y_S) - \cos\omega\sin\phi(Z_M - Z_S)] \end{array} \right\}$$

$$= \frac{f}{\bar{Z}\bar{Z}} \begin{Bmatrix} \bar{Z}\left[-\sin\phi\cos\kappa(X_M - X_S) + r_{11}\sin\omega(Y_M - Y_S) - r_{33}\cos\kappa(Z_M - Z_S)\right] \\ -\bar{X}\left[\cos\phi(X_M - X_S) + \sin\omega\sin\phi(Y_M - Y_S) - \cos\omega\sin\phi(Z_M - Z_S)\right] \end{Bmatrix}$$

$$\frac{\partial x_m}{\partial \omega_S(t)} = f\,\frac{\dfrac{\partial \bar{X}}{\partial \omega_S(t)}\bar{Z} - \dfrac{\partial \bar{Z}}{\partial \omega_S(t)}\bar{X}}{\bar{Z}\bar{Z}}$$

$$= \frac{z_R}{\bar{Z}\bar{Z}} \begin{Bmatrix} \bar{Z}\left[\dfrac{\partial r_{11}}{\partial \omega_S(t)}(X_M - X_S) + \dfrac{\partial r_{12}}{\partial \omega_S(t)}(Y_M - Y_S) + \dfrac{\partial r_{13}}{\partial \omega_S(t)}(Z_M - Z_S)\right] \\ -\bar{X}\left[\dfrac{\partial r_{31}}{\partial \omega_S(t)}(X_M - X_S) + \dfrac{\partial r_{32}}{\partial \omega_S(t)}(Y_M - Y_S) + \dfrac{\partial r_{33}}{\partial \omega_S(t)}(Z_M - Z_S)\right] \end{Bmatrix}$$

$$= \frac{f}{\bar{Z}\bar{Z}} \begin{Bmatrix} \bar{Z}\begin{bmatrix} 0\cdot(X_M - X_S) + (-\sin\omega\sin\kappa + \cos\omega\sin\phi\cos\kappa)(Y_M - Y_S) \\ +(\cos\omega\sin\kappa + \sin\omega\sin\phi\cos\kappa(Z_M - Z_S) \end{bmatrix} \\ -\bar{X}\left[0\cdot(X_M - X_S) - \cos\omega\cos\phi(Y_M - Y_S) - \sin\omega\cos\phi(Z_M - Z_S)\right] \end{Bmatrix}$$

$$= \frac{f}{\bar{Z}\bar{Z}}\left\{\bar{Z}\left[-r_{13}(Y_M - Y_S) + r_{12}(Z_M - Z_S)\right] - \bar{X}\left[-r_{33}(Y_M - Y_S) + r_{32}(Z_M - Z_S)\right]\right\}$$

$$\frac{\partial x_m}{\partial \kappa_S(t)} = f\,\frac{\dfrac{\partial \bar{X}}{\partial \kappa_S(t)}\bar{Z} - \dfrac{\partial \bar{Z}}{\partial \kappa_S(t)}\bar{X}}{\bar{Z}\bar{Z}} = \frac{z_R}{\bar{Z}\bar{Z}} \begin{Bmatrix} \bar{Z}\begin{bmatrix} -\cos\phi\sin\kappa(X_M - X_S) + (\cos\omega\cos\kappa - \sin\omega\sin\phi\sin\kappa) \\ (Y_M - Y_S) + (\sin\omega\cos\kappa + \cos\omega\sin\phi\sin\kappa)(Z_M - Z_S) \end{bmatrix} \\ -\bar{X}\left[0\cdot(X_M - X_S) + 0\cdot(Y_M - Y_S) + 0\cdot(Z_M - Z_S)\right] \end{Bmatrix}$$

$$= \frac{f}{\bar{Z}\bar{Z}}\left\{\bar{Z}\left[r_{21}(X_M - X_S) + r_{22}(Y_M - Y_S) + r_{23}(Z_M - Z_S)\right]\right\}$$

$$= \frac{f}{\bar{Z}}(\bar{Y}) = y_M$$

$$\frac{\partial y_m}{\partial X_S(t)} = f\,\frac{\dfrac{\partial \bar{Y}}{\partial X_S(t)}\bar{Z} - \dfrac{\partial \bar{Z}}{\partial X_S(t)}\bar{Y}}{\bar{Z}\bar{Z}} = f\,\frac{-r_{21}\bar{Z} + r_{31}\bar{Y}}{\bar{Z}\bar{Z}} = \frac{1}{\bar{Z}}(-fr_{21} + r_{31}y_m)$$

$$\frac{\partial y_m}{\partial Y_S(t)} = f\,\frac{\dfrac{\partial \bar{Y}}{\partial Y_S(t)}\bar{Z} - \dfrac{\partial \bar{Z}}{\partial Y_S(t)}\bar{Y}}{\bar{Z}\bar{Z}} = f\,\frac{-r_{22}\bar{Z} + r_{32}\bar{Y}}{\bar{Z}\bar{Z}} = \frac{1}{\bar{Z}}(-fr_{22} + r_{32}y_m)$$

$$\frac{\partial y_m}{\partial Z_S(t)} = f\,\frac{\dfrac{\partial \bar{Y}}{\partial Z_S(t)}\bar{Z} - \dfrac{\partial \bar{Z}}{\partial Z_S(t)}\bar{Y}}{\bar{Z}\bar{Z}} = f\,\frac{-r_{23}\bar{Z} + r_{33}\bar{X}}{\bar{Z}\bar{Z}} = \frac{1}{\bar{Z}}(-fr_{23} + r_{33}y_m)$$

$$\frac{\partial y_m}{\partial \phi_S(t)} = f \frac{\dfrac{\partial \overline{Y}}{\partial \phi_S(t)} \overline{Z} - \dfrac{\partial \overline{Z}}{\partial \phi_S(t)} \overline{Y}}{\overline{Z}\overline{Z}}$$

$$= \frac{z_R}{\overline{Z}\overline{Z}} \left\{ \begin{aligned} &\overline{Z}\left[\frac{\partial r_{21}}{\partial \phi_S(t)}(X_M - X_S) + \frac{\partial r_{22}}{\partial \phi_S(t)}(Y_M - Y_S) + \frac{\partial r_{23}}{\partial \phi_S(t)}(Z_M - Z_S) \right] \\ &-\overline{Y}\left[\frac{\partial r_{31}}{\partial \phi_S(t)}(X_M - X_S) + \frac{\partial r_{32}}{\partial \phi_S(t)}(Y_M - Y_S) + \frac{\partial r_{33}}{\partial \phi_S(t)}(Z_M - Z_S) \right] \end{aligned} \right\}$$

$$= \frac{f}{\overline{Z}\overline{Z}} \left\{ \begin{aligned} &\overline{Z}\left[\sin\phi \sin\kappa (X_M - X_S) - \sin\omega \cos\phi \sin\kappa (Y_M - Y_S) + \cos\omega \cos\phi \sin\kappa (Z_M - Z_S) \right] \\ &-\overline{Y}\left[\cos\phi (X_M - X_S) + \sin\omega \sin\phi (Y_M - Y_S) - \cos\omega \sin\phi (Z_M - Z_S) \right] \end{aligned} \right\}$$

$$= \frac{f}{\overline{Z}\overline{Z}} \left\{ \begin{aligned} &\overline{Z}\left[\sin\phi \sin\kappa (X_M - X_S) + r_{21} \sin\omega (Y_M - Y_S) + r_{33} \sin\kappa (Z_M - Z_S) \right] \\ &-\overline{Y}\left[\cos\phi (X_M - X_S) + \sin\omega \sin\phi (Y_M - Y_S) - \cos\omega \sin\phi (Z_M - Z_S) \right] \end{aligned} \right\}$$

$$\frac{\partial y_m}{\partial \omega_S(t)} = f \frac{\dfrac{\partial \overline{Y}}{\partial \omega_S(t)} \overline{Z} - \dfrac{\partial \overline{Z}}{\partial \omega_S(t)} \overline{Y}}{\overline{Z}\overline{Z}}$$

$$= \frac{z_R}{\overline{Z}\overline{Z}} \left\{ \begin{aligned} &\overline{Z}\left[\frac{\partial r_{21}}{\partial \omega_S(t)} \cdot (X_M - X_S) + \frac{\partial r_{22}}{\partial \omega_S(t)}(Y_M - Y_S) + \frac{\partial r_{23}}{\partial \omega_S(t)}(Z_M - Z_S) \right] \\ &-\overline{Y}\left[\frac{\partial r_{31}}{\partial \phi_S(t)}(X_M - X_S) + \frac{\partial r_{32}}{\partial \phi_S(t)}(Y_M - Y_S) + \frac{\partial r_{33}}{\partial \phi_S(t)}(Z_M - Z_S) \right] \end{aligned} \right\}$$

$$= \frac{f}{\overline{Z}\overline{Z}} \left\{ \begin{aligned} &\overline{Z}\begin{bmatrix} 0 \cdot (X_M - X_S) + (-\sin\omega \cos\kappa - \cos\omega \sin\phi \sin\kappa)(Y_M - Y_S) \\ +(\cos\omega \cos\kappa - \sin\omega \sin\phi \sin\kappa)(Z_M - Z_S) \end{bmatrix} \\ &-\overline{Y}\left[0 \cdot (X_M - X_S) - \cos\omega \cos\phi (Y_M - Y_S) - \sin\omega \cos\phi (Z_M - Z_S) \right] \end{aligned} \right\}$$

$$= \frac{f}{\overline{Z}\overline{Z}} \left\{ \overline{Z}\left[-r_{23}(Y_M - Y_S) + r_{22}(Z_M - Z_S) \right] - \overline{Y}\left[-r_{33}(Y_M - Y_S) + r_{32}(Z_M - Z_S) \right] \right\}$$

$$\frac{\partial y_m}{\partial \kappa_S(t)} = f \frac{\dfrac{\partial \overline{Y}}{\partial \kappa_S(t)} \overline{Z} - \dfrac{\partial \overline{Z}}{\partial \kappa_S(t)} \overline{Y}}{\overline{Z}\overline{Z}}$$

$$= \frac{z_R}{\overline{Z}\overline{Z}} \left\{ \begin{aligned} &\overline{Z}\begin{bmatrix} -\cos\phi \cos\kappa (X_M - X_S) + (-\cos\omega \sin\kappa - \sin\omega \sin\phi \cos\kappa) \\ (Y_M - Y_S) + (-\sin\omega \sin\kappa + \cos\omega \sin\phi \cos\kappa)(Z_M - Z_S) \end{bmatrix} \\ &-\overline{Y}\left[0 \cdot (X_M - X_S) + 0 \cdot (Y_M - Y_S) + 0 \cdot (Z_M - Z_S) \right] \end{aligned} \right\}$$

$$= \frac{f}{\overline{Z}\overline{Z}} \left\{ \overline{Z}\left[-r_{11}(X_M - X_S) - r_{12}(Y_M - Y_S) - r_{13}(Z_M - Z_S) \right] \right\}$$

$$= \frac{f}{\overline{Z}}(-\overline{X}) = -x_m$$

Finally, the coefficients of the observation equations are

$$\gamma_{11} = \frac{\partial x_m}{\partial X_S(t)}, \quad \gamma_{12} = \frac{\partial x_m}{\partial Y_S(t)}, \quad \gamma_{13} = \frac{\partial x_m}{\partial Z_S(t)}, \quad \gamma_{14} = \frac{\partial x_m}{\partial \phi_S(t)}, \quad \gamma_{15} = \frac{\partial x_m}{\partial \omega_S(t)},$$

$$\gamma_{16} = \frac{\partial x_m}{\partial \kappa_S(t)}, \quad \gamma_{17} = t \cdot \gamma_{11}, \quad \gamma_{18} = t \cdot \gamma_{12}, \quad \gamma_{19} = t \cdot \gamma_{13}$$

$$\gamma_{110} = t \cdot \gamma_{14}, \quad \gamma_{111} = t \cdot \gamma_{15}, \quad \gamma_{112} = t \cdot \gamma_{16}, \quad \gamma_{113} = t^2 \cdot \gamma_{11}, \quad \gamma_{114} = t^2 \cdot \gamma_{12}, \quad \gamma_{115} = t^2 \cdot \gamma_{13},$$

$$\gamma_{116} = t^2 \cdot \gamma_{14}, \quad \gamma_{117} = t^2 \cdot \gamma_{15}, \quad \gamma_{118} = t^2 \cdot \gamma_{16}, \quad \gamma_{119} = t^3 \cdot \gamma_{11}, \quad \gamma_{120} = t^3 \cdot \gamma_{12}, \quad \gamma_{121} = t^3 \cdot \gamma_{13},$$

$$\gamma_{122} = t^3 \cdot \gamma_{14}, \quad \gamma_{123} = t^3 \cdot \gamma_{15}, \quad \gamma_{124} = t^3 \cdot \gamma_{16}, \quad \gamma_{126} = -\gamma_{12}, \quad \gamma_{127} = -\gamma_{13}$$

$$\gamma_{21} = \frac{\partial y_m}{\partial X_S(t)}, \quad \gamma_{22} = \frac{\partial y_m}{\partial Y_S(t)}, \quad \gamma_{23} = \frac{\partial y_m}{\partial Z_S(t)}, \quad \gamma_{24} = \frac{\partial y_m}{\partial \phi_S(t)}, \quad \gamma_{25} = \frac{\partial y_m}{\partial \omega_S(t)},$$

$$\gamma_{26} = \frac{\partial y_m}{\partial \kappa_S(t)}, \quad \gamma_{27} = t \cdot \gamma_{21}, \quad \gamma_{28} = t \cdot \gamma_{22}, \quad \gamma_{29} = t \cdot \gamma_{23}$$

$$\gamma_{210} = t \cdot \gamma_{24}, \quad \gamma_{211} = t \cdot \gamma_{25}, \quad \gamma_{212} = t \cdot \gamma_{26}, \quad \gamma_{213} = t^2 \cdot \gamma_{21}, \quad \gamma_{214} = t^2 \cdot \gamma_{22}, \quad \gamma_{215} = t^2 \cdot \gamma_{23},$$

$$\gamma_{216} = t^2 \cdot \gamma_{24}, \quad \gamma_{217} = t^2 \cdot \gamma_{25}, \quad \gamma_{218} = t^2 \cdot \gamma_{26}, \quad \gamma_{219} = t^3 \cdot \gamma_{21}, \quad \gamma_{220} = t^3 \cdot \gamma_{22}, \quad \gamma_{221} = t^3 \cdot \gamma_{23},$$

$$\gamma_{222} = t^3 \cdot \gamma_{24}, \quad \gamma_{223} = t^3 \cdot \gamma_{25}, \quad \gamma_{224} = t^3 \cdot \gamma_{26}, \quad \gamma_{225} = -\gamma_{21}, \quad \gamma_{226} = -\gamma_{22}, \quad \gamma_{227} = -\gamma_{23}$$

A.3 THE STRUCTURE OF OBSERVATION EQUATIONS

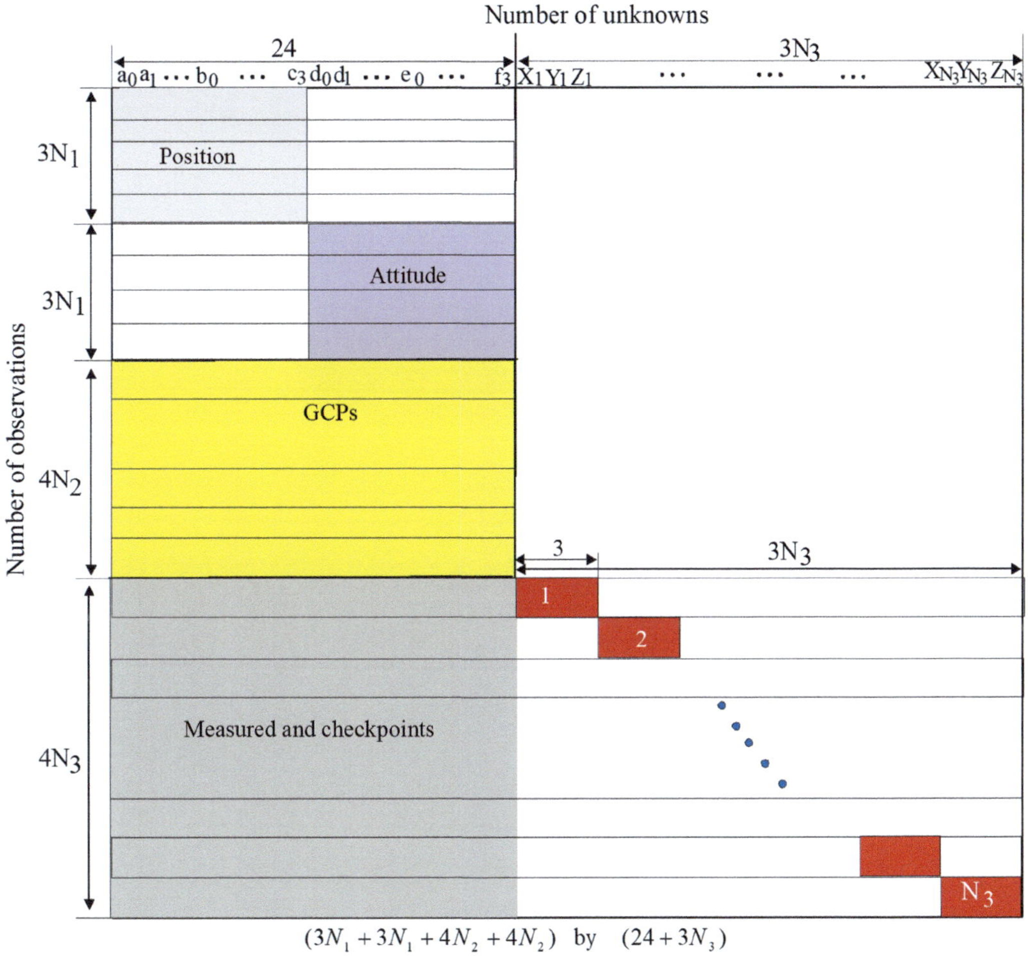

FIGURE A.1 The structure of observation equations.

REFERENCES

Li, R.; Zhou, G. Q., Photogrammetric processing of high-resolution airborne and satellite linear array stereo images for mapping applications. *International Journal of Remote Sensing*, 23, No. 20, 2002, pp. 4451–4473. https://doi.org/10.1080/01431160110107662

Zhou, G.; Li, H.; Song, R.; Wang, Q.; Xu, J.; Song, B., Orthorectification of fisheye image under equidistant projection model. *Remote Sensing*, 14, 2022, p. 4175.

Index

Note: **Bold** page numbers refer to tables and *italic* page numbers refer to figures.

V

Z

For Product Safety Concerns and Information please contact our
EU representative GPSR@taylorandfrancis.com Taylor & Francis
Verlag GmbH, Kaufingerstraße 24, 80331 München, Germany